CISCO™

Course Booklet

Enterprise Networking, Security, and Automation

CCNAv7

Cisco Press

CISCO Networking Academy

Enterprise Networking, Security, and Automation Course Booklet (CCNAv7)

Copyright © 2020 Cisco Systems, Inc.

Published by:
Cisco Press
Hoboken, New Jersey

1 2020

Library of Congress Control Number: 2020936932

ISBN-13: 978-0-13-663473-7

ISBN-10: 0-13-663473-7

Warning and Disclaimer

This book is designed to provide information about Enterprise Networking, Security, and Automation. Every effort has been made to make this book as complete and as accurate as possible, but no warranty or fitness is implied.

The information is provided on an "as is" basis. The authors, Cisco Press, and Cisco Systems, Inc. shall have neither liability nor responsibility to any person or entity with respect to any loss or damages arising from the information contained in this book or from the use of the discs or programs that may accompany it.

The opinions expressed in this book belong to the author and are not necessarily those of Cisco Systems, Inc.

Editor-in-Chief
Mark Taub

**Alliances Manager,
Cisco Press**
Arezou Gol

Product Line Manager
Brett Bartow

Senior Editor
James Manly

Managing Editor
Sandra Schroeder

Senior Project Editor
Tonya Simpson

Editorial Assistant
Cindy Teeters

Cover Designer
Chuti Prasertsith

Composition
codeMantra

Indexer
Cheryl Ann Lenser

CISCO

Trademark Acknowledgments

All terms mentioned in this book that are known to be trademarks or service marks have been appropriately capitalized. Cisco Press or Cisco Systems, Inc., cannot attest to the accuracy of this information. Use of a term in this book should not be regarded as affecting the validity of any trademark or service mark.

Special Sales

For information about buying this title in bulk quantities, or for special sales opportunities (which may include electronic versions; custom cover designs; and content particular to your business, training goals, marketing focus, or branding interests), please contact our corporate sales department at corpsales@pearsoned.com or (800) 382-3419.

For government sales inquiries, please contact governmentsales@pearsoned.com.

For questions about sales outside the U.S., please contact intlcs@pearson.com.

Feedback Information

At Cisco Press, our goal is to create in-depth technical books of the highest quality and value. Each book is crafted with care and precision, undergoing rigorous development that involves the unique expertise of members from the professional technical community.

Readers' feedback is a natural continuation of this process. If you have any comments regarding how we could improve the quality of this book, or otherwise alter it to better suit your needs, you can contact us through email at feedback@ciscopress.com. Please make sure to include the book title and ISBN in your message.

We greatly appreciate your assistance.

Americas Headquarters	Asia Pacific Headquarters	Europe Headquarters
Cisco Systems, Inc.	Cisco Systems (USA) Pte. Ltd.	Cisco Systems International BV Amsterdam,
San Jose, CA	Singapore	The Netherlands

Cisco has more than 200 offices worldwide. Addresses, phone numbers, and fax numbers are listed on the Cisco Website at **www.cisco.com/go/offices**.

Cisco and the Cisco logo are trademarks or registered trademarks of Cisco and/or its affiliates in the U.S. and other countries. To view a list of Cisco trademarks, go to this URL: www.cisco.com/go/trademarks. Third party trademarks mentioned are the property of their respective owners. The use of the word partner does not imply a partnership relationship between Cisco and any other company. (1110R)

Contents at a Glance

Contents

Command Syntax Conventions

The conventions used to present command syntax in this book are the same conventions used in the IOS Command Reference. The Command Reference describes these conventions as follows:

- **Boldface** indicates commands and keywords that are entered literally as shown. In actual configuration examples and output (not general command syntax), boldface indicates commands that are manually input by the user (such as a **show** command).

- *Italic* indicates arguments for which you supply actual values.

- Vertical bars (|) separate alternative, mutually exclusive elements.

- Square brackets ([ ]) indicate an optional element.

- Braces ({ }) indicate a required choice.

- Braces within brackets ([{ }]) indicate a required choice within an optional element.

About This Course Booklet

Your Cisco Networking Academy Course Booklet is designed as a study resource you can easily read, highlight, and review on the go, wherever the Internet is not available or practical:

- The text is extracted directly, word-for-word, from the online course so you can highlight important points and take notes in the "Your Chapter Notes" section.

- Headings with the exact page correlations provide a quick reference to the online course for your classroom discussions and exam preparation.

- An icon system directs you to the online curriculum to take full advantage of the images embedded within the Networking Academy online course interface and reminds you to do the labs, interactive activities, packet tracer activities, watch videos, and take the chapter quizzes.

Refer to **Online Course** for Illustration   Refer to **Lab Activity** for this chapter   Go to the online course to take the quiz and exam.   Refer to **Interactive Graphic** in online course   Refer to **Packet Tracer Activity** for this chapter   Refer to **Video** in online course

The Course Booklet is a basic, economical paper-based resource to help you succeed with the Cisco Networking Academy online course.

Introduction

Networking Academy CCNAv7

Welcome to the final course of the Cisco Networking Academy CCNAv7 curriculum, Enterprise Networking, Security, and Automation (ENSA). This is the third of three courses that are aligned to the CCNA Certification Exam. ENSA contains 14 modules, each with a series of topics.

In Enterprise Networking, Security, and Automation, you will take the skills and knowledge that you learned in ITN and SWRE and apply them to wide area networks (WANs). WANs are large, complex networks that require advanced understanding of network operation and security. ENSA also introduces you to two game-changing areas of networking: virtualization and automation.

By the end of this course you will be able to configure, troubleshoot, and secure enterprise network devices. You will be versed in application programming interfaces (APIs) and the configuration management tools that make network automation possible.

When you have completed ENSA, you will have gained the practical experience you need to prepare for the certification exam. You will also have the skills required for associate-level roles in the Information and Communication Technologies (ICT) industry. Let Cisco Networking Academy help you get where you want to go!

CHAPTER 1

Single-Area OSPFv2 Concepts

Introduction - 1.0

Why should I take this module? - 1.0.1

Welcome to Single-Area OSPFv2 Concepts!

Welcome to the first module in CCNA Enterprise Networking, Security, and Automation v7.0 (ENSA)!

Imagine that it is time for your family to visit your grandparents. You pack your bags and load them into the car. But this takes a bit longer than you planned for and now you are running late. You pull out your map. There are three different routes. One route is no good because there is a lot of construction on the main road and it is temporarily closed. Another route is very scenic, but it takes an additional hour to get to your destination. The third route is not as pretty but it includes a highway, which is much faster. In fact, it is so much faster that you might actually be on time if you take it.

In networking, packets do not need to take the scenic route. The *fastest available* route is always the best. Open Shortest Path First (OSPF) is designed to find the fastest available path for a packet from source to destination. This module covers the basic concepts of single-area OSPFv2. Let's get started!

What will I learn to do in this module? - 1.0.2

Module Title: Single-Area OSPF Concepts

Module Objective: Explain how single-area OSPF operates in both point-to-point and broadcast multiaccess networks.

Topic Title	Topic Objective
OSPF Features and Characteristics	Describe basic OSPF features and characteristics.
OSPF Packets	Describe the OSPF packet types used in single-area OSPF.
OSPF Operation	Explain how single-area OSPF operates.

OSPF Features and Characteristics - 1.1

Introduction to OSPF - 1.1.1

This topic is a brief overview of Open Shortest Path First (OSPF), which includes single-area and multiarea. OSPFv2 is used for IPv4 networks. OSPFv3 is used for IPv6 networks. The primary focus of this entire module is single-area OSPFv2.

OSPF is a link-state routing protocol that was developed as an alternative for the distance vector Routing Information Protocol (RIP). RIP was an acceptable routing protocol in the early days of networking and the internet. However, the RIP reliance on hop count as the only metric for determining best route quickly became problematic. Using hop count does not scale well in larger networks with multiple paths of varying speeds. OSPF has significant advantages over RIP in that it offers faster convergence and scales to much larger network implementations.

OSPF is a link-state routing protocol that uses the concept of areas. A network administrator can divide the routing domain into distinct areas that help control routing update traffic. A link is an interface on a router. A link is also a network segment that connects two routers, or a stub network such as an Ethernet LAN that is connected to a single router. Information about the state of a link is known as a link-state. All link-state information includes the network prefix, prefix length, and cost.

This module covers basic, single-area OSPF implementations and configurations.

Refer to
Interactive Graphic
in online course

Refer to
Online Course
for Illustration

Components of OSPF - 1.1.2

All routing protocols share similar components. They all use routing protocol messages to exchange route information. The messages help build data structures, which are then processed using a routing algorithm.

Click each OSPF component below for more information.

Routing Protocol Messages

Routers running OSPF exchange messages to convey routing information using five types of packets. These packets, as shown in the figure, are as follows:

- Hello packet
- Database description packet
- Link-state request packet
- Link-state update packet
- Link-state acknowledgment packet

These packets are used to discover neighboring routers and also to exchange routing information to maintain accurate information about the network.

Data Structures

OSPF messages are used to create and maintain three OSPF databases, as follows:

- **Adjacency database** - This creates the neighbor table.
- **Link-state database (LSDB)** - This creates the topology table.
- **Forwarding database** - This creates the routing table.

These tables contain a list of neighboring routers to exchange routing information. The tables are kept and maintained in RAM. In the following table, take a particular note of the command used to display each table.

Algorithm

The router builds the topology table using results of calculations based on the Dijkstra shortest-path first (SPF) algorithm. The SPF algorithm is based on the cumulative cost to reach a destination.

The SPF algorithm creates an SPF tree by placing each router at the root of the tree and calculating the shortest path to each node. The SPF tree is then used to calculate the best routes. OSPF places the best routes into the forwarding database, which is used to make the routing table.

Refer to
Interactive Graphic
in online course

Refer to
Online Course
for Illustration

Link-State Operation - 1.1.3

To maintain routing information, OSPF routers complete a generic link-state routing process to reach a state of convergence. The figure shows a five router topology. Each link between routers is labeled with a cost value. In OSPF, cost is used to determine the best path to the destination. The following are the link-state routing steps that are completed by a router:

1. Establish Neighbor Adjacencies

2. Exchange Link-State Advertisements

3. Build the Link State Database

4. Execute the SPF Algorithm

5. Choose the Best Route

Click each button for an illustration of the steps in the link-state routing process that R1 uses to reach convergence.

1. Establish Neighbor Adjacencies

OSPF-enabled routers must recognize each other on the network before they can share information. An OSPF-enabled router sends Hello packets out all OSPF-enabled interfaces to determine if neighbors are present on those links. If a neighbor is present, the OSPF-enabled router attempts to establish a neighbor adjacency with that neighbor.

2. Exchange Link-State Advertisements

After adjacencies are established, routers then exchange link-state advertisements (LSAs). LSAs contain the state and cost of each directly connected link. Routers flood their LSAs to adjacent neighbors. Adjacent neighbors receiving the LSA immediately flood the LSA to other directly connected neighbors, until all routers in the area have all LSAs.

3. Build the Link State Database

After LSAs are received, OSPF-enabled routers build the topology table (LSDB) based on the received LSAs. This database eventually holds all the information about the topology of the area.

4. Execute the SPF Algorithm

Routers then execute the SPF algorithm. The gears in the figure for this step are used to indicate the execution of the SPF algorithm. The SPF algorithm creates the SPF tree.

5. Choose the Best Route

After the SPF tree is built, the best paths to each network are offered to the IP routing table. The route will be inserted into the routing table unless there is a route source to the same network with a lower administrative distance, such as a static route. Routing decisions are made based on the entries in the routing table.

Refer to
Interactive Graphic
in online course

Refer to
Online Course
for Illustration

Single-Area and Multiarea OSPF - 1.1.4

To make OSPF more efficient and scalable, OSPF supports hierarchical routing using areas. An OSPF area is a group of routers that share the same link-state information in their LSDBs. OSPF can be implemented in one of two ways, as follows:

- **Single-Area OSPF** - All routers are in one area. Best practice is to use area 0.

- **Multiarea OSPF** - OSPF is implemented using multiple areas, in a hierarchical fashion. All areas must connect to the backbone area (area 0). Routers interconnecting the areas are referred to as Area Border Routers (ABRs).

The focus of this module is on single-area OSPFv2.

Click each button to compare single-area and multiarea OSPF.

Single-Area OSPF

Multiarea OSPF

Refer to
Online Course
for Illustration

Multiarea OSPF - 1.1.5

With multiarea OSPF, one large routing domain can be divided into smaller areas, to support hierarchical routing. Routing still occurs between the areas (interarea routing), while many of the processor intensive routing operations, such as recalculating the database, are kept within an area.

For instance, any time a router receives new information about a topology change within the area (including the addition, deletion, or modification of a link) the router must rerun the SPF algorithm, create a new SPF tree, and update the routing table. The SPF algorithm is CPU-intensive and the time it takes for calculation depends on the size of the area.

Note: Routers in other areas receive updates regarding topology changes, but these routers only update the routing table, not rerun the SPF algorithm.

Too many routers in one area would make the LSDBs very large and increase the load on the CPU. Therefore, arranging routers into areas effectively partitions a potentially large database into smaller and more manageable databases.

The hierarchical-topology design options with multiarea OSPF can offer the following advantages.

- **Smaller routing tables** - Tables are smaller because there are fewer routing table entries. This is because network addresses can be summarized between areas. Route summarization is not enabled by default.

- **Reduced link-state update overhead** - Designing multiarea OSPF with smaller areas minimizes processing and memory requirements.

- **Reduced frequency of SPF calculations** - Multiarea OSPF localize the impact of a topology change within an area. For instance, it minimizes routing update impact because LSA flooding stops at the area boundary.

For example, in the figure R2 is an ABR for area 51. A topology change in area 51 would cause all area 51 routers to rerun the SPF algorithm, create a new SPF tree and update their IP routing tables. The ABR, R2, would send an LSA to routers in the area 0, which would eventually be flooded to all routers in the OSPF routing domain. This type of LSA does not cause routers in other areas to rerun the SPF algorithm. They only have to update their LSDB and routing table.

Refer to Online Course for Illustration

OSPFv3 - 1.1.6

OSPFv3 is the OSPFv2 equivalent for exchanging IPv6 prefixes. Recall that in IPv6, the network address is referred to as the prefix and the subnet mask is called the prefix-length.

Similar to its IPv4 counterpart, OSPFv3 exchanges routing information to populate the IPv6 routing table with remote prefixes.

Note: With the OSPFv3 Address Families feature, OSPFv3 includes support for both IPv4 and IPv6. OSPF Address Families is beyond the scope of this curriculum.

OSPFv2 runs over the IPv4 network layer, communicating with other OSPF IPv4 peers, and advertising only IPv4 routes.

OSPFv3 has the same functionality as OSPFv2, but uses IPv6 as the network layer transport, communicating with OSPFv3 peers and advertising IPv6 routes. OSPFv3 also uses the SPF algorithm as the computation engine to determine the best paths throughout the routing domain.

OSPFv3 has separate processes from its IPv4 counterpart. The processes and operations are basically the same as in the IPv4 routing protocol, but run independently. OSPFv2 and OSPFv3 each have separate adjacency tables, OSPF topology tables, and IP routing tables, as shown in the figure.

The OSPFv3 configuration and verification commands are similar to those used in OSPFv2.

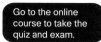
Go to the online
course to take the
quiz and exam.

Check Your Understanding - OSPF Features and Characteristics - 1.1.7

OSPF Packets - 1.2

Refer to **Video**
in online course

Video - OSPF Packets - 1.2.1

Click Play in the figure to view a video about OSPF packets.

Types of OSPF Packets - 1.2.2

Link-state packets are the tools used by OSPF to help determine the fastest available route for a packet. OSPF uses the following link-state packets (LSPs) to establish and maintain neighbor adjacencies and exchange routing updates. Each packet serves a specific purpose in the OSPF routing process, as follows:

- **Type 1: Hello packet** - This is used to establish and maintain adjacency with other OSPF routers.

- **Type 2: Database Description (DBD) packet** - This contains an abbreviated list of the LSDB of the sending router and is used by receiving routers to check against the local LSDB. The LSDB must be identical on all link-state routers within an area to construct an accurate SPF tree.

- **Type 3: Link-State Request (LSR) packet** - Receiving routers can then request more information about any entry in the DBD by sending an LSR.

- **Type 4: Link-State Update (LSU) packet** - This is used to reply to LSRs and to announce new information. LSUs contain several different types of LSAs.

- **Type 5: Link-State Acknowledgment (LSAck) packet** - When an LSU is received, the router sends an LSAck to confirm receipt of the LSU. The LSAck data field is empty.

The table summarizes the five different types of LSPs used by OSPFv2. OSPFv3 has similar packet types.

Type	Packet Name	Description
1	Hello	Discovers neighbors and builds adjacencies between them
2	Database Description (DBD)	Checks for database synchronization between routers
3	Link-State Request (LSR)	Requests specific link-state records from router to router
4	Link-State Update (LSU)	Sends specifically requested link-state records
5	Link-State Acknowledgment (LSAck)	Acknowledges the other packet types

Refer to
Online Course
for Illustration

Link-State Updates - 1.2.3

Routers initially exchange Type 2 DBD packets, which is an abbreviated list of the LSDB of the sending router. It is used by receiving routers to check against the local LSDB.

A Type 3 LSR packet is used by the receiving routers to request more information about an entry in the DBD.

The Type 4 LSU packet is used to reply to an LSR packet.

A Type 5 packet is used to acknowledge the receipt of a Type 4 LSU.

LSUs are also used to forward OSPF routing updates, such as link changes. Specifically, an LSU packet can contain 11 different types of OSPFv2 LSAs, with some of the more common ones shown in the figure. OSPFv3 renamed several of these LSAs and also contains two additional LSAs.

Note: The difference between the LSU and LSA terms can sometimes be confusing because these terms are often used interchangeably. However, an LSU contains one or more LSAs.

Refer to
Online Course
for Illustration

Hello Packet - 1.2.4

The OSPF Type 1 packet is the Hello packet. Hello packets are used to do the following:

- Discover OSPF neighbors and establish neighbor adjacencies.

- Advertise parameters on which two routers must agree to become neighbors.

- Elect the Designated Router (DR) and Backup Designated Router (BDR) on multi-access networks like Ethernet. Point-to-point links do not require DR or BDR.

The figure displays the fields contained in the OSPFv2 Type 1 Hello packet.

Go to the online course to take the quiz and exam.

Check Your Understanding - OSPF Packets - 1.2.5

OSPF Operation - 1.3

Refer to **Video** in online course

Video - OSPF Operation - 1.3.1

Click Play in the figure to view a video about OSPF operation.

OSPF Operational States - 1.3.2

Now that you know about the OSPF link-state packets, this topic explains how they work with OSPF-enabled routers. When an OSPF router is initially connected to a network, it attempts to:

- Create adjacencies with neighbors

- Exchange routing information

- Calculate the best routes
- Reach convergence

The table details the states OSPF progresses through while attempting to reach convergence:

State	Description
Down State	• No Hello packets received = Down. • Router sends Hello packets. • Transition to Init state.
Init State	• Hello packets are received from the neighbor. • They contain the Router ID of the sending router. • Transition to Two-Way state.
Two-Way State	• In this state, communication between the two routers is bidirectional. • On multiaccess links, the routers elect a DR and a BDR. • Transition to ExStart state.
ExStart State	On point-to-point networks, the two routers decide which router will initiate the DBD packet exchange and decide upon the initial DBD packet sequence number.
Exchange State	• Routers exchange DBD packets. • If additional router information is required then transition to Loading; otherwise, transition to the Full state.
Loading State	• LSRs and LSUs are used to gain additional route information. • Routes are processed using the SPF algorithm. • Transition to the Full state.
Full State	The link-state database of the router is fully synchronized.

Refer to
Interactive Graphic
in online course

Refer to
Online Course
for Illustration

Establish Neighbor Adjacencies - 1.3.3

When OSPF is enabled on an interface, the router must determine if there is another OSPF neighbor on the link. To accomplish this, the router sends a Hello packet that contains its router ID out all OSPF-enabled interfaces. The Hello packet is sent to the reserved All OSPF Routers IPv4 multicast address 224.0.0.5. Only OSPFv2 routers will process these packets. The OSPF router ID is used by the OSPF process to uniquely identify each router in the OSPF area. A router ID is a 32-bit number formatted like an IPv4 address and assigned to uniquely identify a router among OSPF peers.

When a neighboring OSPF-enabled router receives a Hello packet with a router ID that is not within its neighbor list, the receiving router attempts to establish an adjacency with the initiating router.

Click each button below to step through the process routers use to establish adjacency on a multiaccess network.

1. Down State to Init State

When OSPFv2 is enabled, the enabled Gigabit Ethernet 0/0 interface transitions from the Down state to the Init state. R1 starts sending Hello packets out all OSPF-enabled interfaces to discover OSPF neighbors to develop adjacencies with.

2. The Init State

R2 receives the Hello packet from R1 and adds the R1 router ID to its neighbor list. R2 then sends a Hello packet to R1. The packet contains the R2 Router ID and the R1 Router ID in its list of neighbors on the same interface.

3. Two-Way State

R1 receives the Hello and adds the R2 Router ID to its list of OSPF neighbors. It also notices its own Router ID in the list of neighbors of the Hello packet. When a router receives a Hello packet with its Router ID listed in the list of neighbors, the router transitions from the Init state to the Two-Way state.

The action performed in Two-Way state depends on the type of interconnection between the adjacent routers, as follows:

- If the two adjacent neighbors are interconnected over a point-to-point link, then they immediately transition from the Two-Way state to the ExStart state.

- If the routers are interconnected over a common Ethernet network, then a designated router DR and a BDR must be elected.

4. Elect the DR and BDR

Because R1 and R2 are interconnected over an Ethernet network, a DR and BDR election takes place. As shown in the figure, R2 becomes the DR and R1 is the BDR. This process only occurs on multiaccess networks such as Ethernet LANs.

Hello packets are continually exchanged to maintain router information.

Refer to
Interactive Graphic
in online course

Refer to
Online Course
for Illustration

Synchronizing OSPF Databases - 1.3.4

After the Two-Way state, routers transition to database synchronization states. While the Hello packet was used to establish neighbor adjacencies, the other four types of OSPF packets are used during the process of exchanging and synchronizing LSDBs. This is a three step process, as follows:

1. Decide first router

2. Exchange DBDs

3. Send an LSR

Click each button below to step through the process routers use to synchronize their LSDBs.

1. Decide First Router

In the ExStart state, the two routers decide which router will send the DBD packets first. The router with the higher router ID will be the first router to send DBD packets during the Exchange state. In the figure, R2 has the higher router ID and sends its DBD packets first.

2. Exchange DBDs

In the Exchange state, the two routers exchange one or more DBD packets. A DBD packet includes information about the LSA entry header that appears in the LSDB of the router. The entries can be about a link or about a network. Each LSA entry header includes information about the link-state type, the address of the advertising router, the cost of the link, and the sequence number. The router uses the sequence number to determine the newness of the received link-state information.

In the figure, R2 sends a DBD packet to R1. When R1 receives the DBD, it performs the following actions:

1. It acknowledges the receipt of the DBD using the LSAck packet.

2. R1 then sends DBD packets to R2.

3. R2 acknowledges R1.

3. Send an LSR

R1 compares the information received with the information it has in its own LSDB. If the DBD packet has a more current link-state entry, the router transitions to the Loading state.

For example, in the figure, R1 sends an LSR regarding network 172.16.6.0 to R2. R2 responds with the complete information about 172.16.6.0 in an LSU packet. Again, when R1 receives an LSU, it sends an LSAck. R1 then adds the new link-state entries into its LSDB.

After all LSRs have been satisfied for a given router, the adjacent routers are considered synchronized and in a full state. Updates (LSUs) are sent only to neighbors in the following conditions:

- When a change is perceived (incremental updates)

- Every 30 minutes

Refer to **Online Course** for Illustration

The Need for a DR - 1.3.5

Why is a DR and BDR election necessary?

Multiaccess networks can create two challenges for OSPF regarding the flooding of LSAs, as follows:

- **Creation of multiple adjacencies** - Ethernet networks could potentially interconnect many OSPF routers over a common link. Creating adjacencies with every router is unnecessary and undesirable. It would lead to an excessive number of LSAs exchanged between routers on the same network.

- **Extensive flooding of LSAs** - Link-state routers flood their LSAs any time OSPF is initialized, or when there is a change in the topology. This flooding can become excessive.

To understand the problem with multiple adjacencies, we must study a formula:

For any number of routers (designated as n) on a multiaccess network, there are $n\,(n-1)\,/\,2$ adjacencies.

For example, the figure shows a simple topology of five routers, all of which are attached to the same multiaccess Ethernet network. Without some type of mechanism to reduce the number of adjacencies, collectively these routers would form 10 adjacencies:

5 (5 − 1) / 2 = 10

This may not seem like much, but as routers are added to the network, the number of adjacencies increases dramatically. For example, a multiaccess network with 20 routers would create 190 adjacencies.

Refer to **Interactive Graphic** in online course

LSA Flooding With a DR - 1.3.6

A dramatic increase in the number of routers also dramatically increases the number of LSAs exchanged between the routers. This flooding of LSAs significantly impacts the operation of OSPF.

Click each button to compare the flooding of LSAs without and with a DR.

Flooding LSAs

To understand the problem of extensive flooding of LSAs, play the animation in the figure. In the animation, R2 sends out an LSA. This event triggers every other router to also send out an LSA. Not shown in the animation are the required acknowledgments sent for every LSA received. If every router in a multiaccess network had to flood and acknowledge all received LSAs to all other routers on that same multiaccess network, the network traffic would become quite chaotic.

LSAs and DR

The solution to managing the number of adjacencies and the flooding of LSAs on a multiaccess network is the DR. On multiaccess networks, OSPF elects a DR to be the collection and distribution point for LSAs sent and received. A BDR is also elected in case the DR fails. All other routers become DROTHERs. A DROTHER is a router that is neither the DR nor the BDR.

Note: The DR is only used for the dissemination of LSAs. The router will still use the best next-hop router indicated in the routing table for the forwarding of all other packets.

Play the animation in the figure to see the role of the DR.

Go to the online course to take the quiz and exam.

Check Your Understanding - OPSF Operation - 1.3.7

Module Practice and Quiz - 1.4

What did I learn in this module? - 1.4.1

OSPF Features and Characteristics

Open Shortest Path First (OSPF) is a link-state routing protocol that was developed as an alternative for the distance vector Routing Information Protocol (RIP). OSPF has significant advantages over RIP in that it offers faster convergence and scales to much larger network implementations. OSPF is a link-state routing protocol that uses the concept of areas for scalability. A link is an interface on a router. A link is also a network segment that connects two routers, or a stub network such as an Ethernet LAN that is connected to a single router. All link-state information includes the network prefix, prefix length, and cost. All routing protocols use routing protocol messages to exchange route information. The messages help build data structures, which are then processed using a routing algorithm. Routers running OSPF exchange messages to convey routing information using five types of packets: the Hello packet, the database description packet, the link-state request packet, the link-state update packet, and the link-state acknowledgment packet. OSPF messages are used to create and maintain three OSPF databases: the adjacency database creates the neighbor table, the link-state database (LSDB) creates the topology table, and the forwarding database creates the routing table. The router builds the topology table using results of calculations based on the Dijkstra SPF (shortest-path first) algorithm. The SPF algorithm is based on the cumulative cost to reach a destination. In OSPF, cost is used to determine the best path to the destination. To maintain routing information, OSPF routers complete a generic link-state routing process to reach a state of convergence:

1. Establish Neighbor Adjacencies

2. Exchange Link-State Advertisements

3. Build the Link State Database

4. Execute the SPF Algorithm

5. Choose the Best Route

With single-area OSPF any number can be used for the area, best practice is to use area 0. Single-area OSPF is useful in smaller networks with few routers. With multiarea OSPF, one large routing domain can be divided into smaller areas, to support hierarchical routing. Routing still occurs between the areas (interarea routing), while many of the processor intensive routing operations, such as recalculating the database, are kept within an area. OSPFv3 is the OSPFv2 equivalent for exchanging IPv6 prefixes. Recall that in IPv6, the network address is referred to as the prefix and the subnet mask is called the prefix-length.

OSPF Packets

OSPF uses the following link-state packets (LSPs) to establish and maintain neighbor adjacencies and exchange routing updates: 1 Hello, 2 DBD, 3 LSR, 4 LSU, and 5 LSAck. LSUs are also used to forward OSPF routing updates, such as link changes. Hello packets are used to:

- Discover OSPF neighbors and establish neighbor adjacencies.

- Advertise parameters on which two routers must agree to become neighbors.

- Elect the Designated Router (DR) and Backup Designated Router (BDR) on multi-access networks like Ethernet. Point-to-point links do not require DR or BDR.

Some important fields in the Hello packet are type, router ID, area ID, network mask, hello interval, router priority, dead interval, DR, BDR and list of neighbors.

OSPF Operation

When an OSPF router is initially connected to a network, it attempts to:

- Create adjacencies with neighbors
- Exchange routing information
- Calculate the best routes
- Reach convergence

The states that OSPF progresses through to do this are down state, init state, two-way state, ExStart state, Exchange state, loading state, and full state. When OSPF is enabled on an interface, the router must determine if there is another OSPF neighbor on the link by sending a Hello packet that contains its router ID out all OSPF-enabled interfaces. The Hello packet is sent to the reserved All OSPF Routers IPv4 multicast address 224.0.0.5. Only OSPFv2 routers will process these packets. When a neighboring OSPF-enabled router receives a Hello packet with a router ID that is not within its neighbor list, the receiving router attempts to establish an adjacency with the initiating router. After the Two-Way state, routers transition to database synchronization states, which is a three step process:

1. Decide First Router
2. Exchange DBDs
3. Send an LSR

Multiaccess networks can create two challenges for OSPF regarding the flooding of LSAs: the creation of multiple adjacencies and extensive flooding of LSAs. A dramatic increase in the number of routers also dramatically increases the number of LSAs exchanged between the routers. This flooding of LSAs significantly impact the operation of OSPF. If every router in a multiaccess network had to flood and acknowledge all received LSAs to all other routers on that same multiaccess network, the network traffic would become quite chaotic. This is why DR and BDR election is necessary. On multiaccess networks, OSPF elects a DR to be the collection and distribution point for LSAs sent and received. A BDR is also elected in case the DR fails.

Go to the online
course to take the
quiz and exam.

Chapter Quiz - Single-Area OSPFv2 Concepts

Your Chapter Notes

CHAPTER 2

Single-Area OSPFv2 Configuration

Introduction - 2.0

Why should I take this module? - 2.0.1

Welcome to Single-Area OSPFv2 Configuration!

Now that you know about single-area OSPFv2, you can probably think of all the ways it could benefit your own network. As a link-state protocol, OSPF is designed to not only find the fastest available route, it is designed to *create* fast, available routes. If you prefer a bit more control over some areas of your network, OSPF gives you several ways to manually override the DR election process and create your own preferred routes. With OSPF, your network can combine the automated processes with your own choices to make a network that you could troubleshoot in your sleep! You know you want to learn how to do this!

What will I learn to do in this module? - 2.0.2

Module Title: Single-Area OSPFv2 Configuration

Module Objective: Implement single-area OSPFv2 in both point-to-point and broadcast multi-access networks.

Topic Title	Topic Objective
OSPF Router ID	Configure an OSPFv2 router ID.
Point-to-Point OSPF Networks	Configure single-area OSPFv2 in a point-to-point network.
Multiaccess OSPF Networks	Configure the OSPF interface priority to influence the DR/BDR election in a multiaccess network.
Modify Single-Area OSPFv2	Implement modifications to change the operation of single-area OSPFv2.
Default Route Propagation	Configure OSPF to propagate a default route.
Verify Single-Area OSPFv2	Verify a single-area OSPFv2 implementation.

OSPF Router ID - 2.1

OSPF Reference Topology - 2.1.1

Refer to
Online Course
for Illustration

To get you started, this topic discusses the foundation on which OSPF bases its entire process, the OSPF router ID.

The figure shows the topology used for configuring OSPFv2 in this module. The routers in the topology have a starting configuration, including interface addresses. There is currently no static routing or dynamic routing configured on any of the routers. All interfaces on R1, R2, and R3 (except the loopback 1 on R2) are within the OSPF backbone area. The ISP router is used as the gateway to the internet of the routing domain.

Note: In this topology the loopback interface is used to simulate the WAN link to the Internet and a LAN connected to each router. This is done to allow this topology to be duplicated for demonstration purposes on routers that only have two Gigabit Ethernet interfaces.

Router Configuration Mode for OSPF - 2.1.2

OSPFv2 is enabled using the **router ospf** *process-id* global configuration mode command, as shown in the command window for R1. The *process-id* value represents a number between 1 and 65,535 and is selected by the network administrator. The *process-id* value is locally significant, which means that it does not have to be the same value on the other OSPF routers to establish adjacencies with those neighbors. It is considered best practice to use the same *process-id* on all OSPF routers.

After entering the **router ospf** *process-id* command, the router enters router configuration mode, as indicated by the **R1(config-router)#** prompt. Enter a question mark (**?**), to view all the commands available in this mode. The list of commands shown here has been altered to display only the commands that are relevant to this module.

```
R1(config)# router ospf 10

R1(config-router)# ?

  area                   OSPF area parameters
  auto-cost              Calculate OSPF interface cost according to
                         bandwidth
  default-information    Control distribution of default information
  distance               Define an administrative distance
  exit                   Exit from routing protocol configuration mode
  log-adjacency-changes  Log changes in adjacency state
  neighbor               Specify a neighbor router
  network                Enable routing on an IP network
  no                     Negate a command or set its defaults
  passive-interface      Suppress routing updates on an interface
  redistribute           Redistribute information from another routing
                         protocol
  router-id              router-id for this OSPF process
R1(config-router)#
```

Router IDs - 2.1.3

An OSPF router ID is a 32-bit value, represented as an IPv4 address. The router ID is used to uniquely identify an OSPF router. All OSPF packets include the router ID of the originating router. Every router requires a router ID to participate in an OSPF domain. The router ID can be defined by an administrator or automatically assigned by the router. The router ID is used by an OSPF-enabled router to do the following:

- **Participate in the synchronization of OSPF databases** - During the Exchange State, the router with the highest router ID will send their database descriptor (DBD) packets first.

- **Participate in the election of the designated router (DR)** - In a multiaccess LAN environment, the router with the highest router ID is elected the DR. The routing device with the second highest router ID is elected the backup designated router (BDR).

Note: The DR and BDR election process is discussed in more detail later in this module.

Refer to
Online Course
for Illustration

Router ID Order of Precedence - 2.1.4

But how does the router determine the router ID? As illustrated in the figure, Cisco routers derive the router ID based on one of three criteria, in the following preferential order:

1. The router ID is explicitly configured using the OSPF **router-id** *rid* router configuration mode command. The *rid* value is any 32-bit value expressed as an IPv4 address. This is the recommended method to assign a router ID.

2. If the router ID is not explicitly configured, the router chooses the highest IPv4 address of any of configured loopback interfaces. This is the next best alternative to assigning a router ID.

3. If no loopback interfaces are configured, then the router chooses the highest active IPv4 address of any of its physical interfaces. This is the least recommended method because it makes it more difficult for administrators to distinguish between specific routers.

Configure a Loopback Interface as the Router ID - 2.1.5

In the reference topology, only the physical interfaces are configured and active. The loopback interfaces have not been configured. When OSPF routing is enabled on the router, the routers would pick the following highest active configured IPv4 address as the router ID.

- R1: 10.1.1.14 (G0/0/1)

- R2: 10.1.1.9 (G0/0/1)

- R3: 10.1.1.13 (G0/0/0)

Note: OSPF does not need to be enabled on an interface for that interface to be chosen as the router ID.

Instead of relying on physical interface, the router ID can be assigned to a loopback interface. Typically, the IPv4 address for this type of loopback interface should be configured using a 32-bit subnet mask (255.255.255.255). This effectively creates a host route. A 32-bit host route would not get advertised as a route to other OSPF routers.

The example shows how to configure a loopback interface on R1. Assuming the router ID was not explicitly configured or previously learned, R1 will use IPv4 address 1.1.1.1 as its router ID. Assume R1 has not yet learned a router ID.

```
R1(config-if)# interface Loopback 1
R1(config-if)# ip address 1.1.1.1 255.255.255.255
R1(config-if)# end
R1# show ip protocols | include Router ID
   Router ID 1.1.1.1
R1#
```

Refer to
Online Course
for Illustration

Explicitly Configure a Router ID - 2.1.6

In the figure, the topology has been updated to show the router ID for each router:

- R1 uses router ID 1.1.1.1
- R2 uses router ID 2.2.2.2
- R3 uses router ID 3.3.3.3

Use the **router-id** *rid* router configuration mode command to manually assign a router ID. In the example, the router ID 1.1.1.1 is assigned to R1. Use the **show ip protocols** command to verify the router ID.

```
R1(config)# router ospf 10
R1(config-router)# router-id 1.1.1.1
R1(config-router)# end
*May 23 19:33:42.689: %SYS-5-CONFIG_I: Configured from console by console
R1# show ip protocols | include Router ID
   Router ID 1.1.1.1
R1#
```

Modify a Router ID - 2.1.7

After a router selects a router ID, an active OSPF router does not allow the router ID to be changed until the router is reloaded or the OSPF process is reset.

In example for R1, the configured router ID has been removed and the router reloaded. Notice that the current router ID is 10.10.1.1, which is the Loopback 0 IPv4 address. The router ID should be 1.1.1.1. Therefore, R1 is configured with the command **router-id 1.1.1.1**.

Notice how an informational message appears stating that the OSPF process must be cleared or that the router must be reloaded. The reason is because R1 already has adjacencies with other neighbors using the router ID 10.10.1.1. Those adjacencies must be renegotiated using the new router ID 1.1.1.1. Use the **clear ip ospf process** command to reset the adjacencies. You can then verify that R1 is using the new router ID command with the **show ip protocols** command piped to display only the router ID section.

Clearing the OSPF process is the preferred method to reset the router ID.

```
R1# show ip protocols | include Router ID
  Router ID 10.10.1.1
R1# conf t
Enter configuration commands, one per line.  End with CNTL/Z.
R1(config)# router ospf 10
R1(config-router)# router-id 1.1.1.1
% OSPF: Reload or use "clear ip ospf process" command, for this to take
effect
R1(config-router)# end
R1# clear ip ospf process
Reset ALL OSPF processes? [no]: y

*Jun  6 01:09:46.975: %OSPF-5-ADJCHG: Process 10, Nbr 3.3.3.3 on
GigabitEthernet0/0/1 from FULL to DOWN, Neighbor Down: Interface down or
detached

*Jun  6 01:09:46.975: %OSPF-5-ADJCHG: Process 10, Nbr 2.2.2.2 on
GigabitEthernet0/0/0 from FULL to DOWN, Neighbor Down: Interface down or
detached

*Jun  6 01:09:46.981: %OSPF-5-ADJCHG: Process 10, Nbr 3.3.3.3 on
GigabitEthernet0/0/1 from LOADING to FULL, Loading Done

*Jun  6 01:09:46.981: %OSPF-5-ADJCHG: Process 10, Nbr 2.2.2.2 on
GigabitEthernet0/0/0 from LOADING to FULL, Loading Done

R1# show ip protocols | include Router ID
  Router ID 1.1.1.1
R1#
```

Note: The **router-id** command is the preferred method. However, some older versions of the IOS do not recognize the **router-id** command; therefore, the best way to set the router ID on those routers is by using a loopback interface.

Refer to **Interactive Graphic** in online course

Syntax Checker - Configure R2 and R3 Router IDs - 2.1.8

Use the Syntax Checker to configure R2 and R3 with router IDs.

Go to the online course to take the quiz and exam.

Check Your Understanding - OSPF Router ID - 2.1.9

Point-to-Point OSPF Networks - 2.2

The network Command Syntax - 2.2.1

One type of network that uses OSPF is the point-to-point network. You can specify the interfaces that belong to a point-to-point network by configuring the **network** command. You can also configure OSPF directly on the interface with the **ip ospf** command, as we will see later.

Both commands are used to determine which interfaces participate in the routing process for an OSPFv2 area. The basic syntax for the **network** command is as follows:

```
Router(config-router)# network network-address wildcard-mask area area-id
```

- The *network-address wildcard-mask* syntax is used to enable OSPF on interfaces. Any interfaces on a router that match the network address in the **network** command are enabled to send and receive OSPF packets.

- The **area** *area-id* syntax refers to the OSPF area. When configuring single-area OSPFv2, the **network** command must be configured with the same *area-id* value on all routers. Although any area ID can be used, it is good practice to use an area ID of 0 with single-area OSPFv2. This convention makes it easier if the network is later altered to support multiarea OSPFv2.

Refer to
Online Course
for Illustration

The Wildcard Mask - 2.2.2

The wildcard mask is typically the inverse of the subnet mask configured on that interface. In a subnet mask, binary 1 is equal to a match and binary 0 is not a match. In a wildcard mask, the reverse is true, as shown in here:

- **Wildcard mask bit 0** - Matches the corresponding bit value in the address.

- **Wildcard mask bit 1** - Ignores the corresponding bit value in the address.

The easiest method for calculating a wildcard mask is to subtract the network subnet mask from 255.255.255.255, as shown for /24 and /26 subnet masks in the figure.

Refer to
Interactive Graphic
in online course

Check Your Understanding - The Wildcard Masks - 2.2.3

Calculate the subnet mask and wildcard mask required to advertise the specified network address in OSPF. Type your answers in the fields provided. Click Check to verify your answers. Click Show Me to see the correct answer.

Click New Problem to continue the activity.

Refer to
Online Course
for Illustration

Configure OSPF Using the network Command - 2.2.4

Within routing configuration mode, there are two ways to identify the interfaces that will participate in the OSPFv2 routing process. The figure shows the reference topology.

In the first example, the wildcard mask identifies the interface based on the network addresses. Any active interface that is configured with an IPv4 address belonging to that network will participate in the OSPFv2 routing process.

```
R1(config)# router ospf 10
R1(config-router)# network 10.10.1.0 0.0.0.255 area 0
R1(config-router)# network 10.1.1.4 0.0.0.3 area 0
R1(config-router)# network 10.1.1.12 0.0.0.3 area 0
R1(config-router)#
```

Note: Some IOS versions allow the subnet mask to be entered instead of the wildcard mask. The IOS then converts the subnet mask to the wildcard mask format.

As an alternative, the second example shows how OSPFv2 can be enabled by specifying the exact interface IPv4 address using a quad zero wildcard mask. Entering **network 10.1.1.5 0.0.0.0 area 0** on R1 tells the router to enable interface Gigabit Ethernet 0/0/0 for the routing process. As a result, the OSPFv2 process will advertise the network that is on this interface (10.1.1.4/30).

```
R1(config)# router ospf 10
R1(config-router)# network 10.10.1.1 0.0.0.0 area 0
R1(config-router)# network 10:1.1.5 0.0.0.0 area 0
R1(config-router)# network 10.1.1.14 0.0.0.0 area 0
R1(config-router)#
```

The advantage of specifying the interface is that the wildcard mask calculation is not necessary. Notice that in all cases, the **area** argument specifies area 0.

Refer to
Interactive Graphic
in online course

Syntax Checker - Configure R2 and R3 Using the network Command - 2.2.5

Use the Syntax Checker to advertise the networks connected to R2 and R3.

Note: While completing the syntax checker, observe the informational messages describing the adjacency between R1 (1.1.1.1) and R2 (2.2.2.2). The IPv4 addressing scheme used for the router ID makes it easy to identify the neighbor.

Configure OSPF Using the ip ospf Command - 2.2.6

You can also configure OSPF directly on the interface instead of using the **network** command. To configure OSPF directly on the interface, use the **ip ospf** interface configuration mode command. The syntax is as follows:

```
Router(config-if)# ip ospf process-id area area-id
```

For R1, remove the network commands by using the **no** form of the **network** commands. And then go to each interface and configure the **ip ospf** command, as shown in the command window.

```
R1(config)# router ospf 10

R1(config-router)# no network 10.10.1.1 0.0.0.0 area 0

R1(config-router)# no network 10.1.1.5 0.0.0.0 area 0

R1(config-router)# no network 10.1.1.14 0.0.0.0 area 0

R1(config-router)# interface GigabitEthernet 0/0/0

R1(config-if)# ip ospf 10 area 0

R1(config-if)# interface GigabitEthernet 0/0/1

R1(config-if)# ip ospf 10 area 0

R1(config-if)# interface Loopback 0

R1(config-if)# ip ospf 10 area 0

R1(config-if)#
```

Refer to
Interactive Graphic
in online course

Syntax Checker - Configure R2 and R3 Using the ip ospf Command - 2.2.7

Use the Syntax Checker to advertise the networks by configuring the interfaces for OSPF on R2 and R3.

Refer to
Online Course
for Illustration

Passive Interface - 2.2.8

By default, OSPF messages are forwarded out all OSPF-enabled interfaces. However, these messages really only need to be sent out interfaces that are connecting to other OSPF-enabled routers.

Refer to the topology in the figure. OSPFv2 messages are forwarded out the three loopback interfaces even though no OSPFv2 neighbor exists on these simulated LANs. In a production network, these loopbacks would be physical interfaces to networks with users and traffic. Sending out unneeded messages on a LAN affects the network in three ways, as follows:

- **Inefficient Use of Bandwidth** - Available bandwidth is consumed transporting unnecessary messages.

- **Inefficient Use of Resources** - All devices on the LAN must process and eventually discard the message.

- **Increased Security Risk** - Without additional OSPF security configurations, OSPF messages can be intercepted with packet sniffing software. Routing updates can be modified and sent back to the router, corrupting the routing table with false metrics that misdirect traffic.

Configure Passive Interfaces - 2.2.9

Use the **passive-interface** router configuration mode command to prevent the transmission of routing messages through a router interface, but still allow that network to be advertised to other routers. The configuration example identifies the R1 Loopback 0/0/0 interface as passive.

The **show ip protocols** command is then used to verify that the Loopback 0 interface is listed as passive. The interface is still listed under the heading, "Routing on Interfaces Configured Explicitly (Area 0)", which means that this network is still included as a route entry in OSPFv2 updates that are sent to R2 and R3.

```
R1(config)# router ospf 10

R1(config-router)# passive-interface loopback 0

R1(config-router)# end

R1#

*May 23 20:24:39.309: %SYS-5-CONFIG_I: Configured from console by
console

R1# show ip protocols

*** IP Routing is NSF aware ***

(output omitted)

Routing Protocol is "ospf 10"

  Outgoing update filter list for all interfaces is not set

  Incoming update filter list for all interfaces is not set

  Router ID 1.1.1.1

  Number of areas in this router is 1. 1 normal 0 stub 0 nssa

  Maximum path: 4

  Routing for Networks:

  Routing on Interfaces Configured Explicitly (Area 0):

    Loopback0

    GigabitEthernet0/0/1

    GigabitEthernet0/0/0

  Passive Interface(s):

    Loopback0

  Routing Information Sources:

    Gateway         Distance      Last Update

    3.3.3.3              110      01:01:48

    2.2.2.2              110      01:01:38

  Distance: (default is 110)

R1#
```

Refer to
Interactive Graphic
in online course

Syntax Checker - Configure R2 and R3 Passive Interfaces - 2.2.10

Use the Syntax Checker to configure the Loopback interfaces on R2 as a passive. As an alternative, all interfaces can be made passive using the **passive-interface default** command. Interfaces that should not be passive can be re-enabled using the **no passive-interface** command. Configure R3 with the **passive-interface default** command and then re-enable the Gigabit Ethernet interfaces.

OSPF Point-to-Point Networks - 2.2.11

By default, Cisco routers elect a DR and BDR on Ethernet interfaces, even if there is only one other device on the link. You can verify this with the **show ip ospf interface** command, as shown in the example for G0/0/0 of R1.

```
R1# show ip ospf interface GigabitEthernet 0/0/0

GigabitEthernet0/0/0 is up, line protocol is up

  Internet Address 10.1.1.5/30, Area 0, Attached via Interface Enable

  Process ID 10, Router ID 1.1.1.1, Network Type BROADCAST, Cost: 1

  Topology-MTID    Cost    Disabled    Shutdown       Topology Name

        0           1         no          no              Base

  Enabled by interface config, including secondary ip addresses

  Transmit Delay is 1 sec, State BDR, Priority 1

  Designated Router (ID) 2.2.2.2, Interface address 10.1.1.6

  Backup Designated router (ID) 1.1.1.1, Interface address 10.1.1.5

  Timer intervals configured, Hello 10, Dead 40, Wait 40, Retransmit 5

    oob-resync timeout 40

    Hello due in 00:00:08

  Supports Link-local Signaling (LLS)

  Cisco NSF helper support enabled

  IETF NSF helper support enabled

  Index 1/2/2, flood queue length 0

  Next 0x0(0)/0x0(0)/0x0(0)

  Last flood scan length is 1, maximum is 1

  Last flood scan time is 0 msec, maximum is 0 msec

  Neighbor Count is 1, Adjacent neighbor count is 1

    Adjacent with neighbor 2.2.2.2   (Designated Router)

  Suppress hello for 0 neighbor(s)

R1#
```

R1 is the BDR and R2 is the DR. The DR/ BDR election process is unnecessary as there can only be two routers on the point-to-point network between R1 and R2. Notice in the output that the router has designated the network type as BROADCAST. To change this to a point-to-point network, use the interface configuration command **ip ospf network point-to-point** on all interfaces where you want to disable the DR/BDR election process. The example below shows this configuration for R1. The OSPF neighbor adjacency status will go down for a few milliseconds.

```
R1(config)# interface GigabitEthernet 0/0/0

R1(config-if)# ip ospf network point-to-point

*Jun  6 00:44:05.208: %OSPF-5-ADJCHG: Process 10, Nbr 2.2.2.2 on
GigabitEthernet0/0/0 from FULL to DOWN, Neighbor Down: Interface down or
detached

*Jun  6 00:44:05.211: %OSPF-5-ADJCHG: Process 10, Nbr 2.2.2.2 on
GigabitEthernet0/0/0 from LOADING to FULL, Loading Done

R1(config-if)# interface GigabitEthernet 0/0/1

R1(config-if)# ip ospf network point-to-point

*Jun  6 00:44:45.532: %OSPF-5-ADJCHG: Process 10, Nbr 3.3.3.3 on
GigabitEthernet0/0/1 from FULL to DOWN, Neighbor Down: Interface down or
detached

*Jun  6 00:44:45.535: %OSPF-5-ADJCHG: Process 10, Nbr 3.3.3.3 on
GigabitEthernet0/0/1 from LOADING to FULL, Loading Done

R1(config-if)# end

R1# show ip ospf interface GigabitEthernet 0/0/0

GigabitEthernet0/0/0 is up, line protocol is up
   Internet Address 10.1.1.5/30, Area 0, Attached via Interface Enable
   Process ID 10, Router ID 1.1.1.1, Network Type POINT_TO_POINT, Cost: 1
   Topology-MTID    Cost    Disabled    Shutdown      Topology Name
        0             1        no          no            Base
   Enabled by interface config, including secondary ip addresses
   Transmit Delay is 1 sec, State POINT_TO_POINT
   Timer intervals configured, Hello 10, Dead 40, Wait 40, Retransmit 5
     oob-resync timeout 40
     Hello due in 00:00:04
   Supports Link-local Signaling (LLS)
   Cisco NSF helper support enabled
   IETF NSF helper support enabled
   Index 1/2/2, flood queue length 0
   Next 0x0(0)/0x0(0)/0x0(0)
   Last flood scan length is 1, maximum is 2
```

```
Last flood scan time is 0 msec, maximum is 1 msec

Neighbor Count is 1, Adjacent neighbor count is 1

  Adjacent with neighbor 2.2.2.2

Suppress hello for 0 neighbor(s)
R1#
```

Notice that the Gigabit Ethernet 0/0/0 interface now lists the network type as POINT_ TO_POINT and that there is no DR or BDR on the link.

Loopbacks and Point-to-Point Networks - 2.2.12

We use loopbacks to provide additional interfaces for a variety of purposes. In this case, we are using loopbacks to simulate more networks than the equipment can support. By default, loopback interfaces are advertised as /32 host routes. For example, R1 would advertise the 10.10.1.0/24 network as 10.10.1.1/32 to R2 and R3.

```
R2# show ip route | include 10.10.1

O        10.10.1.1/32 [110/2] via 10.1.1.5, 00:03:05,
GigabitEthernet0/0/0
```

To simulate a real LAN, the Loopback 0 interface is configured as a point-to-point network so that R1 will advertise the full 10.10.1.0/24 network to R2 and R3.

```
R1(config-if)# interface Loopback 0

R1(config-if)# ip ospf network point-to-point
```

Now R2 receives the more accurate, simulated LAN network address of 10.10.1.0/24.

```
R2# show ip route | include 10.10.1

O        10.10.1.0/24 [110/2] via 10.1.1.5, 00:00:30,
GigabitEthernet0/0/0
```

Note: At the time of this writing, Packet Tracer does not support the **ip ospf network point-to-point** command on Gigabit Ethernet interfaces. However, it is supported on Loopback interfaces.

Refer to **Packet Tracer Activity** for this chapter

Packet Tracer - Point-to-Point Single-Area OSPFv2 Configuration - 2.2.13

In this Packet Tracer activity, you will configure the single-area OSPFv2 with the following:

■ Explicitly configure router IDs.

■ Configure the **network** command on R1 using wildcard mask based on the subnet mask.

■ Configure the **network** command on R2 using a quad-zero wildcard mask.

- Configure the **ip ospf** interface command on R3.
- Configure passive interfaces.
- Verify OSPF operation using the **show ip protocols** and **show ip route** commands.

Multiaccess OSPF Networks - 2.3

OSPF Network Types - 2.3.1

Refer to
Online Course
for Illustration

Another type of network that uses OSPF is the multiaccess OSPF network. Multiaccess OSPF networks are unique in that one router controls the distribution of LSAs. The router that is elected for this role should be determined by the network administrator through proper configuration.

OSPF may include additional processes depending on the type of network. The previous topology used point-to-point links between the routers. However, routers can be connected to the same switch to form a multiaccess network, as shown in the figure. Ethernet LANs are the most common example of broadcast multiaccess networks. In broadcast networks, all devices on the network see all broadcast and multicast frames.

OSPF Designated Router - 2.3.2

Refer to
Interactive Graphic
in online course

Recall that, in multiaccess networks, OSPF elects a DR and BDR as a solution to manage the number of adjacencies and the flooding of link-state advertisements (LSAs). The DR is responsible for collecting and distributing LSAs sent and received. The DR uses the multicast IPv4 address 224.0.0.5 which is meant for all OSPF routers.

A BDR is also elected in case the DR fails. The BDR listens passively and maintains a relationship with all the routers. If the DR stops producing Hello packets, the BDR promotes itself and assumes the role of DR.

All other routers become a DROTHER (a router that is neither the DR nor the BDR). DROTHERs use the multiaccess address 224.0.0.6 (all designated routers) to send OSPF packets to the DR and BDR. Only the DR and BDR listen for 224.0.0.6.

In the figure, R1, R5, and R4 are DROTHERs. Click play to see the animation of R2 acting as DR. Notice that only the DR and the BDR process the LSA sent by R1 using the multicast IPv4 address 224.0.0.6. The DR then sends out the LSA to all OSPF routers using the multicast IPv4 address 224.0.0.5.

OSPF Multiaccess Reference Topology - 2.3.3

Refer to
Online Course
for Illustration

In the multiaccess topology shown in the figure, there are three routers interconnected over a common Ethernet multiaccess network, 192.168.1.0/24. Each router is configured with the indicated IPv4 address on the Gigabit Ethernet 0/0/0 interface.

Because the routers are connected over a common multiaccess network, OSPF has automatically elected a DR and BDR. In this example, R3 has been elected as the DR because its router ID is 3.3.3.3, which is the highest in this network. R2 is the BDR because it has the second highest router ID in the network.

Refer to **Interactive Graphic** in online course

Verify OSPF Router Roles - 2.3.4

To verify the roles of the OSPFv2 router, use the **show ip ospf interface** command.

Click each button see the output for the show ip ospf interface command on each router.

R1 DROTHER

The output generated by R1 confirms that the following:

1. R1 is not the DR or BDR, but is a DROTHER with a default priority of 1. (Line 7)

2. The DR is R3 with router ID 3.3.3.3 at IPv4 address 192.168.1.3, while the BDR is R2 with router ID 2.2.2.2 at IPv4 address 192.168.1.2. (Lines 8 and 9)

3. R1 has two adjacencies: one with the BDR and one with the DR. (Lines 20-22)

```
R1# show ip ospf interface GigabitEthernet 0/0/0
GigabitEthernet0/0/0 is up, line protocol is up
   Internet Address 192.168.1.1/24, Area 0, Attached via Interface Enable
   Process ID 10, Router ID 1.1.1.1, Network Type BROADCAST, Cost: 1
   Topology-MTID    Cost    Disabled    Shutdown    Topology Name
        0            1         no          no           Base
   Enabled by interface config, including secondary ip addresses
   Transmit Delay is 1 sec, State DROTHER, Priority 1
   Designated Router (ID) 3.3.3.3, Interface address 192.168.1.3
   Backup Designated router (ID) 2.2.2.2, Interface address 192.168.1.2
   Timer intervals configured, Hello 10, Dead 40, Wait 40, Retransmit 5
      oob-resync timeout 40
      Hello due in 00:00:07
   Supports Link-local Signaling (LLS)
   Cisco NSF helper support enabled
   IETF NSF helper support enabled
   Index 1/1/1, flood queue length 0
   Next 0x0(0)/0x0(0)/0x0(0)
   Last flood scan length is 0, maximum is 1
   Last flood scan time is 0 msec, maximum is 1 msec
   Neighbor Count is 2, Adjacent neighbor count is 2
      Adjacent with neighbor 2.2.2.2   (Backup Designated Router)
      Adjacent with neighbor 3.3.3.3   (Designated Router)
   Suppress hello for 0 neighbor(s)
R1#
```

R2 BDR

The output generated by R2 confirms that:

1. R2 is the BDR with a default priority of 1. (Line 7)

2. The DR is R3 with router ID 3.3.3.3 at IPv4 address 192.168.1.3, while the BDR is R2 with router ID 2.2.2.2 at IPv4 address 192.168.1.2. (Lines 8 and 9)

3. R2 has two adjacencies; one with a neighbor with router ID 1.1.1.1 (R1) and the other with the DR. (Lines 20-22)

```
R2# show ip ospf interface GigabitEthernet 0/0/0
GigabitEthernet0/0/0 is up, line protocol is up
  Internet Address 192.168.1.2/24, Area 0, Attached via Interface Enable
  Process ID 10, Router ID 2.2.2.2, Network Type BROADCAST, Cost: 1
  Topology-MTID    Cost    Disabled    Shutdown    Topology Name
        0           1        no          no           Base
  Enabled by interface config, including secondary ip addresses
  Transmit Delay is 1 sec, State BDR, Priority 1
  Designated Router (ID) 3.3.3.3, Interface address 192.168.1.3
  Backup Designated router (ID) 2.2.2.2, Interface address 192.168.1.2
  Timer intervals configured, Hello 10, Dead 40, Wait 40, Retransmit 5
    oob-resync timeout 40
    Hello due in 00:00:01
  Supports Link-local Signaling (LLS)
  Cisco NSF helper support enabled
  IETF NSF helper support enabled
  Index 1/1, flood queue length 0
  Next 0x0(0)/0x0(0)
  Last flood scan length is 0, maximum is 1
  Last flood scan time is 0 msec, maximum is 0 msec
  Neighbor Count is 2, Adjacent neighbor count is 2
    Adjacent with neighbor 1.1.1.1
    Adjacent with neighbor 3.3.3.3   (Designated Router)
  Suppress hello for 0 neighbor(s)
R2#
```

R3 DR

The output generated by R3 confirms that:

1. R3 is the DR with a default priority of 1. (Line 7)

2. The DR is R3 with router ID 3.3.3.3 at IPv4 address 192.168.1.3, while the BDR is R2 with router ID 2.2.2.2 at IPv4 address 192.168.1.2. (Lines 8 and 9)

3. R3 has two adjacencies: one with a neighbor with router ID 1.1.1.1 (R1) and the other with the BDR. (Lines 20-22)

```
R3# show ip ospf interface GigabitEthernet 0/0/0
GigabitEthernet0/0/0 is up, line protocol is up
  Internet Address 192.168.1.3/24, Area 0, Attached via Interface Enable
  Process ID 10, Router ID 3.3.3.3, Network Type BROADCAST, Cost: 1
  Topology-MTID    Cost    Disabled    Shutdown       Topology Name
        0            1        no          no              Base
  Enabled by interface config, including secondary ip addresses
  Transmit Delay is 1 sec, State DR, Priority 1
  Designated Router (ID) 3.3.3.3, Interface address 192.168.1.3
  Backup Designated router (ID) 2.2.2.2, Interface address 192.168.1.2
  Timer intervals configured, Hello 10, Dead 40, Wait 40, Retransmit 5
    oob-resync timeout 40
    Hello due in 00:00:06
  Supports Link-local Signaling (LLS)
  Cisco NSF helper support enabled
  IETF NSF helper support enabled
  Index 1/1/1, flood queue length 0
  Next 0x0(0)/0x0(0)/0x0(0)
  Last flood scan length is 2, maximum is 2
  Last flood scan time is 0 msec, maximum is 0 msec
  Neighbor Count is 2, Adjacent neighbor count is 2
    Adjacent with neighbor 1.1.1.1
    Adjacent with neighbor 2.2.2.2   (Backup Designated Router)
  Suppress hello for 0 neighbor(s)
R3#
```

> Refer to
> **Interactive Graphic**
> in online course

Verify DR/BDR Adjacencies - 2.3.5

To verify the OSPFv2 adjacencies, use the **show ip ospf neighbor** command, as shown in the example for R1. The state of neighbors in multiaccess networks can be as follows:

■ **FULL/DROTHER** - This is a DR or BDR router that is fully adjacent with a non-DR or BDR router. These two neighbors can exchange Hello packets, updates, queries, replies, and acknowledgments.

- **FULL/DR** - The router is fully adjacent with the indicated DR neighbor. These two neighbors can exchange Hello packets, updates, queries, replies, and acknowledgments.

- **FULL/BDR** - The router is fully adjacent with the indicated BDR neighbor. These two neighbors can exchange Hello packets, updates, queries, replies, and acknowledgments.

- **2-WAY/DROTHER** - The non-DR or BDR router has a neighbor relationship with another non-DR or BDR router. These two neighbors exchange Hello packets.

The normal state for an OSPF router is usually FULL. If a router is stuck in another state, it is an indication that there are problems in forming adjacencies. The only exception to this is the 2-WAY state, which is normal in a multiaccess broadcast network. For examples, DROTHERs will form a 2-WAY neighbor adjacency with any DROTHERs that join the network. When this happens, the neighbor state displays as 2-WAY/DROTHER.

Click each button see the output for the show ip ospf neighbor command on each router.

R1 Adjacencies

The output generated by R1 confirms that R1 has adjacencies with the following routers:

- R2 with router ID 2.2.2.2 is in a Full state and the role of R2 is BDR.

- R3 with router ID 3.3.3.3 is in a Full state and the role of R3 is DR.

```
R1# show ip ospf neighbor

Neighbor ID   Pri   State      Dead Time   Address       Interface

2.2.2.2        1    FULL/BDR   00:00:31    192.168.1.2   GigabitEthernet0/0/0

3.3.3.3        1    FULL/DR    00:00:39    192.168.1.3   GigabitEthernet0/0/0

R1#
```

R2 Adjacencies

The output generated by R2 confirms that R2 has adjacencies with the following routers:

- R1 with router ID 1.1.1.1 is in a Full state and R1 is neither the DR nor BDR.

- R3 with router ID 3.3.3.3 is in a Full state and the role of R3 is DR.

```
R2# show ip ospf neighbor

Neighbor ID Pri State       Dead Time Address       Interface

1.1.1.1      1  FULL/DROTHER 00:00:31  192.168.1.1   GigabitEthernet0/0/0

3.3.3.3      1  FULL/DR      00:00:34  192.168.1.3   GigabitEthernet0/0/0

R2#
```

R3 Adjacencies

The output generated by R3 confirms that R3 has adjacencies with the following routers:

- R1 with router ID 1.1.1.1 is in a Full state and R1 is neither the DR nor BDR.

- R2 with router ID 2.2.2.2 is in a Full state and the role of R2 is BDR.

```
R3# show ip ospf neighbor

Neighbor ID    Pri  State       Dead Time  Address      nterface
1.1.1.1         1   FULL/DROTHER 00:00:37  192.168.1.1  GigabitEthernet0/0/0
2.2.2.2         1   FULL/BDR     00:00:33  192.168.1.2  GigabitEthernet0/0/0
R3#
```

Refer to
Online Course
for Illustration

Default DR/BDR Election Process - 2.3.6

How do the DR and BDR get elected? The OSPF DR and BDR election decision is based on the following criteria, in sequential order:

1. The routers in the network elect the router with the highest interface priority as the DR. The router with the second highest interface priority is elected as the BDR. The priority can be configured to be any number between 0 – 255. If the interface priority value is set to 0, that interface cannot be elected as DR nor BDR. The default priority of multiaccess broadcast interfaces is 1. Therefore, unless otherwise configured, all routers have an equal priority value and must rely on another tie breaking method during the DR/BDR election.

2. If the interface priorities are equal, then the router with the highest router ID is elected the DR. The router with the second highest router ID is the BDR.

Recall that the router ID is determined in one of the following three ways:

1. The router ID can be manually configured.

2. If no router IDs are configured, the router ID is determined by the highest loopback IPv4 address.

3. If no loopback interfaces are configured, the router ID is determined by the highest active IPv4 address.

In the figure, all Ethernet router interfaces have a default priority of 1. As a result, based on the selection criteria listed above, the OSPF router ID is used to elect the DR and BDR. R3 with the highest router ID becomes the DR; and R2, with the second highest router ID, becomes the BDR.

The DR and BDR election process takes place as soon as the first router with an OSPF-enabled interface is active on the multiaccess network. This can happen when the preconfigured OSPF routers are powered on, or when OSPF is activated on the interface. The election process only takes a few seconds. If all of the routers on the multiaccess network have not finished booting, it is possible that a router with a lower router ID becomes the DR.

OSPF DR and BDR elections are not pre-emptive. If a new router with a higher priority or higher router ID is added to the network after the DR and BDR election, the newly added router does not take over the DR or the BDR role. This is because those roles have already been assigned. The addition of a new router does not initiate a new election process.

Refer to
Interactive Graphic
in online course

DR Failure and Recovery - 2.3.7

Refer to
Online Course
for Illustration

After the DR is elected, it remains the DR until one of the following events occurs:

- The DR fails.
- The OSPF process on the DR fails or is stopped.
- The multiaccess interface on the DR fails or is shutdown.

If the DR fails, the BDR is automatically promoted to DR. This is the case even if another DROTHER with a higher priority or router ID is added to the network after the initial DR/BDR election. However, after a BDR is promoted to DR, a new BDR election occurs and the DROTHER with the highest priority or router ID is elected as the new BDR.

Click each button for an illustration of various scenarios relating to the DR and BDR election process.

R3 Fails

In this scenario, the current DR (R3) fails. Therefore, the pre-elected BDR (R2) assumes the role of DR. Subsequently, an election is held to choose a new BDR. Because R1 is the only DROTHER, it is elected as the BDR.

R3 Re-Joins Network

In this scenario, R3 has re-joined the network after several minutes of being unavailable. Because the DR and BDR already exist, R3 does not take over either role. Instead, it becomes a DROTHER.

R4 Joins Network

In this scenario, a new router (R4) with a higher router ID is added to the network. DR (R2) and BDR (R1) retain the DR and BDR roles. R4 automatically becomes a DROTHER.

R2 Fails

In this scenario, R2 has failed. The BDR (R1) automatically becomes the DR and an election process selects R4 as the BDR because it has the higher router ID.

The ip ospf priority Command - 2.3.8

If the interface priorities are equal on all routers, the router with the highest router ID is elected the DR. It is possible to configure the router ID to manipulate the DR/BDR election. However, this process only works if there is a stringent plan for setting the router ID on all routers. Configuring the router ID can help control this. However, in large networks this can be cumbersome.

Instead of relying on the router ID, it is better to control the election by setting interface priorities. This also allows a router to be the DR in one network and a DROTHER in another. To set the priority of an interface, use the command **ip ospf priority** *value*, where *value* is 0 to 255. A value of 0 does not become a DR or a BDR. A value of 1 to 255 on the interface makes it more likely that the router becomes the DR or the BDR.

Configure OSPF Priority - 2.3.9

In the topology, the **ip ospf priority** command will be used to change the DR and BDR as follows:

- R1 should be the DR and will be configured with a priority of 255.

- R2 should be the BDR and will be left with the default priority of 1.

- R3 should never be a DR or BDR and will be configured with a priority of 0.

Change the R1 G0/0/0 interface priority from 1 to 255.

```
R1(config)# interface GigabitEthernet 0/0/0

R1(config-if)# ip ospf priority 255

R1(config-if)# end

R1#
```

Change the R3 G0/0/0 interface priority from 1 to 0.

```
R3(config)# interface GigabitEthernet 0/0/0

R3(config-if)# ip ospf priority 0

R3(config-if)# end

R3#
```

The following example, shows how to clear the OSPF process on R1. The **clear ip ospf process** command also must be entered on R2 and R3 (not shown). Notice the OSPF state information that is generated.

```
R1# clear ip ospf process

Reset ALL OSPF processes? [no]: y

R1#

*Jun  5 03:47:41.563: %OSPF-5-ADJCHG: Process 10, Nbr 2.2.2.2 on
GigabitEthernet0/0/0 from FULL to DOWN, Neighbor Down: Interface down
or detached

*Jun  5 03:47:41.563: %OSPF-5-ADJCHG: Process 10, Nbr 3.3.3.3 on
GigabitEthernet0/0/0 from FULL to DOWN, Neighbor Down: Interface down
or detached

*Jun  5 03:47:41.569: %OSPF-5-ADJCHG: Process 10, Nbr 2.2.2.2 on
GigabitEthernet0/0/0 from LOADING to FULL, Loading Done

*Jun  5 03:47:41.569: %OSPF-5-ADJCHG: Process 10, Nbr 3.3.3.3 on
GigabitEthernet0/0/0 from LOADING to FULL, Loading Done
```

The output from the **show in ospf interface g0/0/0** command on R1 confirms that R1 is now the DR with a priority of 255 and identifies the new neighbor adjacencies of R1.

```
R1# show ip ospf interface GigabitEthernet 0/0/0
GigabitEthernet0/0/0 is up, line protocol is up
  Internet Address 192.168.1.1/24, Area 0, Attached via Interface Enable
  Process ID 10, Router ID 1.1.1.1, Network Type BROADCAST, Cost: 1
  Topology-MTID    Cost    Disabled    Shutdown      Topology Name
       0            1         no          no             Base
  Enabled by interface config, including secondary ip addresses
  Transmit Delay is 1 sec, State DR, Priority 255
  Designated Router (ID) 1.1.1.1, Interface address 192.168.1.1
  Backup Designated router (ID) 2.2.2.2, Interface address 192.168.1.2
  Timer intervals configured, Hello 10, Dead 40, Wait 40, Retransmit 5
    oob-resync timeout 40
    Hello due in 00:00:00
  Supports Link-local Signaling (LLS)
  Cisco NSF helper support enabled
  IETF NSF helper support enabled
  Index 1/1/1, flood queue length 0
  Next 0x0(0)/0x0(0)/0x0(0)
  Last flood scan length is 1, maximum is 2
  Last flood scan time is 0 msec, maximum is 1 msec
  Neighbor Count is 2, Adjacent neighbor count is 2
    Adjacent with neighbor 2.2.2.2   (Backup Designated Router)
    Adjacent with neighbor 3.3.3.3
  Suppress hello for 0 neighbor(s)
R1#
```

Refer to **Interactive Graphic** in online course

Syntax Checker - Configure OSPF Priority - 2.3.10

Use the Syntax Checker to configure a different OSPF priority scenario for R1, R2, and R3.

Refer to **Packet Tracer Activity** for this chapter

Packet Tracer - Determine the DR and BDR - 2.3.11

In this activity, you will complete the following:

- Examine DR and BDR roles and watch the roles change when there is a change in the network.

- Modify the priority to control the roles and force a new election.

- Verify routers are filling the desired roles.

Modify Single-Area OSPFv2 - 2.4

Refer to
Online Course
for Illustration

Cisco OSPF Cost Metric - 2.4.1

Recall that a routing protocol uses a metric to determine the best path of a packet across a network. A metric gives indication of the overhead that is required to send packets across a certain interface. OSPF uses cost as a metric. A lower cost indicates a better path than a higher cost.

The Cisco cost of an interface is inversely proportional to the bandwidth of the interface. Therefore, a higher bandwidth indicates a lower cost. The formula used to calculate the OSPF cost is:

Cost = reference bandwidth / interface bandwidth

The default reference bandwidth is 10^8 (100,000,000); therefore, the formula is:

Cost = 100,000,000 bps / interface bandwidth in bps

Refer to the table for a breakdown of the cost calculation. Because the OSPF cost value must be an integer, FastEthernet, Gigabit Ethernet, and 10 GigE interfaces share the same cost. To correct this situation, you can:

■ Adjust the reference bandwidth with the **auto-cost reference-bandwidth** command on each OSPF router.

■ Manually set the OSPF cost value with the **ip ospf cost** command on necessary interfaces.

Adjust the Reference Bandwidth - 2.4.2

The cost value must be an integer. If something less than an integer is calculated, OSPF rounds up to the nearest integer. Therefore, the OSPF cost assigned to a Gigabit Ethernet interface with the default reference bandwidth of 100,000,000 bps would equal 1, because the nearest integer for 0.1 is 0 instead of 1.

Cost = 100,000,000 bps / 1,000,000,000 = 1

For this reason, all interfaces faster than Fast Ethernet will have the same cost value of 1 as a Fast Ethernet interface. To assist OSPF in making the correct path determination, the reference bandwidth must be changed to a higher value to accommodate networks with links faster than 100 Mbps.

Changing the reference bandwidth does not actually affect the bandwidth capacity on the link; rather, it simply affects the calculation used to determine the metric. To adjust the reference bandwidth, use the **auto-cost reference-bandwidth** *Mbps* router configuration command.

```
Router(config-router)# auto-cost reference-bandwidth Mbps
```

This command must be configured on every router in the OSPF domain. Notice that the value is expressed in Mbps; therefore, to adjust the costs for Gigabit Ethernet, use the command **auto-cost reference-bandwidth 1000**. For 10 Gigabit Ethernet, use the command **auto-cost reference-bandwidth 10000**.

To return to the default reference bandwidth, use the **auto-cost reference-bandwidth 100** command.

Whichever method is used, it is important to apply the configuration to all routers in the OSPF routing domain. The table shows the OSPF cost if the reference bandwidth is adjusted to accommodate 10 Gigabit Ethernet links. The reference bandwidth should be adjusted anytime there are links faster than FastEthernet (100 Mbps).

Interface Type	Reference Bandwidth in bps	Default Bandwidth in bps	Cost
10 Gigabit Ethernet 10 Gbps	10,000,000,000	÷ 10,000,000,000	1
Gigabit Ethernet 1 Gbps	10,000,000,000	÷ 1,000,000,000	10
Fast Ethernet 100 Mbps	10,000,000,000	÷ 100,000,000	100
Ethernet 10 Mbps	10,000,000,000	÷ 10,000,000	1000

Use the **show ip ospf interface g0/0/0** command to verify the current OSPFv2 cost assigned to the R1 GigabitEthernet 0/0/0 interface. Notice how it displays a cost of 1. Then, after adjusting the reference bandwidth, the cost is now 10. This will allow for scaling to 10 Gigabit Ethernet interfaces in the future without having adjust the reference bandwidth again.

Note: The **auto-cost reference-bandwidth** command must be configured consistently on all routers in the OSPF domain to ensure accurate route calculations.

```
R1# show ip ospf interface gigabitethernet0/0/0
GigabitEthernet0/0/0 is up, line protocol is up
   Internet Address 10.1.1.5/30, Area 0, Attached via Interface Enable
   Process ID 10, Router ID 1.1.1.1, Network Type POINT_TO_POINT, Cost: 1
(output omitted)
R1# config t
Enter configuration commands, one per line.  End with CNTL/Z.
R1(config)# router ospf 10
R1(config-router)# auto-cost reference-bandwidth 10000
% OSPF: Reference bandwidth is changed.
        Please ensure reference bandwidth is consistent across all
routers.
R1(config-router)# do show ip ospf interface gigabitethernet0/0/0
GigabitEthernet0/0 is up, line protocol is up
   Internet address is 172.16.1.1/24, Area 0
   Process ID 10, Router ID 1.1.1.1, Network Type BROADCAST, Cost: 10
   Transmit Delay is 1 sec, State DR, Priority 1
(output omitted)
```

Refer to
Online Course
for Illustration

OSPF Accumulates Costs - 2.4.3

The cost of an OSPF route is the accumulated value from one router to the destination network. Assuming the **auto-cost reference-bandwidth 10000** command has been configured on all three routers, the cost of the links between each router is now 10. The loopback interfaces have a default cost of 1, as shown in the figure.

Therefore, we can calculate the cost for each router to reach each network. For example, the total cost for R1 to reach the 10.10.2.0/24 network is 11. This is because the link to R2 cost = 10 and the loopback default cost = 1. 10 + 1 = 11.

The routing table of R1 in Figure 2 confirms that the metric to reach the R2 LAN is a cost of 11.

```
R1# show ip route | include 10.10.2.0

O       10.10.2.0/24 [110/11] via 10.1.1.6, 01:05:02,
GigabitEthernet0/0/0

R1# show ip route 10.10.2.0

Routing entry for 10.10.2.0/24

  Known via "ospf 10", distance 110, metric 11, type intra area

  Last update from 10.1.1.6 on GigabitEthernet0/0/0, 01:05:13 ago

  Routing Descriptor Blocks:

  * 10.1.1.6, from 2.2.2.2, 01:05:13 ago, via GigabitEthernet0/0/0

      Route metric is 11, traffic share count is 1

R1#
```

Refer to
Online Course
for Illustration

Manually Set OSPF Cost Value - 2.4.4

OSPF cost values can be manipulated to influence the route chosen by OSPF. For example, in the current configuration, R1 is load balancing to the 10.1.1.8/30 network. It will send some traffic to R2 and some traffic to R3. You can see this in the routing table.

```
R1# show ip route ospf | begin 10

      10.0.0.0/8 is variably subnetted, 9 subnets, 3 masks

O       10.1.1.8/30 [110/20] via 10.1.1.13, 00:54:50,
GigabitEthernet0/0/1

                  [110/20] via 10.1.1.6, 00:55:14,
GigabitEthernet0/0/0

(output omitted)

R1#
```

The administrator may want traffic to go through R2 and use R3 as a backup route in case the link between R1 and R2 goes down.

Another reason to change the cost value is because other vendors may calculate OSPF in a different manner. By manipulating the cost value, the administrator can make sure the route costs shared between OSPF multivendor routers are accurately reflected in routing tables.

To change the cost value reported by the local OSPF router to other OSPF routers, use the interface configuration command **ip ospf cost** *value*. In the figure, we need to change cost of the loopback interfaces to 10 to simulate Gigabit Ethernet speeds. In addition, we will change the cost of the link between R2 and R3 to 30 so that this link is used as a backup link.

The following example is the configuration for R1.

```
R1(config)# interface g0/0/1

R1(config-if)# ip ospf cost 30

R1(config-if)# interface lo0

R1(config-if)# ip ospf cost 10

R1(config-if)# end

R1#
```

Assuming OSPF costs for R2 and R3 have been configured to match the topology in the above figure, the OSPF routes for R1 would have the following cost values. Notice that R1 is no longer load balancing to the 10.1.1.8/30 network. In fact, all routes go through R2 as desired by the network administrator.

```
R1# show ip route ospf | begin 10

      10.0.0.0/8 is variably subnetted, 9 subnets, 3 masks

O       10.1.1.8/30 [110/20] via 10.1.1.6, 01:18:25,
GigabitEthernet0/0/0

O       10.10.2.0/24 [110/20] via 10.1.1.6, 00:04:31,
GigabitEthernet0/0/0

O       10.10.3.0/24 [110/30] via 10.1.1.6, 00:03:21,
GigabitEthernet0/0/0

R1#
```

Note: Although using the **ip ospf cost** command is the recommended method to manipulate the OSPF cost values, an administrator could also do this by using the interface configuration **bandwidth** *kbps* command. However, that would only work if all the routers are Cisco routers.

Test Failover to Backup Route - 2.4.5

What happens if the link between R1 and R2 goes down? We can simulate that by shutting down the Gigabit Ethernet 0/0/0 interface and verifying the routing table is updated to use R3 as the next-hop router. Notice that R1 can now reach the 10.1.1.4/30 network through R3 with a cost value of 50.

```
R1(config)# interface g0/0/0

R1(config-if)# shutdown

*Jun  7 03:41:34.866: %OSPF-5-ADJCHG: Process 10, Nbr 2.2.2.2 on
GigabitEthernet0/0/0 from FULL to DOWN, Neighbor Down: Interface down or
detached
```

```
*Jun  7 03:41:36.865: %LINK-5-CHANGED: Interface GigabitEthernet0/0/0,
changed state to administratively down

*Jun  7 03:41:37.865: %LINEPROTO-5-UPDOWN: Line protocol on Interface
GigabitEthernet0/0/0, changed state to down

R1(config-if)# end

R1# show ip route ospf | begin 10
      10.0.0.0/8 is variably subnetted, 8 subnets, 3 masks
O      10.1.1.4/30 [110/50] via 10.1.1.13, 00:00:14, GigabitEthernet0/0/1
O      10.1.1.8/30 [110/40] via 10.1.1.13, 00:00:14, GigabitEthernet0/0/1
O      10.10.2.0/24 [110/50] via 10.1.1.13, 00:00:14, GigabitEthernet0/0/1
O      10.10.3.0/24 [110/40] via 10.1.1.13, 00:00:14, GigabitEthernet0/0/1
R1#
```

Refer to
Interactive Graphic
in online course

Syntax Checker - Modify the Cost Values for R2 and R3 - 2.4.6

Use the Syntax Checker to modify the cost values for R2 and R3.

Refer to
Online Course
for Illustration

Hello Packet Intervals - 2.4.7

As shown in the figure, OSPFv2 Hello packets are transmitted to multicast address 224.0.0.5 (all OSPF routers) every 10 seconds. This is the default timer value on multi-access and point-to-point networks.

Note: Hello packets are not sent on the simulated the LAN interfaces because those interfaces were set to passive by using the router configuration **passive-interface** command.

The Dead interval is the period that the router waits to receive a Hello packet before declaring the neighbor down. If the Dead interval expires before the routers receive a Hello packet, OSPF removes that neighbor from its link-state database (LSDB). The router floods the LSDB with information about the down neighbor out all OSPF-enabled interfaces. Cisco uses a default of 4 times the Hello interval. This is 40 seconds on multiaccess and point-to-point networks.

Note: On non-broadcast multiaccess (NBMA) networks, the default Hello interval is 30 seconds and the default dead interval is 120 seconds. NBMA networks are beyond the scope of this module.

Verify Hello and Dead Intervals - 2.4.8

The OSPF Hello and Dead intervals are configurable on a per-interface basis. The OSPF intervals must match or a neighbor adjacency does not occur. To verify the currently configured OSPFv2 interface intervals, use the **show ip ospf interface** command, as shown in the example. The Gigabit Ethernet 0/0/0 Hello and Dead intervals are set to the default 10 seconds and 40 seconds respectively.

```
R1# show ip ospf interface g0/0/0
GigabitEthernet0/0/0 is up, line protocol is up
   Internet Address 10.1.1.5/30, Area 0, Attached via Interface Enable
   Process ID 10, Router ID 1.1.1.1, Network Type POINT_TO_POINT, Cost: 10
   Topology-MTID    Cost    Disabled    Shutdown    Topology Name
        0            10        no          no           Base
   Enabled by interface config, including secondary ip addresses
   Transmit Delay is 1 sec, State POINT_TO_POINT
   Timer intervals configured, Hello 10, Dead 40, Wait 40, Retransmit 5
     oob-resync timeout 40
     Hello due in 00:00:06
   Supports Link-local Signaling (LLS)
   Cisco NSF helper support enabled
   IETF NSF helper support enabled
   Index 1/2/2, flood queue length 0
   Next 0x0(0)/0x0(0)/0x0(0)
   Last flood scan length is 1, maximum is 1
   Last flood scan time is 0 msec, maximum is 0 msec
   Neighbor Count is 1, Adjacent neighbor count is 1
     Adjacent with neighbor 2.2.2.2
   Suppress hello for 0 neighbor(s)
R1#
```

Use the **show ip ospf neighbor** command to see the Dead Time counting down from 40 seconds, as shown in the following example. By default, this value is refreshed every 10 seconds when R1 receives a Hello from the neighbor.

```
R1# show ip ospf neighbor
Neighbor ID     Pri    State       Dead Time   Address    Interface
3.3.3.3          0     FULL/  -    00:00:35    10.1.1.13  GigabitEthernet0/0/1
2.2.2.2          0     FULL/  -    00:00:31    10.1.1.6   GigabitEthernet0/0/0
R1#
```

Modify OSPFv2 Intervals - 2.4.9

It may be desirable to change the OSPF timers so that routers detect network failures in less time. Doing this increases traffic, but sometimes the need for quick convergence is more important than the extra traffic it creates.

Note: The default Hello and Dead intervals are based on best practices and should only be altered in rare situations.

OSPFv2 Hello and Dead intervals can be modified manually using the following interface configuration mode commands:

```
Router(config-if)# ip ospf hello-interval seconds
```

```
Router(config-if)# ip ospf dead-interval seconds
```

Use the **no ip ospf hello-interval** and **no ip ospf dead-interval** commands to reset the intervals to their default.

In the example, the Hello interval for the link between R1 and R2 is changed to 5 seconds. Immediately after changing the Hello interval, the Cisco IOS automatically modifies the Dead interval to four times the Hello interval. However, you can document the new dead interval in the configuration by manually setting it to 20 seconds, as shown.

As displayed by the highlighted OSPFv2 adjacency message, when the Dead Timer on R1 expires, R1 and R2 lose adjacency. The reason is because the R1 and R2 must be configured with the same Hello interval. Use the **show ip ospf neighbor** command on R1 to verify the neighbor adjacencies. Notice that the only neighbor listed is the 3.3.3.3 (R3) router and that R1 is no longer adjacent with the 2.2.2.2 (R2) neighbor.

```
R1(config)# interface g0/0/0

R1(config-if)# ip ospf hello-interval 5

R1(config-if)# ip ospf dead-interval 20

R1(config-if)#

*Jun  7 04:56:07.571: %OSPF-5-ADJCHG: Process 10, Nbr 2.2.2.2 on
GigabitEthernet0/0/0 from FULL to DOWN, Neighbor Down: Dead timer
expired

R1(config-if)# end

R1# show ip ospf neighbor

Neighbor ID    Pri    State      Dead Time    Address      Interface

3.3.3.3          0    FULL/  -   00:00:37     10.1.1.13    GigabitEthernet0/0/1

R1#
```

To restore adjacency between R1 and R2, the R2 Gigabit Ethernet 0/0/0 interface Hello interval is set to 5 seconds, as shown in the following example. Almost immediately, the IOS displays a message that adjacency has been established with a state of FULL. Verify the interface intervals using the **show ip ospf interface** command. Notice that the Hello time is 5 seconds and that the Dead Time was automatically set to 20 seconds instead of the default 40 seconds.

```
R2(config)# interface g0/0/0

R2(config-if)# ip ospf hello-interval 5

*Jun  7 15:08:30.211: %OSPF-5-ADJCHG: Process 10, Nbr 1.1.1.1 on
GigabitEthernet0/0/0 from LOADING to FULL, Loading Done

R2(config-if)# end

R2# show ip ospf interface g0/0/0 | include Timer

  Timer intervals configured, Hello 5, Dead 20, Wait 20, Retransmit 5

R2# show ip ospf neighbor

Neighbor ID   Pri   State       Dead Time   Address      Interface

3.3.3.3         0   FULL/  -     00:00:38   10.1.1.10    GigabitEthernt0/0/1

1.1.1.1         0   FULL/  -     00:00:17   10.1.1.5     GigabitEthernet0/0/0

R2#
```

Refer to **Interactive Graphic** in online course

Syntax Checker - Modifying Hello and Dead Intervals on R3 - 2.4.10

The Hello and Dead intervals are set to 5 and 20, respectively, on R1 and R2. Use the Syntax Checker to modify the Hello and Dead intervals on R3 and verify adjacencies are re-established with R1 and R2.

Refer to **Packet Tracer Activity** for this chapter

Packet Tracer - Modify Single-Area OSPFv2 - 2.4.11

In this Packet Tracer activity, you will complete the following:

- Adjust the reference bandwidth to account for gigabit and faster speeds
- Modify the OSPF cost value
- Modify the OSPF Hello timers
- Verify the modifications are accurately reflected in the routers.

Default Route Propagation - 2.5

Refer to **Online Course** for Illustration

Propagate a Default Static Route in OSPFv2 - 2.5.1

Your network users will need to send packets out of your network to non-OSPF networks, such as the internet. This is where you will need to have a default static route that they can use. In the topology in the figure, R2 is connected to the internet and should propagate a default route to R1 and R3. The router connected to the internet is sometimes called the edge router or the gateway router. However, in OSPF terminology, the router located between an OSPF routing domain and a non-OSPF network is called the autonomous system boundary router (ASBR).

All that is required for R2 to reach the internet is a default static route to the service provider.

Note: In this example, a loopback interface with IPv4 address 64.100.0.1 is used to simulate the connection to the service provider.

To propagate a default route, the edge router (R2) must be configured with the following:

■ A default static route using the **ip route 0.0.0.0 0.0.0.0** [*next-hop-address* | *exit-intf*] command.

■ The **default-information originate** router configuration command. This instructs R2 to be the source of the default route information and propagate the default static route in OSPF updates.

In the following example, R2 is configured with a loopback to simulate a connection to the internet. Then a default route is configured and propagated to all other OSPF routers in the routing domain.

Note: When configuring static routes, best practice is to use the next-hop IP address. However, when simulating a connection to the internet, there is no next-hop IP address. Therefore, we use the *exit-intf* argument

```
R2(config)# interface lo1

R2(config-if)# ip address 64.100.0.1 255.255.255.252

R2(config-if)# exit

R2(config)# ip route 0.0.0.0 0.0.0.0 loopback 1

%Default route without gateway, if not a point-to-point interface, may
impact performance

R2(config)# router ospf 10

R2(config-router)# default-information originate

R2(config-router)# end

R2#
```

Refer to
Interactive Graphic
in online course

Verify the Propagated Default Route - 2.5.2

You can verify the default route settings on R2 using the **show ip route** command. You can also verify that R1 and R3 received a default route.

Notice that the route source on R1 and R3 is O*E2, signifying that it was learned using OSPFv2. The asterisk identifies this as a good candidate for the default route. The E2 designation identifies that it is an external route. The meaning of E1 and E2 is beyond the scope of this module.

Click each button see the output for the show ip route command on each router.

R2 Routing Table

```
R2# show ip route | begin Gateway
Gateway of last resort is 0.0.0.0 to network 0.0.0.0
S*    0.0.0.0/0 is directly connected, Loopback1
      10.0.0.0/8 is variably subnetted, 9 subnets, 3 masks
C        10.1.1.4/30 is directly connected, GigabitEthernet0/0/0
L        10.1.1.6/32 is directly connected, GigabitEthernet0/0/0
C        10.1.1.8/30 is directly connected, GigabitEthernet0/0/1
L        10.1.1.9/32 is directly connected, GigabitEthernet0/0/1
O        10.1.1.12/30 [110/40] via 10.1.1.10, 00:48:42,
GigabitEthernet0/0/1
                      [110/40] via 10.1.1.5, 00:59:30,
GigabitEthernet0/0/0
O        10.10.1.0/24 [110/20] via 10.1.1.5, 00:59:30, GigabitEthernet0/0/0
C        10.10.2.0/24 is directly connected, Loopback0
L        10.10.2.1/32 is directly connected, Loopback0
O        10.10.3.0/24 [110/20] via 10.1.1.10, 00:48:42,
GigabitEthernet0/0/1
      64.0.0.0/8 is variably subnetted, 2 subnets, 2 masks
C        64.100.0.0/30 is directly connected, Loopback1
L        64.100.0.1/32 is directly connected, Loopback1
R2#
```

R1 Routing Table

```
R1# show ip route | begin Gateway
Gateway of last resort is 10.1.1.6 to network 0.0.0.0
O*E2  0.0.0.0/0 [110/1] via 10.1.1.6, 00:11:08, GigabitEthernet0/0/0
      10.0.0.0/8 is variably subnetted, 9 subnets, 3 masks
C        10.1.1.4/30 is directly connected, GigabitEthernet0/0/0
L        10.1.1.5/32 is directly connected, GigabitEthernet0/0/0
O        10.1.1.8/30 [110/20] via 10.1.1.6, 00:58:59, GigabitEthernet0/0/0
C        10.1.1.12/30 is directly connected, GigabitEthernet0/0/1
L        10.1.1.14/32 is directly connected, GigabitEthernet0/0/1
C        10.10.1.0/24 is directly connected, Loopback0
L        10.10.1.1/32 is directly connected, Loopback0
```

```
O         10.10.2.0/24 [110/20] via 10.1.1.6, 00:58:59,
GigabitEthernet0/0/0

O         10.10.3.0/24 [110/30] via 10.1.1.6, 00:48:11,
GigabitEthernet0/0/0

R1#
```

R3 Routing Table

```
R3# show ip route | begin Gateway

Gateway of last resort is 10.1.1.9 to network 0.0.0.0

O*E2   0.0.0.0/0 [110/1] via 10.1.1.9, 00:12:04, GigabitEthernet0/0/1

         10.0.0.0/8 is variably subnetted, 9 subnets, 3 masks

O         10.1.1.4/30 [110/20] via 10.1.1.9, 00:49:08,
GigabitEthernet0/0/1

C         10.1.1.8/30 is directly connected, GigabitEthernet0/0/1

L         10.1.1.10/32 is directly connected, GigabitEthernet0/0/1

C         10.1.1.12/30 is directly connected, GigabitEthernet0/0/0

L         10.1.1.13/32 is directly connected, GigabitEthernet0/0/0

O         10.10.1.0/24 [110/30] via 10.1.1.9, 00:49:08,
GigabitEthernet0/0/1

O         10.10.2.0/24 [110/20] via 10.1.1.9, 00:49:08,
GigabitEthernet0/0/1

C         10.10.3.0/24 is directly connected, Loopback0

L         10.10.3.1/32 is directly connected, Loopback0

R3#
```

Refer to **Packet Tracer Activity** for this chapter

Packet Tracer - Propagate a Default Route in OSPFv2 - 2.5.3

In this activity, you will configure an IPv4 default route to the internet and propagate that default route to other OSPF routers. You will then verify the default route is in downstream routing tables and that hosts can now access a web server on the internet.

Verify Single-Area OSPFv2 - 2.6

Refer to **Online Course** for Illustration

Verify OSPF Neighbors - 2.6.1

If you have configured single-area OSPFv2, you will need to verify your configurations. This topic details the many commands that you can use to verify OSPF.

As you know, the following two commands are particularly useful for verifying routing:

- **show ip interface brief** - This verifies that the desired interfaces are active with correct IP addressing.

- **show ip route** - This verifies that the routing table contains all the expected routes.

Additional commands for determining that OSPF is operating as expected include the following:

- **show ip ospf neighbor**
- **show ip protocols**
- **show ip ospf**
- **show ip ospf interface**

The figure shows the OSPF reference topology used to demonstrate these commands.

Use the **show ip ospf neighbor** command to verify that the router has formed an adjacency with its neighboring routers. If the router ID of the neighboring router is not displayed, or if it does not show as being in a state of FULL, the two routers have not formed an OSPFv2 adjacency.

If two routers do not establish adjacency, link-state information is not exchanged. Incomplete LSDBs can cause inaccurate SPF trees and routing tables. Routes to destination networks may not exist, or may not be the most optimum path.

Note: A non-DR or BDR router that has a neighbor relationship with another non-DR or BDR router will display a two-way adjacency instead of full.

The following command output displays the neighbor table of R1.

```
R1# show ip ospf neighbor

Neighbor ID  Pri   State      Dead Time   Address     Interface

3.3.3.3        0   FULL/   -  00:00:19    10.1.1.13   GigabitEthernet0/0/1

2.2.2.2        0   FULL/   -  00:00:18    10.1.1.6    abitEthernet0/0/0

R1#
```

For each neighbor, this command displays the following:

- **Neighbor ID** - This is the router ID of the neighboring router.

- **Pri** - This is the OSPFv2 priority of the interface. This value is used in the DR and BDR election.

- **State** - This is the OSPFv2 state of the interface. FULL state means that the router and its neighbor have identical OSPFv2 LSDBs. On multiaccess networks, such as Ethernet, two routers that are adjacent may have their states displayed as 2WAY. The dash indicates that no DR or BDR is required because of the network type.

- **Dead Time** - This is the amount of time remaining that the router waits to receive an OSPFv2 Hello packet from the neighbor before declaring the neighbor down. This value is reset when the interface receives a Hello packet.

- **Address** - This is the IPv4 address of the interface of the neighbor to which this router is directly connected.

- **Interface** - This is the interface on which this router has formed adjacency with the neighbor.

Two routers may not form an OSPFv2 adjacency if the following occurs:

■ The subnet masks do not match, causing the routers to be on separate networks.

■ The OSPFv2 Hello or Dead Timers do not match.

■ The OSPFv2 Network Types do not match.

■ There is a missing or incorrect OSPFv2 network command.

Verify OSPF Protocol Settings - 2.6.2

The **show ip protocols** command is a quick way to verify vital OSPF configuration information, as shown in the following command output. This includes the OSPFv2 process ID, the router ID, interfaces explicitly configured to advertise OSPF routes, the neighbors the router is receiving updates from, and the default administrative distance, which is 110 for OSPF.

```
R1# show ip protocols
*** IP Routing is NSF aware ***
(output omitted)
Routing Protocol is "ospf 10"
  Outgoing update filter list for all interfaces is not set
  Incoming update filter list for all interfaces is not set
  Router ID 1.1.1.1
  Number of areas in this router is 1. 1 normal 0 stub 0 nssa
  Maximum path: 4
  Routing for Networks:
  Routing on Interfaces Configured Explicitly (Area 0):
    Loopback0
    GigabitEthernet0/0/1
    GigabitEthernet0/0/0
  Routing Information Sources:
    Gateway         Distance      Last Update
    3.3.3.3             110       00:09:30
    2.2.2.2             110       00:09:58
  Distance: (default is 110)
R1#
```

Verify OSPF Process Information - 2.6.3

The **show ip ospf** command can also be used to examine the OSPFv2 process ID and router ID, as shown in the following command output. This command displays the OSPFv2 area information and the last time the SPF algorithm was executed.

```
R1# show ip ospf
 Routing Process "ospf 10" with ID 1.1.1.1
 Start time: 00:01:47.390, Time elapsed: 00:12:32.320
 Supports only single TOS(TOS0) routes
 Supports opaque LSA
 Supports Link-local Signaling (LLS)
 Supports area transit capability
 Supports NSSA (compatible with RFC 3101)
 Supports Database Exchange Summary List Optimization (RFC 5243)
 Event-log enabled, Maximum number of events: 1000, Mode: cyclic
 Router is not originating router-LSAs with maximum metric
 Initial SPF schedule delay 5000 msecs
 Minimum hold time between two consecutive SPFs 10000 msecs
 Maximum wait time between two consecutive SPFs 10000 msecs
 Incremental-SPF disabled
 Minimum LSA interval 5 secs
 Minimum LSA arrival 1000 msecs
 LSA group pacing timer 240 secs
 Interface flood pacing timer 33 msecs
 Retransmission pacing timer 66 msecs
 EXCHANGE/LOADING adjacency limit: initial 300, process maximum 300
 Number of external LSA 1. Checksum Sum 0x00A1FF
 Number of opaque AS LSA 0. Checksum Sum 0x000000
 Number of DCbitless external and opaque AS LSA 0
 Number of DoNotAge external and opaque AS LSA 0
 Number of areas in this router is 1. 1 normal 0 stub 0 nssa
 Number of areas transit capable is 0
 External flood list length 0
 IETF NSF helper support enabled
 Cisco NSF helper support enabled
 Reference bandwidth unit is 10000 mbps
    Area BACKBONE(0)
        Number of interfaces in this area is 3
        Area has no authentication
        SPF algorithm last executed 00:11:31.231 ago
        SPF algorithm executed 4 times
```

```
        Area ranges are

        Number of LSA 3. Checksum Sum 0x00E77E

        Number of opaque link LSA 0. Checksum Sum 0x000000

        Number of DCbitless LSA 0

        Number of indication LSA 0

        Number of DoNotAge LSA 0

        Flood list length 0

R1#
```

Verify OSPF Interface Settings - 2.6.4

The **show ip ospf interface** command provides a detailed list for every OSPFv2-enabled interface. Specify an interface to display the settings of just that interface, as shown in the following output for Gigabit Ethernet 0/0/0. This command shows the process ID, the local router ID, the type of network, OSPF cost, DR and BDR information on multiaccess links (not shown), and adjacent neighbors.

```
R1# show ip ospf interface GigabitEthernet 0/0/0

GigabitEthernet0/0/0 is up, line protocol is up

   Internet Address 10.1.1.5/30, Area 0, Attached via Interface Enable

   Process ID 10, Router ID 1.1.1.1, Network Type POINT_TO_POINT, Cost: 10

   Topology-MTID    Cost    Disabled    Shutdown    Topology Name

          0          10       no          no          Base

   Enabled by interface config, including secondary ip addresses

   Transmit Delay is 1 sec, State POINT_TO_POINT

   Timer intervals configured, Hello 5, Dead 20, Wait 20, Retransmit 5

      oob-resync timeout 40

      Hello due in 00:00:01

   Supports Link-local Signaling (LLS)

   Cisco NSF helper support enabled

   IETF NSF helper support enabled

   Index 1/2/2, flood queue length 0

   Next 0x0(0)/0x0(0)/0x0(0)

   Last flood scan length is 1, maximum is 1

   Last flood scan time is 0 msec, maximum is 0 msec

   Neighbor Count is 1, Adjacent neighbor count is 1

      Adjacent with neighbor 2.2.2.2

   Suppress hello for 0 neighbor(s)

R1#
```

To get a quick summary of OSPFv2-enabled interfaces, use the **show ip ospf interface brief** command, as shown in the following command output. This command is useful for seeing important information including the following:

- Interfaces are participating in OSPF

- Networks that are being advertised (IP Address/Mask)

- Cost of each link

- Network state

- Number of neighbors on each link

```
R1# show ip ospf interface brief
Interface    PID    Area    IP Address/Mask    Cost    State Nbrs F/C
Lo0          10     0       10.10.1.1/24       10      P2P   0/0
Gi0/0/1      10     0       10.1.1.14/30       30      P2P   1/1
Gi0/0/0      10     0       10.1.1.5/30        10      P2P   1/1
R1#
```

Refer to
Interactive Graphic
in online course

Syntax Checker - Verify Single-Area OSPFv2 - 2.6.5

Use the Syntax Checker to verify single-area OSPFv2 configuration on R2 and R3.

Refer to **Packet Tracer Activity** for this chapter

Packet Tracer - Verify Single-Area OSPFv2 - 2.6.6

In this Packet Tracer activity, you will use a variety of commands to verify the single-area OSPFv2 configuration.

Module Practice and Quiz - 2.7

Refer to **Packet Tracer Activity** for this chapter

Packet Tracer - Single-Area OSPFv2 Configuration - 2.7.1

You are helping a network engineer test an OSPF set up by building the network in the lab where you work. You have interconnected the devices and configured the interfaces and have connectivity within the local LANs. Your job is to complete the OSPF configuration according to the requirements left by the engineer.

In this Packet Tracer activity, use the information provided and the list of requirements to configure the test network. When the task has been successfully completed, all hosts should be able to ping the internet server.

Refer to
Lab Activity
for this chapter

Lab - Single-Area OSPFv2 Configuration - 2.7.2

In this lab, you will complete the following objectives:

- Part 1: Build the network and configure basic device settings

- Part 2: Configure and verify single-area OSPFv2 for basic operation

- Part 3: Optimize and verify the single-area OSPFv2 configuration

What did I learn in this module? - 2.7.3

OSPF Router ID

OSPFv2 is enabled using the **router ospf** *process-id* global configuration mode command. The *process-id* value represents a number between 1 and 65,535 and is selected by the network administrator. An OSPF router ID is a 32-bit value, represented as an IPv4 address. The router ID is used by an OSPF-enabled router to synchronize OSPF databases and participate in the election of the DR and BDR. Cisco routers derive the router ID based on one of three criteria, in the following preferential order:

1. The router ID is explicitly configured using the OSPF **router-id** *rid* router configuration mode command. The *rid* value is any 32-bit value expressed as an IPv4 address.

2. If the router ID is not explicitly configured, the router chooses the highest IPv4 address of any of configured loopback interfaces.

3. If no loopback interfaces are configured, then the router chooses the highest active IPv4 address of any of its physical interfaces.

The router ID can be assigned to a loopback interface. The IPv4 address for this type of loopback interface should be configured using a 32-bit subnet mask (255.255.255.255), creating a host route. A 32-bit host route would not get advertised as a route to other OSPF routers. After a router selects a router ID, an active OSPF router does not allow the router ID to be changed until the router is reloaded or the OSPF process is reset. Use the **clear ip ospf process** command to reset the adjacencies. You can then verify that R1 is using the new router ID command with the **show ip protocols** command piped to display only the router ID section.

Point-to-Point OSPF Networks

The network command is used to determine which interfaces participate in the routing process for an OSPFv2 area. The basic syntax for the **network** command is **network** *network-address wildcard-mask* **area** *area-id*. Any interfaces on a router that match the network address in the **network** command can send and receive OSPF packets. When configuring single-area OSPFv2, the **network** command must be configured with the same *area-id* value on all routers. The wildcard mask is typically the inverse of the subnet mask configured on that interface. In a wildcard mask:

- **Wildcard mask bit 0** - Matches the corresponding bit value in the address
- **Wildcard mask bit 1** - Ignores the corresponding bit value in the address

Within routing configuration mode, there are two ways to identify the interfaces that will participate in the OSPFv2 routing process. One way is when the wildcard mask identifies the interface based on the network addresses. Any active interface that is configured with an IPv4 address belonging to that network will participate in the OSPFv2 routing process. The other way is OSPFv2 can be enabled by specifying the exact interface IPv4 address using a quad zero wildcard mask. To configure OSPF directly on the interface, use the **ip ospf** interface configuration mode command. The syntax is **ip ospf** *process-id* **area** *area-id*. Sending out unneeded messages on a LAN affects the network through inefficient use of bandwidth and resources, and creates an increased security risk. Use the **passive-interface** router configuration mode command to stop transmitting routing messages through a router interface, but still allow that network to be advertised to other routers. The **show ip protocols** command is then used to verify that the Loopback 0 interface is listed as passive. The DR/ BDR election process is unnecessary as there can only be two routers on the point-to-point network between R1 and R2. Use the interface configuration command **ip ospf network point-to-point** on all interfaces where you want to disable the DR/BDR election process. Use loopbacks to simulate more networks than the equipment can support. By default, loopback interfaces are advertised as /32 host routes. To simulate a real LAN, the Loopback 0 interface is configured as a point-to-point network.

OSPF Network Types

Routers can be connected to the same switch to form a multiaccess network. Ethernet LANs are the most common example of broadcast multiaccess networks. In broadcast networks, all devices on the network see all broadcast and multicast frames. The DR is responsible for collecting and distributing LSAs . The DR uses the multicast IPv4 address 224.0.0.5 which is meant for all OSPF routers. If the DR stops producing Hello packets, the BDR promotes itself and assumes the role of DR. All other routers become a DROTHER. DROTHERs use the multiaccess address 224.0.0.6 (all designated routers) to send OSPF packets to the DR and BDR. Only the DR and BDR listen for 224.0.0.6. To verify the roles of the OSPFv2 router, use the **show ip ospf interface** command. To verify the OSPFv2 adjacencies, use the **show ip ospf neighbor** command. The state of neighbors in multiaccess networks can be:

- **FULL/DROTHER** - This is a DR or BDR router that is fully adjacent with a non-DR or BDR router.

- **FULL/DR** - The router is fully adjacent with the indicated DR neighbor.

- **FULL/BDR** - The router is fully adjacent with the indicated BDR neighbor.

- **2-WAY/DROTHER** - The non-DR or BDR router has a neighbor relationship with another non-DR or BDR router.

The OSPF DR and BDR election decision is based on the following criteria, in sequential order:

1. The routers in the network elect the router with the highest interface priority as the DR. The router with the second highest interface priority is elected as the BDR. The priority can be configured to be any number between 0 – 255. If the interface priority value is set to 0, that interface cannot be elected as DR nor BDR. The default priority of multiaccess broadcast interfaces is 1. Therefore, unless otherwise configured, all routers have an equal priority value and must rely on another tie breaking method during the DR/BDR election.

2. If the interface priorities are equal, then the router with the highest router ID is elected the DR. The router with the second highest router ID is the BDR.

OSPF DR and BDR elections are not pre-emptive. If the DR fails, the BDR is automatically promoted to DR. This is the case even if another DROTHER with a higher priority or router ID is added to the network after the initial DR/BDR election. However, after a BDR is promoted to DR, a new BDR election occurs and the DROTHER with the highest priority or router ID is elected as the new BDR. To set the priority of an interface, use the command **ip ospf priority** *value*, where value is 0 to 255. If the value is 0, the router will not become a DR or BDR. If the value is 1 to 255, then the router with the higher priority value will more likely become the DR or BDR on the interface.

Modify Single-Area OSPFv2

OSPF uses cost as a metric. A lower cost indicates a better path than a higher cost. The Cisco cost of an interface is inversely proportional to the bandwidth of the interface. Therefore, a higher bandwidth indicates a lower cost. The formula used to calculate the OSPF cost is: Cost = reference bandwidth / interface bandwidth. Because the OSPF cost value must be an integer, FastEthernet, Gigabit Ethernet, and 10 GigE interfaces share the same cost. To correct this situation, you can adjust the reference bandwidth with the **auto-cost reference-bandwidth** command on each OSPF router, or manually set the OSPF cost value with the **ip ospf cost** command. To adjust the reference bandwidth, use the **auto-cost reference-bandwidth** *Mbps* router configuration command. The cost of an OSPF route is the accumulated value from one router to the destination network. OSPF cost values can be manipulated to influence the route chosen by OSPF. To change the cost value report by the local OSPF router to other OSPF routers, use the interface configuration command **ip ospf cost** *value*. If the Dead interval expires before the routers receive a Hello packet, OSPF removes that neighbor from its link-state database (LSDB). The router floods the LSDB with information about the down neighbor out all OSPF-enabled interfaces. Cisco uses a default of 4 times the Hello interval or 40 seconds on multiaccess and point-to-point networks. To verify the OSPFv2 interface intervals, use the **show ip ospf interface** command. OSPFv2 Hello and Dead intervals can be modified manually using the following interface configuration mode commands: **ip ospf hello-interval** *seconds* and **ip ospf dead-interval** *seconds*.

Default Route Propagation

In OSPF terminology, the router located between an OSPF routing domain and a non-OSPF network is called the ASBR. To propagate a default route, the ASBR must be configured with a default static route using the **ip route 0.0.0.0 0.0.0.0** *[next-hop-address | exit-intf]* command, and the **default-information originate** router configuration command. This instructs the ASBR to be the source of the default route information and propagate the default static route in OSPF updates. Verify the default route settings on the ASBR using the **show ip route** command.

Verify Single-Area OSPFv2

The following two commands are used to verify routing:

- **show ip interface brief** - Used to verify that the desired interfaces are active with correct IP addressing.

- **show ip route** - Used to verify that the routing table contains all the expected routes.

Additional commands for determining that OSPF is operating as expected include: **show ip ospf neighbor, show ip protocols, show ip ospf,** and **show ip ospf interface**.

Use the **show ip ospf neighbor** command to verify that the router has formed an adjacency with its neighboring routers. For each neighbor, this command displays:

- **Neighbor ID** - The router ID of the neighboring router.

- **Pri** - The OSPFv2 priority of the interface. This value is used in the DR and BDR election.

- **State** - The OSPFv2 state of the interface. FULL state means that the router and its neighbor have identical OSPFv2 LSDBs. On multiaccess networks, such as Ethernet, two routers that are adjacent may have their states displayed as 2WAY. The dash indicates that no DR or BDR is required because of the network type.

- **Dead Time** - The amount of time remaining that the router waits to receive an OSPFv2 Hello packet from the neighbor before declaring the neighbor down. This value is reset when the interface receives a Hello packet.

- **Address** - The IPv4 address of the neighbor's interface to which this router is directly connected.

- **Interface** - The interface on which this router has formed adjacency with the neighbor.

The **show ip protocols** command is a quick way to verify vital OSPF configuration information such as the OSPFv2 process ID, the router ID, interfaces explicitly configured to advertise OSPF routes, the neighbors the router is receiving updates from, and the default administrative distance, which is 110 for OSPF. Use the **show ip ospf** command to examine the OSPFv2 process ID and router ID. This command displays the OSPFv2 area information and the last time the SPF algorithm was executed. The **show ip ospf interface** command provides a detailed list for every OSPFv2-enabled interface. Specify an interface for just one interface to display the process ID, the local router ID, the type of network, OSPF cost, DR and BDR information on multiaccess links, and adjacent neighbors.

Go to the online course to take the quiz and exam.

Chapter Quiz - Single-Area OSPFv2 Configuration

Your Chapter Notes

CHAPTER 3

Network Security Concepts

Introduction - 3.0

Why should I take this module? - 3.0.1

Welcome to Network Security Concepts!

Perhaps you've heard one of the hundreds of news stories about a data security breach within a large corporation or even a government. Was your credit card number exposed by a breach? Your private health information? Would you like to know how to prevent these data breaches? The field of network security is growing every day. This module provides a detailed landscape of the types of cybercrime and the many ways we have to fight back against cyber-criminals. Let's get started!

What will I learn in this module? - 3.0.2

Module Title: Network Security Concepts

Module Objective: Explain how vulnerabilities, threats, and exploits can be mitigated to enhance network security.

Topic	Topic Title
Current State of Cybersecurity	Describe the current state of cybersecurity and vectors of data loss.
Threat Actors	Describe tools used by threat actors to exploit networks.
Malware	Describe malware types.
Common Network Attacks	Describe common network attacks.
IP Vulnerabilities and Threats	Explain how IP vulnerabilities are exploited by threat actors.
TCP and UDP Vulnerabilities	Explain how TCP and UDP vulnerabilities are exploited by threat actors.
IP Services	Explain how IP services are exploited by threat actors.
Network Security Best Practices	Describe best practices for protecting a network.
Cryptography	Describe common cryptographic processes used to protect data in transit.

Refer to
Online Course
for Illustration

Ethical Hacking Statement - 3.0.3

In this module, learners may be exposed to tools and techniques in a "sandboxed", virtual machine environment to demonstrate various types of cyber attacks. Experimentation with these tools, techniques, and resources is at the discretion of the instructor and local institution. If the learner is considering using attack tools for educational purposes, they should contact their instructor prior to any experimentation.

Unauthorized access to data, computer, and network systems is a crime in many jurisdictions and often is accompanied by severe consequences, regardless of the perpetrator's motivations. It is the learner's responsibility, as the user of this material, to be cognizant of and compliant with computer use laws.

Current State of Cybersecurity - 3.1

Current State of Affairs - 3.1.1

Cyber criminals now have the expertise and tools necessary to take down critical infrastructure and systems. Their tools and techniques continue to evolve.

Cyber criminals are taking malware to unprecedented levels of sophistication and impact. They are becoming more adept at using stealth and evasion techniques to hide their activity. Lastly, cyber criminals are exploiting undefended gaps in security.

Network security breaches can disrupt e-commerce, cause the loss of business data, threaten people's privacy, and compromise the integrity of information. These breaches can result in lost revenue for corporations, theft of intellectual property, lawsuits, and can even threaten public safety.

Maintaining a secure network ensures the safety of network users and protects commercial interests. Organizations need individuals who can recognize the speed and scale at which adversaries are amassing and refining their cyber weaponry. All users should be aware of security terms in the table.

Security Terms	Description
Assets	An asset is anything of value to the organization. It includes people, equipment, resources, and data.
Vulnerability	A vulnerability is a weakness in a system, or its design, that could be exploited by a threat.
Threat	A threat is a potential danger to a company's assets, data, or network functionality.
Exploit	An exploit is a mechanism that takes advantage of a vulnerability.
Mitigation	Mitigation is the counter-measure that reduces the likelihood or severity of a potential threat or risk. Network security involves multiple mitigation techniques.
Risk	Risk is the likelihood of a threat to exploit the vulnerability of an asset, with the aim of negatively affecting an organization. Risk is measured using the probability of the occurrence of an event and its consequences.

Assets must be identified and protected. Vulnerabilities must be addressed before they become a threat and are exploited. Mitigation techniques are required before, during, and after an attack.

Vectors of Network Attacks - 3.1.2

Refer to
Online Course
for Illustration

An attack vector is a path by which a threat actor can gain access to a server, host, or network. Attack vectors originate from inside or outside the corporate network, as shown in the figure. For example, threat actors may target a network through the internet, to disrupt network operations and create a denial of service (DoS) attack.

Note: A DoS attack occurs when a network device or application is incapacitated and no longer capable of supporting requests from legitimate users.

An internal user, such as an employee, can accidentally or intentionally:

- Steal and copy confidential data to removable media, email, messaging software, and other media.

- Compromise internal servers or network infrastructure devices.

- Disconnect a critical network connection and cause a network outage.

- Connect an infected USB drive into a corporate computer system.

Internal threats have the potential to cause greater damage than external threats because internal users have direct access to the building and its infrastructure devices. Employees may also have knowledge of the corporate network, its resources, and its confidential data.

Network security professionals must implement tools and apply techniques for mitigating both external and internal threats.

Data Loss - 3.1.3

Data is likely to be an organization's most valuable asset. Organizational data can include research and development data, sales data, financial data, human resource and legal data, employee data, contractor data, and customer data.

Data loss or data exfiltration is when data is intentionally or unintentionally lost, stolen, or leaked to the outside world. The data loss can result in:

- Brand damage and loss of reputation

- Loss of competitive advantage

- Loss of customers

- Loss of revenue

- Litigation/legal action resulting in fines and civil penalties

- Significant cost and effort to notify affected parties and recover from the breach

Common data loss vectors are displayed in the table.

Data Loss Vectors	Description
Email/Social Networking	Intercepted email or IM messages could be captured and reveal confidential information.
Unencrypted Devices	If the data is not stored using an encryption algorithm, then the thief can retrieve valuable confidential data.
Cloud Storage Devices	Sensitive data can be lost if access to the cloud is compromised due to weak security settings.

Data Loss Vectors	Description
Removable Media	One risk is that an employee could perform an unauthorized transfer of data to a USB drive. Another risk is that a USB drive containing valuable corporate data could be lost.
Hard Copy	Confidential data should be shredded when no longer required.
Improper Access Control	Passwords or weak passwords which have been compromised can provide a threat actor with easy access to corporate data.

Network security professionals must protect the organization's data. Various Data Loss Prevention (DLP) controls must be implemented which combine strategic, operational and tactical measures.

Go to the online course to take the quiz and exam.

Check Your Understanding - Current State of Cybersecurity - 3.1.4

Threat Actors - 3.2

The Hacker - 3.2.1

In the previous topic, you gained a high-level look at the current landscape of cybersecurity, including the types of threats and vulnerabilities that plague all network administrators and architects. In this topic, you will learn more details about particular types of threat actors.

Hacker is a common term used to describe a threat actor. As shown in the table, the terms white hat hacker, black hat hacker, and gray hat hacker are often used to describe a type of hacker.

Hacker Type	Description
White Hat Hackers	These are ethical hackers who use their programming skills for good, ethical, and legal purposes. White hat hackers may perform network penetration tests in an attempt to compromise networks and systems by using their knowledge of computer security systems to discover network vulnerabilities. Security vulnerabilities are reported to developers for them to fix before the vulnerabilities can be exploited.
Gray Hat Hackers	These are individuals who commit crimes and do arguably unethical things, but not for personal gain or to cause damage. Gray hat hackers may disclose a vulnerability to the affected organization after having compromised their network.
Black Hat Hackers	These are unethical criminals who compromise computer and network security for personal gain, or for malicious reasons, such as attacking networks.

Note: In this course, we will not use the term hacker outside of this module. We will use the term threat actor. The term threat actor includes hackers. But threat actor also includes any device, person, group, or nation state that is, intentionally or unintentionally, the source of an attack.

Evolution of Hackers - 3.2.2

Hacking started in the 1960s with phone freaking, or phreaking, which refers to using audio frequencies to manipulate phone systems. At that time, telephone switches used various tones to indicate different functions. Early hackers realized that by mimicking a tone using a whistle, they could exploit the phone switches to make free long-distance calls.

In the mid-1980s, computer dial-up modems were used to connect computers to networks. Hackers wrote "war dialing" programs which dialed each telephone number in a given area in search of computers. When a computer was found, password-cracking programs were used to gain access.

The table displays modern hacking terms and a brief description of each.

Hacking Term	Description
Script Kiddies	These are teenagers or inexperienced hackers running existing scripts, tools, and exploits, to cause harm, but typically not for profit.
Vulnerability Broker	These are usually gray hat hackers who attempt to discover exploits and report them to vendors, sometimes for prizes or rewards.
Hacktivists	These are gray hat hackers who publicly protest organizations or governments by posting articles, videos, leaking sensitive information, and performing network attacks.
Cyber criminals	These are black hat hackers who are either self-employed or working for large cybercrime organizations.
State-Sponsored	These are either white hat or black hat hackers who steal government secrets, gather intelligence, and sabotage networks. Their targets are foreign governments, terrorist groups, and corporations. Most countries in the world participate to some degree in state-sponsored hacking.

Cyber Criminals - 3.2.3

It is estimated that cyber criminals steal billions of dollars from consumers and businesses. Cyber criminals operate in an underground economy where they buy, sell, and trade attack toolkits, zero day exploit code, botnet services, banking Trojans, keyloggers, and much more. They also buy and sell the private information and intellectual property they steal. Cyber criminals target small businesses and consumers, as well as large enterprises and entire industries.

Hacktivists - 3.2.4

Two examples of hacktivist groups are Anonymous and the Syrian Electronic Army. Although most hacktivist groups are not well organized, they can cause significant problems for governments and businesses. Hacktivists tend to rely on fairly basic, freely available tools.

State-Sponsored Hackers - 3.2.5

State-sponsored hackers create advanced, customized attack code, often using previously undiscovered software vulnerabilities called zero-day vulnerabilities. An example of a state-sponsored attack involves the Stuxnet malware that was created to damage Iran's nuclear enrichment capabilities.

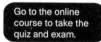
Go to the online course to take the quiz and exam.

Check Your Understanding - Threat Actors - 3.2.6

Threat Actor Tools - 3.3

Refer to **Video** in online course

Video - Threat Actor Tools - 3.3.1

As you learned in the previous topic, there are different types of hackers with different motivations for what they do. In this topic, you will learn about some of the tools these individuals use.

Click Play in the figure to view a video about threat actor tools.

Refer to **Interactive Graphic** in online course

Introduction to Attack Tools - 3.3.2

To exploit a vulnerability, a threat actor must have a technique or tool. Over the years, attack tools have become more sophisticated, and highly automated. These new tools require less technical knowledge to implement.

In the figure, drag the white circle across the timeline to view the relationship between the sophistication of attack tools versus the technical knowledge required to use them.

Evolution of Security Tools - 3.3.3

Ethical hacking involves many different types of tools used to test the network and keep its data secure. To validate the security of a network and its systems, many network penetration testing tools have been developed. It is unfortunate that many of these tools can be used by black hat hackers for exploitation.

Black hat hackers have also created many hacking tools. These tools are created explicitly for nefarious reasons. White hat hackers must also know how to use these tools when performing network penetration tests.

The table highlights categories of common penetration testing tools. Notice how some tools are used by white hats and black hats. Keep in mind that the list is not exhaustive as new tools are always being developed.

Penetration Testing Tool	Description
Password Crackers	Password cracking tools are often referred to as password recovery tools and can be used to crack or recover a password. This is accomplished either by removing the original password, after bypassing the data encryption, or by outright discovery of the password. Password crackers repeatedly make guesses in order to crack the password. Examples of password cracking tools include John the Ripper, Ophcrack, L0phtCrack, THC Hydra, RainbowCrack, and Medusa.
Wireless Hacking Tools	Wireless hacking tools are used to intentionally hack into a wireless network to detect security vulnerabilities. Examples of wireless hacking tools include Aircrack-ng, Kismet, InSSIDer, KisMAC, Firesheep, and NetStumbler.

Penetration Testing Tool	Description
Network Scanning and Hacking Tools	Network scanning tools are used to probe network devices, servers, and hosts for open TCP or UDP ports. Examples of scanning tools include Nmap, SuperScan, Angry IP Scanner, and NetScanTools.
Packet Crafting Tools	These tools are used to probe and test a firewall's robustness using specially crafted forged packets. Examples include Hping, Scapy, Socat, Yersinia, Netcat, Nping, and Nemesis.
Packet Sniffers	These tools are used to capture and analyze packets within traditional Ethernet LANs or WLANs. Tools include Wireshark, Tcpdump, Ettercap, Dsniff, EtherApe, Paros, Fiddler, Ratproxy, and SSLstrip.
Rootkit Detectors	This is a directory and file integrity checker used by white hats to detect installed root kits. Example tools include AIDE, Netfilter, and PF: OpenBSD Packet Filter.
Fuzzers to Search Vulnerabilities	Fuzzers are tools used by threat actors to discover a computer's security vulnerabilities. Examples of fuzzers include Skipfish, Wapiti, and W3af.
Forensic Tools	These tools are used by white hat hackers to sniff out any trace of evidence existing in a computer. Example of tools include Sleuth Kit, Helix, Maltego, and Encase.
Debuggers	These tools are used by black hats to reverse engineer binary files when writing exploits. They are also used by white hats when analyzing malware. Debugging tools include GDB, WinDbg, IDA Pro, and Immunity Debugger.
Hacking Operating Systems	These are specially designed operating systems preloaded with tools optimized for hacking. Examples of specially designed hacking operating systems include Kali Linux, Knoppix, BackBox Linux.
Encryption Tools	Encryption tools use algorithm schemes to encode the data to prevent unauthorized access to the encrypted data. Examples of these tools include VeraCrypt, CipherShed, OpenSSH, OpenSSL, Tor, OpenVPN, and Stunnel.
Vulnerability Exploitation Tools	These tools identify whether a remote host is vulnerable to a security attack. Examples of vulnerability exploitation tools include Metasploit, Core Impact, Sqlmap, Social Engineer Toolkit, and Netsparker.
Vulnerability Scanners	These tools scan a network or system to identify open ports. They can also be used to scan for known vulnerabilities and scan VMs, BYOD devices, and client databases. Examples of tools include Nipper, Secunia PSI, Core Impact, Nessus v6, SAINT, and Open VAS.

Note: Many of these tools are UNIX or Linux based; therefore, a security professional should have a strong UNIX and Linux background.

Attack Types - 3.3.4

Threat actors can use the previously mentioned attack tools, or a combination of tools, to create attacks. The table displays common types of attacks. However, the list of attacks is not exhaustive as new attack vulnerabilities are constantly being discovered.

Attack Type	Description
Eavesdropping Attack	This is when a threat actor captures and "listens" to network traffic. This attack is also referred to as sniffing or snooping.
Data Modification Attack	If threat actors have captured enterprise traffic, they can alter the data in the packet without the knowledge of the sender or receiver.
IP Address Spoofing Attack	A threat actor constructs an IP packet that appears to originate from a valid address inside the corporate intranet.
Password-Based Attacks	If threat actors discover a valid user account, the threat actors have the same rights as the real user. Threat actors could use that valid account to obtain lists of other users, network information, change server and network configurations, and modify, reroute, or delete data.
Denial of Service Attack	A DoS attack prevents normal use of a computer or network by valid users. A DoS attack can flood a computer or the entire network with traffic until a shutdown occurs because of the overload. A DoS attack can also block traffic, which results in a loss of access to network resources by authorized users.
Man-in-the-Middle Attack	This attack occurs when threat actors have positioned themselves between a source and destination. They can now actively monitor, capture, and control the communication transparently.
Compromised-Key Attack	If a threat actor obtains a secret key, that key is referred to as a compromised key. A compromised key can be used to gain access to a secured communication without the sender or receiver being aware of the attack.
Sniffer Attack	A sniffer is an application or device that can read, monitor, and capture network data exchanges and read network packets. If the packets are not encrypted, a sniffer provides a full view of the data inside the packet.

Go to the online course to take the quiz and exam.

Check Your Understanding - Threat Actor Tools - 3.3.5

Malware - 3.4

Refer to Interactive Graphic in online course

Overview of Malware - 3.4.1

Now that you know about the tools that hacker use, this topic introduces you to different types of malware that hackers use to gain access to end devices.

End devices are particularly prone to malware attacks. It is important to know about malware because threat actors rely on users to install malware to help exploit the security gaps.

Click Play to view an animation of the three most common types of malware.

Viruses and Trojan Horses - 3.4.2

The first and most common type of computer malware is a virus. Viruses require human action to propagate and infect other computers. For example, a virus can infect a computer when a victim opens an email attachment, opens a file on a USB drive, or downloads a file.

The virus hides by attaching itself to computer code, software, or documents on the computer. When opened, the virus executes and infects the computer.

Viruses can:

- Alter, corrupt, delete files, or erase entire drives.
- Cause computer booting issues, and corrupt applications.
- Capture and send sensitive information to threat actors.
- Access and use email accounts to spread.
- Lay dormant until summoned by the threat actor.

Modern viruses are developed for specific intent such as those listed in the table.

Types of Viruses	Description
Boot sector virus	Virus attacks the boot sector, file partition table, or file system.
Firmware virus	Virus attacks the device firmware.
Macro virus	Virus uses the MS Office or other applications macro feature maliciously.
Program virus	Virus inserts itself in another executable program.
Script virus	Virus attacks the OS interpreter which is used to execute scripts.

Threat actors use Trojan horses to compromise hosts. A Trojan horse is a program that looks useful but also carries malicious code. Trojan horses are often provided with free online programs such as computer games. Unsuspecting users download and install the game, along with the Trojan horse.

There are several types of Trojan horses as described in the table.

Type of Trojan Horse	Description
Remote-access	Trojan horse enables unauthorized remote access.
Data-sending	Trojan horse provides the threat actor with sensitive data, such as passwords.
Destructive	Trojan horse corrupts or deletes files.
Proxy	Trojan horse will use the victim's computer as the source device to launch attacks and perform other illegal activities.
FTP	Trojan horse enables unauthorized file transfer services on end devices.
Security software disabler	Trojan horse stops antivirus programs or firewalls from functioning.
Denial of Service (DoS)	Trojan horse slows or halts network activity.
Keylogger	Trojan horse actively attempts to steal confidential information, such as credit card numbers, by recording key strokes entered into a web form.

Viruses and Trojan horses are only two types of malware that threat actors use. There are many other types of malware that have been designed for specific purposes.

Other Types of Malware - 3.4.3

The table shows details about many different types of malware.

Malware	Description
Adware	• Adware is usually distributed by downloading online software. • Adware can display unsolicited advertising using pop-up web browser windows, new toolbars, or unexpectedly redirect a webpage to a different website. • Pop-up windows may be difficult to control as new windows can pop-up faster than the user can close them.
Ransomware	• Ransomware typically denies a user access to their files by encrypting the files and then displaying a message demanding a ransom for the decryption key. • Users without up-to-date backups must pay the ransom to decrypt their files. • Payment is usually made using wire transfer or crypto currencies such as Bitcoin.
Rootkit	• Rootkits are used by threat actors to gain administrator account-level access to a computer. • They are very difficult to detect because they can alter firewall, antivirus protection, system files, and even OS commands to conceal their presence. • They can provide a backdoor to threat actors giving them access to the PC, and allowing them to upload files, and install new software to be used in a DDoS attack. • Special rootkit removal tools must be used to remove them, or a complete OS re-install may be required.
Spyware	• Similar to adware, but used to gather information about the user and send to threat actors without the user's consent. • Spyware can be a low threat, gathering browsing data, or it can be a high threat capturing personal and financial information.
Worm	• A worm is a self-replicating program that propagates automatically without user actions by exploiting vulnerabilities in legitimate software. • It uses the network to search for other victims with the same vulnerability. • The intent of a worm is usually to slow or disrupt network operations.

Go to the online course to take the quiz and exam.

Check Your Understanding - Malware - 3.4.4

Common Network Attacks - 3.5

Overview of Network Attacks - 3.5.1

As you have learned, there are many types of malware that hackers can use. But these are not the only ways that they can attack a network, or even an organization.

When malware is delivered and installed, the payload can be used to cause a variety of network related attacks.

To mitigate attacks, it is useful to understand the types of attacks. By categorizing network attacks, it is possible to address types of attacks rather than individual attacks.

Networks are susceptible to the following types of attacks:

■ Reconnaissance Attacks

■ Access Attacks

■ DoS Attacks

Refer to **Video**
in online course

Video - Reconnaissance Attacks - 3.5.2

Click Play in the figure to view a video about reconnaissance attacks.

Refer to
Interactive Graphic
in online course

Reconnaissance Attacks - 3.5.3

Reconnaissance is information gathering. It is analogous to a thief surveying a neighborhood by going door-to-door pretending to sell something. What the thief is actually doing is looking for vulnerable homes to break into, such as unoccupied residences, residences with easy-to-open doors or windows, and those residences without security systems or security cameras.

Threat actors use reconnaissance (or recon) attacks to do unauthorized discovery and mapping of systems, services, or vulnerabilities. Recon attacks precede access attacks or DoS attacks.

Some of the techniques used by malicious threat actors to conduct reconnaissance attacks are described in the table.

Technique	Description
Perform an information query of a target	The threat actor is looking for initial information about a target. Various tools can be used, including the Google search, organizations website, whois, and more.
Initiate a ping sweep of the target network	The information query usually reveals the target's network address. The threat actor can now initiate a ping sweep to determine which IP addresses are active.
Initiate a port scan of active IP addresses	This is used to determine which ports or services are available. Examples of port scanners include Nmap, SuperScan, Angry IP Scanner, and NetScanTools.
Run vulnerability scanners	This is to query the identified ports to determine the type and version of the application and operating system that is running on the host. Examples of tools include Nipper, Secuna PSI, Core Impact, Nessus v6, SAINT, and Open VAS.
Run exploitation tools	The threat actor now attempts to discover vulnerable services that can be exploited. A variety of vulnerability exploitation tools exist including Metasploit, Core Impact, Sqlmap, Social Engineer Toolkit, and Netsparker.

Click each button to view the progress of a reconnaissance attack from information query, to ping sweep, to port scan.

Internet Information Queries

Click Play in the figure to view an animation of a threat actor using the whois command to find information about a target.

Performing Ping Sweeps

Click Play in the figure to view an animation of a threat actor doing a ping sweep of the target's network address to discover live and active IP addresses.

Performing Port Scans

Click Play in the figure to view an animation of a threat actor performing a port scan on the discovered active IP addresses using Nmap.

Refer to **Video**
in online course

Video - Access and Social Engineering Attacks - 3.5.4

Click Play in the figure to view a video about access and social engineering attacks.

Refer to
Interactive Graphic
in online course

Access Attacks - 3.5.5

Refer to
Online Course
for Illustration

Access attacks exploit known vulnerabilities in authentication services, FTP services, and web services. The purpose of these types of attacks is to gain entry to web accounts, confidential databases, and other sensitive information.

Threat actors use access attacks on network devices and computers to retrieve data, gain access, or to escalate access privileges to administrator status.

Password Attacks

In a password attack, the threat actor attempts to discover critical system passwords using various methods. Password attacks are very common and can be launched using a variety of password cracking tools.

Spoofing Attacks

In spoofing attacks, the threat actor device attempts to pose as another device by falsifying data. Common spoofing attacks include IP spoofing, MAC spoofing, and DHCP spoofing. These spoofing attacks will be discussed in more detail later in this module

Other Access attacks include:

- Trust exploitations
- Port redirections
- Man-in-the-middle attacks
- Buffer overflow attacks

Click each button to view an illustration and explanation of these access attacks.

Trust Exploitation Example

In a trust exploitation attack, a threat actor uses unauthorized privileges to gain access to a system, possibly compromising the target. Click Play in the figure to view an example of trust exploitation.

Port Redirection Example

In a port redirection attack, a threat actor uses a compromised system as a base for attacks against other targets. The example in the figure shows a threat actor using SSH (port 22) to connect to a compromised Host A. Host A is trusted by Host B and, therefore, the threat actor can use Telnet (port 23) to access it.

Man-in-the-Middle Attack Example

In a man-in-the-middle attack, the threat actor is positioned in between two legitimate entities in order to read or modify the data that passes between the two parties. The figure displays an example of a man-in-the-middle attack.

Buffer Overflow Attack

In a buffer overflow attack, the threat actor exploits the buffer memory and overwhelms it with unexpected values. This usually renders the system inoperable, creating a DoS attack. The figure shows that the threat actor is sending many packets to the victim in an attempt to overflow the victim's buffer.

Refer to **Online Course** for Illustration

Social Engineering Attacks - 3.5.6

Social engineering is an access attack that attempts to manipulate individuals into performing actions or divulging confidential information. Some social engineering techniques are performed in-person while others may use the telephone or internet.

Social engineers often rely on people's willingness to be helpful. They also prey on people's weaknesses. For example, a threat actor could call an authorized employee with an urgent problem that requires immediate network access. The threat actor could appeal to the employee's vanity, invoke authority using name-dropping techniques, or appeal to the employee's greed.

Information about social engineering techniques is shown in the table.

Social Engineering Attack	Description
Pretexting	A threat actor pretends to need personal or financial data to confirm the identity of the recipient.
Phishing	A threat actor sends fraudulent email which is disguised as being from a legitimate, trusted source to trick the recipient into installing malware on their device, or to share personal or financial information.
Spear phishing	A threat actor creates a targeted phishing attack tailored for a specific individual or organization.
Spam	Also known as junk mail, this is unsolicited email which often contains harmful links, malware, or deceptive content.
Something for Something	Sometimes called "Quid pro quo", this is when a threat actor requests personal information from a party in exchange for something such as a gift.
Baiting	A threat actor leaves a malware infected flash drive in a public location. A victim finds the drive and unsuspectingly inserts it into their laptop, unintentionally installing malware.
Impersonation	This type of attack is where a threat actor pretends to be someone they are not to gain the trust of a victim.
Tailgating	This is where a threat actor quickly follows an authorized person into a secure location to gain access to a secure area.
Shoulder surfing	This is where a threat actor inconspicuously looks over someone's shoulder to steal their passwords or other information.
Dumpster diving	This is where a threat actor rummages through trash bins to discover confidential documents.

The Social Engineering Toolkit (SET) was designed to help white hat hackers and other network security professionals create social engineering attacks to test their own networks.

Enterprises must educate their users about the risks of social engineering, and develop strategies to validate identities over the phone, via email, or in person.

The figure shows recommended practices that should be followed by all users.

Refer to **Lab Activity** for this chapter

Lab - Social Engineering - 3.5.7

In this lab, you will research examples of social engineering and identify ways to recognize and prevent it.

Refer to **Video** in online course

Video - Denial of Service Attacks - 3.5.8

Click Play in the figure to view a video about denial of service attacks.

Refer to **Interactive Graphic** in online course

DoS and DDoS Attacks - 3.5.9

A Denial of Service (DoS) attack creates some sort of interruption of network services to users, devices, or applications. There are two major types of DoS attacks:

- **Overwhelming Quantity of Traffic** - The threat actor sends an enormous quantity of data at a rate that the network, host, or application cannot handle. This causes transmission and response times to slow down. It can also crash a device or service.

- **Maliciously Formatted Packets** - The threat actor sends a maliciously formatted packet to a host or application and the receiver is unable to handle it. This causes the receiving device to run very slowly or crash.

Click each button for an illustration and explanation of DoS and DDoS attacks.

DoS Attack

DoS attacks are a major risk because they interrupt communication and cause significant loss of time and money. These attacks are relatively simple to conduct, even by an unskilled threat actor.

Click Play in the figure to view the animation of a DoS attack.

DDoS Attack

A Distributed DoS Attack (DDoS) is similar to a DoS attack, but it originates from multiple, coordinated sources. For example, A threat actor builds a network of infected hosts, known as zombies. The threat actor uses a command and control (CnC) system to send control messages to the zombies. The zombies constantly scan and infect more hosts with bot malware. The bot malware is designed to infect a host, making it a zombie that can communicate with the CnC system. The collection of zombies is called a botnet. When ready, the threat actor instructs the CnC system to make the botnet of zombies carry out a DDoS attack.

Click Play in the figure to view the animations of a DDoS attack.

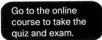

Go to the online
course to take the
quiz and exam.

Check Your Understanding - Common Network Attacks - 3.5.10

IP Vulnerabilities and Threats - 3.6

Refer to **Video**
in online course

Video - Common IP and ICMP Attacks - 3.6.1

There are even more types of attacks than the ones discussed in the previous topics. Some specifically target IP vulnerabilities, as you will learn in this topic.

Click Play in the figure to view a video about common IP and ICMP attacks.

IPv4 and IPv6 - 3.6.2

IP does not validate whether the source IP address contained in a packet actually came from that source. For this reason, threat actors can send packets using a spoofed source IP address. Threat actors can also tamper with the other fields in the IP header to carry out their attacks. Security analysts must understand the different fields in both the IPv4 and IPv6 headers.

Some of the more common IP related attacks are shown in the table.

IP Attack Techniques	Description
ICMP attacks	Threat actors use Internet Control Message Protocol (ICMP) echo packets (pings) to discover subnets and hosts on a protected network, to generate DoS flood attacks, and to alter host routing tables.
Amplification and reflection attacks	Threat actors attempt to prevent legitimate users from accessing information or services using DoS and DDoS attacks.
Address spoofing attacks	Threat actors spoof the source IP address in an IP packet to perform blind spoofing or non-blind spoofing.
Man-in-the-middle attack (MITM)	Threat actors position themselves between a source and destination to transparently monitor, capture, and control the communication. They could eavesdrop by inspecting captured packets, or alter packets and forward them to their original destination.
Session hijacking	Threat actors gain access to the physical network, and then use an MITM attack to hijack a session.

ICMP Attacks - 3.6.3

Threat actors use ICMP for reconnaissance and scanning attacks. They can launch information-gathering attacks to map out a network topology, discover which hosts are active (reachable), identify the host operating system (OS fingerprinting), and determine the state of a firewall. Threat actors also use ICMP for DoS attacks.

Note: ICMP for IPv4 (ICMPv4) and ICMP for IPv6 (ICMPv6) are susceptible to similar types of attacks.

Networks should have strict ICMP access control list (ACL) filtering on the network edge to avoid ICMP probing from the internet. Security analysts should be able to detect ICMP-related attacks by looking at captured traffic and log files. In the case of large networks, security devices such as firewalls and intrusion detection systems (IDS) detect such attacks and generate alerts to the security analysts.

Common ICMP messages of interest to threat actors are listed in the table.

ICMP Messages used by Hackers	Description
ICMP echo request and echo reply	This is used to perform host verification and DoS attacks.
ICMP unreachable	This is used to perform network reconnaissance and scanning attacks.
ICMP mask reply	This is used to map an internal IP network.
ICMP redirects	This is used to lure a target host into sending all traffic through a compromised device and create a MITM attack.
ICMP router discovery	This is used to inject bogus route entries into the routing table of a target host.

Refer to **Video** in online course

Video - Amplification, Reflection, and Spoofing Attacks - 3.6.4

Click Play in the figure to view a video about amplification, reflection, and spoofing attacks.

Refer to **Online Course** for Illustration

Amplification and Reflection Attacks - 3.6.5

Threat actors often use amplification and reflection techniques to create DoS attacks. The example in the figure illustrates how an amplification and reflection technique called a Smurf attack is used to overwhelm a target host.

Note: Newer forms of amplification and reflection attacks such as DNS-based reflection and amplification attacks and Network Time Protocol (NTP) amplification attacks are now being used.

Threat actors also use resource exhaustion attacks. These attacks consume the resources of a target host to either to crash it or to consume the resources of a network.

Refer to **Online Course** for Illustration

Address Spoofing Attacks - 3.6.6

IP address spoofing attacks occur when a threat actor creates packets with false source IP address information to either hide the identity of the sender, or to pose as another legitimate user. The threat actor can then gain access to otherwise inaccessible data or circumvent security configurations. Spoofing is usually incorporated into another attack such as a Smurf attack.

Spoofing attacks can be non-blind or blind:

- **Non-blind spoofing** - The threat actor can see the traffic that is being sent between the host and the target. The threat actor uses non-blind spoofing to inspect the reply packet from the target victim. Non-blind spoofing determines the state of a firewall and sequence-number prediction. It can also hijack an authorized session.

- **Blind spoofing** - The threat actor cannot see the traffic that is being sent between the host and the target. Blind spoofing is used in DoS attacks.

MAC address spoofing attacks are used when threat actors have access to the internal network. Threat actors alter the MAC address of their host to match another known MAC address of a target host, as shown in the figure. The attacking host then sends a frame throughout the network with the newly-configured MAC address. When the switch receives the frame, it examines the source MAC address.

The switch overwrites the current CAM table entry and assigns the MAC address to the new port, as shown in the figure. It then forwards frames destined for the target host to the attacking host.

Application or service spoofing is another spoofing example. A threat actor can connect a rogue DHCP server to create an MITM condition.

Go to the online course to take the quiz and exam.

Check Your Understanding - IP Vulnerabilities and Threats - 3.6.7

TCP and UDP Vulnerabilities - 3.7

Refer to **Online Course** for Illustration

TCP Segment Header - 3.7.1

While some attacks target IP, this topic discusses attacks that target TCP and UDP.

TCP segment information appears immediately after the IP header. The fields of the TCP segment and the flags for the Control Bits field are displayed in the figure.

Refer to **Online Course** for Illustration

TCP Services - 3.7.2

TCP provides these services:

- **Reliable delivery** - TCP incorporates acknowledgments to guarantee delivery, instead of relying on upper-layer protocols to detect and resolve errors. If a timely acknowledgment is not received, the sender retransmits the data. Requiring acknowledgments of received data can cause substantial delays. Examples of application layer protocols that make use of TCP reliability include HTTP, SSL/TLS, FTP, DNS zone transfers, and others.

- **Flow control** - TCP implements flow control to address this issue. Rather than acknowledge one segment at a time, multiple segments can be acknowledged with a single acknowledgment segment.

- **Stateful communication** - TCP stateful communication between two parties occurs during the TCP three-way handshake. Before data can be transferred using TCP, a three-way handshake opens the TCP connection, as shown in the figure. If both sides agree to the TCP connection, data can be sent and received by both parties using TCP.

Refer to
Online Course
for Illustration

TCP Attacks - 3.7.3

Network applications use TCP or UDP ports. Threat actors conduct port scans of target devices to discover which services they offer.

TCP SYN Flood Attack

The TCP SYN Flood attack exploits the TCP three-way handshake. The figure shows a threat actor continually sending TCP SYN session request packets with a randomly spoofed source IP address to a target. The target device replies with a TCP SYN-ACK packet to the spoofed IP address and waits for a TCP ACK packet. Those responses never arrive. Eventually the target host is overwhelmed with half-open TCP connections, and TCP services are denied to legitimate users.

TCP Reset Attack

A TCP reset attack can be used to terminate TCP communications between two hosts. The figure displays how TCP uses a four-way exchange to close the TCP connection using a pair of FIN and ACK segments from each TCP endpoint. A TCP connection terminates when it receives an RST bit. This is an abrupt way to tear down the TCP connection and inform the receiving host to immediately stop using the TCP connection. A threat actor could do a TCP reset attack and send a spoofed packet containing a TCP RST to one or both endpoints.

TCP Session Hijacking

TCP session hijacking is another TCP vulnerability. Although difficult to conduct, a threat actor takes over an already-authenticated host as it communicates with the target. The threat actor must spoof the IP address of one host, predict the next sequence number, and send an ACK to the other host. If successful, the threat actor could send, but not receive, data from the target device.

Refer to
Online Course
for Illustration

UDP Segment Header and Operation - 3.7.4

UDP is commonly used by DNS, TFTP, NFS, and SNMP. It is also used with real-time applications such as media streaming or VoIP. UDP is a connectionless transport layer protocol. It has much lower overhead than TCP because it is not connection-oriented and does not offer the sophisticated retransmission, sequencing, and flow control mechanisms that provide reliability. The UDP segment structure, shown in the figure, is much smaller than TCP's segment structure.

Although UPD is normally called unreliable, in contrast to TCP's reliability, this does not mean that applications that use UDP are always unreliable, nor does it mean that UDP is an inferior protocol. It means that these functions are not provided by the transport layer protocol and must be implemented elsewhere if required.

The low overhead of UDP makes it very desirable for protocols that make simple request and reply transactions. For example, using TCP for DHCP would introduce unnecessary network traffic. If no response is received, the device resends the request.

UDP Attacks - 3.7.5

UDP is not protected by any encryption. You can add encryption to UDP, but it is not available by default. The lack of encryption means that anyone can see the traffic, change it, and send it on to its destination. Changing the data in the traffic will alter the 16-bit checksum, but the checksum is optional and is not always used. When the checksum is used, the threat actor can create a new checksum based on the new data payload, and then record it in the header as a new checksum. The destination device will find that the checksum matches the data without knowing that the data has been altered. This type of attack is not widely used.

UDP Flood Attacks

You are more likely to see a UDP flood attack. In a UDP flood attack, all the resources on a network are consumed. The threat actor must use a tool like UDP Unicorn or Low Orbit Ion Cannon. These tools send a flood of UDP packets, often from a spoofed host, to a server on the subnet. The program will sweep through all the known ports trying to find closed ports. This will cause the server to reply with an ICMP port unreachable message. Because there are many closed ports on the server, this creates a lot of traffic on the segment, which uses up most of the bandwidth. The result is very similar to a DoS attack.

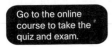
Go to the online course to take the quiz and exam.

Check Your Understanding - TCP and UDP Vulnerabilities - 3.7.6

IP Services - 3.8

Refer to **Interactive Graphic** in online course

ARP Vulnerabilities - 3.8.1

Earlier in this module you learned about vulnerabilities with IP, TCP and UDP. The TCP/IP protocol suite was never built for security. Therefore, the services that IP uses for addressing functions such as ARP, DNS, and DHCP, are also not secure, as you will learn in this topic.

Hosts broadcast an ARP Request to other hosts on the segment to determine the MAC address of a host with a particular IP address. All hosts on the subnet receive and process the ARP Request. The host with the matching IP address in the ARP Request sends an ARP Reply.

Click Play in the figure to see the ARP process at work.

Any client can send an unsolicited ARP Reply called a "gratuitous ARP." This is often done when a device first boots up to inform all other devices on the local network of the new device's MAC address. When a host sends a gratuitous ARP, other hosts on the subnet store the MAC address and IP address contained in the gratuitous ARP in their ARP tables.

This feature of ARP also means that any host can claim to be the owner of any IP or MAC. A threat actor can poison the ARP cache of devices on the local network, creating an MITM attack to redirect traffic. The goal is to target a victim host, and have it change its default gateway to the threat actor's device. This positions the threat actor in between the victim and all other systems outside of the local subnet.

Refer to
Interactive Graphic
in online course

Refer to
Online Course
for Illustration

ARP Cache Poisoning - 3.8.2

ARP cache poisoning can be used to launch various man-in-the-middle attacks.

Click each button for an illustration and an explanation of the ARP cache poisoning process.

ARP Request

The figure shows how ARP cache poisoning works. PC-A requires the MAC address of its default gateway (R1); therefore, it sends an ARP Request for the MAC address of 192.168.10.1.

ARP Reply

In this figure, R1 updates its ARP cache with the IP and MAC addresses of PC-A. R1 sends an ARP Reply to PC-A, which then updates its ARP cache with the IP and MAC addresses of R1.

Spoofed Gratuitous ARP Replies

In the figure, the threat actor sends two spoofed gratuitous ARP Replies using its own MAC address for the indicated destination IP addresses. PC-A updates its ARP cache with its default gateway which is now pointing to the threat actor's host MAC address. R1 also updates its ARP cache with the IP address of PC-A pointing to the threat actor's MAC address.

The threat actor's host is executing an ARP poisoning attack. The ARP poisoning attack can be passive or active. Passive ARP poisoning is where threat actors steal confidential information. Active ARP poisoning is where threat actors modify data in transit, or inject malicious data.

Note: There are many tools available on the internet to create ARP MITM attacks including dsniff, Cain & Abel, ettercap, Yersinia, and others.

Refer to **Video**
in online course

Video - ARP Spoofing - 3.8.3

Click Play in the figure to view a video about ARP Spoofing.

DNS Attacks - 3.8.4

The Domain Name Service (DNS) protocol defines an automated service that matches resource names, such as www.cisco.com, with the required numeric network address, such as the IPv4 or IPv6 address. It includes the format for queries, responses, and data and uses resource records (RR) to identify the type of DNS response.

Securing DNS is often overlooked. However, it is crucial to the operation of a network and should be secured accordingly.

DNS attacks include the following:

- DNS open resolver attacks
- DNS stealth attacks

- DNS domain shadowing attacks
- DNS tunneling attacks

DNS Open Resolver Attacks

Many organizations use the services of publicly open DNS servers such as GoogleDNS (8.8.8.8) to provide responses to queries. This type of DNS server is called an open resolver. A DNS open resolver answers queries from clients outside of its administrative domain. DNS open resolvers are vulnerable to multiple malicious activities described in the table.

DNS Resolver Vulnerabilities	Description
DNS cache poisoning attacks	Threat actors send spoofed, falsified record resource (RR) information to a DNS resolver to redirect users from legitimate sites to malicious sites. DNS cache poisoning attacks can all be used to inform the DNS resolver to use a malicious name server that is providing RR information for malicious activities.
DNS amplification and reflection attacks	Threat actors use DoS or DDoS attacks on DNS open resolvers to increase the volume of attacks and to hide the true source of an attack. Threat actors send DNS messages to the open resolvers using the IP address of a target host. These attacks are possible because the open resolver will respond to queries from anyone asking a question.
DNS resource utilization attacks	A DoS attack that consumes the resources of the DNS open resolvers. This DoS attack consumes all the available resources to negatively affect the operations of the DNS open resolver. The impact of this DoS attack may require the DNS open resolver to be rebooted or services to be stopped and restarted.

DNS Stealth Attacks

To hide their identity, threat actors also use the DNS stealth techniques described in the table to carry out their attacks.

DNS Stealth Techniques	Description
Fast Flux	Threat actors use this technique to hide their phishing and malware delivery sites behind a quickly-changing network of compromised DNS hosts. The DNS IP addresses are continuously changed within minutes. Botnets often employ Fast Flux techniques to effectively hide malicious servers from being detected.
Double IP Flux	Threat actors use this technique to rapidly change the hostname to IP address mappings and to also change the authoritative name server. This increases the difficulty of identifying the source of the attack.
Domain Generation Algorithms	Threat actors use this technique in malware to randomly generate domain names that can then be used as rendezvous points to their command and control (C&C) servers.

DNS Domain Shadowing Attacks

Domain shadowing involves the threat actor gathering domain account credentials in order to silently create multiple sub-domains to be used during the attacks. These subdomains typically point to malicious servers without alerting the actual owner of the parent domain.

DNS Tunneling - 3.8.5

Threat actors who use DNS tunneling place non-DNS traffic within DNS traffic. This method often circumvents security solutions when a threat actor wishes to communicate with bots inside a protected network, or exfiltrate data from the organization, such as a password database. When the threat actor uses DNS tunneling, the different types of DNS records are altered. This is how DNS tunneling works for CnC commands sent to a botnet:

1. The command data is split into multiple encoded chunks.

2. Each chunk is placed into a lower level domain name label of the DNS query.

3. Because there is no response from the local or networked DNS for the query, the request is sent to the ISP's recursive DNS servers.

4. The recursive DNS service will forward the query to the threat actor's authoritative name server.

5. The process is repeated until all the queries containing the chunks of are sent.

6. When the threat actor's authoritative name server receives the DNS queries from the infected devices, it sends responses for each DNS query, which contain the encapsulated, encoded CnC commands.

7. The malware on the compromised host recombines the chunks and executes the commands hidden within the DNS record.

To stop DNS tunneling, the network administrator must use a filter that inspects DNS traffic. Pay close attention to DNS queries that are longer than average, or those that have a suspicious domain name. DNS solutions, like Cisco OpenDNS, block much of the DNS tunneling traffic by identifying suspicious domains.

Refer to
Online Course
for Illustration

DHCP - 3.8.6

DHCP servers dynamically provide IP configuration information to clients. The figure shows the typical sequence of a DHCP message exchange between client and server.

Refer to
Interactive Graphic
in online course

DHCP Attacks - 3.8.7

Refer to
Online Course
for Illustration

DHCP Spoofing Attack

A DHCP spoofing attack occurs when a rogue DHCP server is connected to the network and provides false IP configuration parameters to legitimate clients. A rogue server can provide a variety of misleading information:

- **Wrong default gateway** - Threat actor provides an invalid gateway, or the IP address of its host to create a MITM attack. This may go entirely undetected as the intruder intercepts the data flow through the network.

- **Wrong DNS server** - Threat actor provides an incorrect DNS server address pointing the user to a malicious website.

- **Wrong IP address** - Threat actor provides an invalid IP address, invalid default gateway IP address, or both. The threat actor then creates a DoS attack on the DHCP client.

Assume a threat actor has successfully connected a rogue DHCP server to a switch port on the same subnet as the target clients. The goal of the rogue server is to provide clients with false IP configuration information.

Click each button for an illustration and explanation of the steps in a DHCP spoofing attack.

1. Client Broadcasts DHCP Discovery Messages

In the figure, a legitimate client connects to the network and requires IP configuration parameters. The client broadcasts a DHCP Discover request looking for a response from a DHCP server. Both servers receive the message.

2. DHCP Servers Respond with Offers

The figure shows how the legitimate and rogue DHCP servers each respond with valid IP configuration parameters. The client replies to the first offer received.

3. Client Accepts Rogue DHCP Request

In this scenario, the client received the rogue offer first. It broadcasts a DHCP request accepting the parameters from the rogue server, as shown in the figure. The legitimate and rogue server each receive the request.

4. Rogue DHCP Acknowledges the Request

However, only the rogue server unicasts a reply to the client to acknowledge its request, as shown in the figure. The legitimate server stops communicating with the client because the request has already been acknowledged.

Refer to
Lab Activity
for this chapter

Lab - Explore DNS Traffic - 3.8.8

In this lab, you will complete the following objectives:

- Capture DNS Traffic

- Explore DNS Query Traffic

- Explore DNS Response Traffic

Network Security Best Practices - 3.9

Refer to **Online Course** for Illustration

Confidentiality, Integrity, and Availability - 3.9.1

It is true that the list of network attack types is long. But there are many best practices that you can use to defend your network, as you will learn in this topic.

Network security consists of protecting information and information systems from unauthorized access, use, disclosure, disruption, modification, or destruction.

Most organizations follow the CIA information security triad:

- **Confidentiality** - Only authorized individuals, entities, or processes can access sensitive information. It may require using cryptographic encryption algorithms such as AES to encrypt and decrypt data.

- **Integrity** - Refers to protecting data from unauthorized alteration. It requires the use of cryptographic hashing algorithms such as SHA.

- **Availability** - Authorized users must have uninterrupted access to important resources and data. It requires implementing redundant services, gateways, and links.

Refer to **Online Course** for Illustration

The Defense-in-Depth Approach - 3.9.2

To ensure secure communications across both public and private networks, you must secure devices including routers, switches, servers, and hosts. Most organizations employ a defense-in-depth approach to security. This is also known as a layered approach. It requires a combination of networking devices and services working together. Consider the network in the figure.

All network devices including the router and switches are hardened, which means that they have been secured to prevent threat actors from gaining access and tampering with the devices.

Next, you must secure the data as it travels across various links. This may include internal traffic, but it is more important to protect the data that travels outside of the organization to branch sites, telecommuter sites, and partner sites.

Refer to **Interactive Graphic** in online course

Firewalls - 3.9.3

A firewall is a system, or group of systems, that enforces an access control policy between networks. Click Play in the figure to view an animation of how a firewall operates.

All firewalls share some common properties:

- Firewalls are resistant to network attacks.

- Firewalls are the only transit points between internal corporate networks and external networks because all traffic flows through the firewall.

- Firewalls enforce the access control policy.

There are several benefits of using a firewall in a network:

- They prevent the exposure of sensitive hosts, resources, and applications to untrusted users.

- They sanitize protocol flow, which prevents the exploitation of protocol flaws.

- They block malicious data from servers and clients.

- They reduce security management complexity by off-loading most of the network access control to a few firewalls in the network.

Firewalls also present some limitations:

- A misconfigured firewall can have serious consequences for the network, such as becoming a single point of failure.

- The data from many applications cannot be passed through firewalls securely.

- Users might proactively search for ways around the firewall to receive blocked material, which exposes the network to potential attack.

- Network performance can slow down.

- Unauthorized traffic can be tunneled or hidden so that it appears as legitimate traffic through the firewall.

IPS - 3.9.4

Refer to **Online Course** for Illustration

To defend against fast-moving and evolving attacks, you may need cost-effective detection and prevention systems, such as intrusion detection systems (IDS), or the more scalable intrusion prevention systems (IPS). The network architecture integrates these solutions into the entry and exit points of the network.

IDS and IPS technologies share several characteristics, as shown in the figure. IDS and IPS technologies are both deployed as sensors. An IDS or IPS sensor can be in the form of several different devices:

- A router configured with Cisco IOS IPS software

- A device specifically designed to provide dedicated IDS or IPS services

- A network module installed in an adaptive security appliance (ASA), switch, or router

IDS and IPS technologies detect patterns in network traffic using signatures. A signature is a set of rules that an IDS or IPS uses to detect malicious activity. Signatures can be used to detect severe breaches of security, to detect common network attacks, and to gather information. IDS and IPS technologies can detect atomic signature patterns (single-packet) or composite signature patterns (multi-packet).

Content Security Appliances - 3.9.5

Refer to **Online Course** for Illustration

Content security appliances include fine-grained control over email and web browsing for an organization's users.

Cisco Email Security Appliance (ESA)

The Cisco Email Security Appliance (ESA) is a special device designed to monitor Simple Mail Transfer Protocol (SMTP). The Cisco ESA is constantly updated by real-time feeds from the Cisco Talos, which detects and correlates threats and solutions by using a world-wide database monitoring system. This threat intelligence data is pulled by the Cisco ESA every three to five minutes.

In the figure, a threat actor sends a phishing email.

Cisco Web Security Appliance (WSA)

The Cisco Web Security Appliance (WSA) is a mitigation technology for web-based threats. It helps organizations address the challenges of securing and controlling web traffic. The Cisco WSA combines advanced malware protection, application visibility and control, acceptable use policy controls, and reporting.

Cisco WSA provides complete control over how users access the internet. Certain features and applications, such as chat, messaging, video and audio, can be allowed, restricted with time and bandwidth limits, or blocked, according to the organization's requirements. The WSA can perform blacklisting of URLs, URL-filtering, malware scanning, URL categorization, web application filtering, and encryption and decryption of web traffic.

In the figure, a corporate user attempts to connect to a known blacklisted site.

Go to the online course to take the quiz and exam.

Check Your Understanding - Network Security Best Practices - 3.9.6

Cryptography - 3.10

Refer to Video in online course

Video - Cryptography - 3.10.1

Early in the previous topic, cryptography is mentioned as part of the CIA information security triad. In this topic you will get a deeper dive into the many types of cryptography and how they are used to secure the network.

Click Play in the figure to view a video about cryptography.

Securing Communications - 3.10.2

Organizations must provide support to secure the data as it travels across links. This may include internal traffic, but it is even more important to protect the data that travels outside of the organization to branch sites, telecommuter sites, and partner sites.

These are the four elements of secure communications:

- **Data Integrity** - Guarantees that the message was not altered. Any changes to data in transit will be detected. Integrity is ensured by implementing either Message Digest version 5 (MD5) or Secure Hash Algorithm (SHA) hash-generating algorithms.

- **Origin Authentication** - Guarantees that the message is not a forgery and does actually come from whom it states. Many modern networks ensure authentication with protocols, such as hash message authentication code (HMAC).

- **Data Confidentiality** - Guarantees that only authorized users can read the message. If the message is intercepted, it cannot be deciphered within a reasonable amount of time. Data confidentiality is implemented using symmetric and asymmetric encryption algorithms.

- **Data Non-Repudiation** - Guarantees that the sender cannot repudiate, or refute, the validity of a message sent. Nonrepudiation relies on the fact that only the sender has the unique characteristics or signature for how that message is treated.

Cryptography can be used almost anywhere that there is data communication. In fact, the trend is toward all communication being encrypted.

Refer to **Online Course** for Illustration

Data Integrity - 3.10.3

Hash functions are used to ensure the integrity of a message. They guarantee that message data has not changed accidentally or intentionally.

In the figure, the sender is sending a $100 money transfer to Alex.

Refer to **Online Course** for Illustration

Hash Functions - 3.10.4

There are three well-known hash functions.

MD5 with 128-bit Digest

MD5 is a one-way function that produces a 128-bit hashed message, as shown in the figure. MD5 is a legacy algorithm that should only be used when no better alternatives are available. Use SHA-2 instead.

SHA Hashing Algorithm

SHA-1 is very similar to the MD5 hash functions, as shown in the figure. Several versions exist. SHA-1 creates a 160-bit hashed message and is slightly slower than MD5. SHA-1 has known flaws and is a legacy algorithm. Use SHA-2 when possible.

SHA-2

This includes SHA-224 (224 bit), SHA-256 (256 bit), SHA-384 (384 bit), and SHA-512 (512 bit). SHA-256, SHA-384, and SHA-512 are next-generation algorithms and should be used whenever possible.

While hashing can be used to detect accidental changes, it cannot be used to guard against deliberate changes. There is no unique identifying information from the sender in the hashing procedure. This means that anyone can compute a hash for any data, if they have the correct hash function.

For example, when the message traverses the network, a potential threat actor could intercept the message, change it, recalculate the hash, and append it to the message. The receiving device will only validate against whatever hash is appended.

Therefore, hashing is vulnerable to man-in-the-middle attacks and does not provide security to transmitted data. To provide integrity and origin authentication, something more is required.

Refer to
Interactive Graphic
in online course

Refer to
Online Course
for Illustration

Origin Authentication - 3.10.5

To add authentication to integrity assurance, use a keyed-hash message authentication code (HMAC). HMAC uses an additional secret key as input to the hash function.

Click each button for an illustration and explanation about origin authentication using HMAC.

HMAC Hashing Algorithm

As shown in the figure, an HMAC is calculated using any cryptographic algorithm that combines a cryptographic hash function with a secret key. Hash functions are the basis of the protection mechanism of HMACs.

Only the sender and the receiver know the secret key, and the output of the hash function now depends on the input data and the secret key. Only parties who have access to that secret key can compute the digest of an HMAC function. This defeats man-in-the-middle attacks and provides authentication of the data origin.

If two parties share a secret key and use HMAC functions for authentication, a properly constructed HMAC digest of a message that a party has received indicates that the other party was the originator of the message. This is because the other party possesses the secret key.

Creating the HMAC Value

As shown in the figure, the sending device inputs data (such as Terry Smith's pay of $100 and the secret key) into the hashing algorithm and calculates the fixed-length HMAC digest. This authenticated digest is then attached to the message and sent to the receiver.

Verifying the HMAC Value

In the figure, the receiving device removes the digest from the message and uses the plaintext message with its secret key as input into the same hashing function. If the digest that is calculated by the receiving device is equal to the digest that was sent, the message has not been altered. Additionally, the origin of the message is authenticated because only the sender possesses a copy of the shared secret key. The HMAC function has ensured the authenticity of the message.

Cisco Router HMAC Example

The figure shows how HMACs are used by Cisco routers that are configured to use Open Shortest Path First (OSPF) routing authentication.

R1 is sending a link state update (LSU) regarding a route to network 10.2.0.0/16:

1. R1 calculates the hash value using the LSU message and the secret key.

2. The resulting hash value is sent with the LSU to R2.

3. R2 calculates the hash value using the LSU and its secret key. R2 accepts the update if the hash values match. If they do not match, R2 discards the update.

Refer to **Online Course** for Illustration

Data Confidentiality - 3.10.6

There are two classes of encryption used to provide data confidentiality. These two classes differ in how they use keys.

Symmetric encryption algorithms such as (DES), 3DES, and Advanced Encryption Standard (AES) are based on the premise that each communicating party knows the pre-shared key. Data confidentiality can also be ensured using asymmetric algorithms, including Rivest, Shamir, and Adleman (RSA) and the public key infrastructure (PKI).

The figure highlights some differences between each encryption algorithm method.

Refer to **Online Course** for Illustration

Symmetric Encryption - 3.10.7

Symmetric algorithms use the same pre-shared key to encrypt and decrypt data. A pre-shared key, also called a secret key, is known by the sender and receiver before any encrypted communications can take place.

To help illustrate how symmetric encryption works, consider an example where Alice and Bob live in different locations and want to exchange secret messages with one another through the mail system. In this example, Alice wants to send a secret message to Bob.

In the figure, Alice and Bob have identical keys to a single padlock. These keys were exchanged prior to sending any secret messages. Alice writes a secret message and puts it in a small box that she locks using the padlock with her key. She mails the box to Bob. The message is safely locked inside the box as the box makes its way through the post office system. When Bob receives the box, he uses his key to unlock the padlock and retrieve the message. Bob can use the same box and padlock to send a secret reply to Alice.

Today, symmetric encryption algorithms are commonly used with VPN traffic. This is because symmetric algorithms use less CPU resources than asymmetric encryption algorithms. Encryption and decryption of data is fast when using a VPN. When using symmetric encryption algorithms, like any other type of encryption, the longer the key, the longer it will take for someone to discover the key. Most encryption keys are between 112 and 256 bits. To ensure that the encryption is safe, use a minimum key length of 128 bits. Use a longer key for more secure communications.

Well-known symmetric encryption algorithms are described in the table.

Symmetric Encryption Algorithms	Description
Data Encryption Standard (DES)	This is a legacy symmetric encryption algorithm. It can be used in stream cipher mode but usually operates in block mode by encrypting data in 64-bit block size. A stream cipher encrypts one byte or one bit at a time.
3DES (Triple DES)	This is a newer version of DES, but it repeats the DES algorithm process three times. The basic algorithm has been well tested in the field for more than 35 years. It is considered very trustworthy when implemented using very short key lifetimes.

Symmetric Encryption Algorithms	Description
Advanced Encryption Standard (AES)	AES is a secure and more efficient algorithm than 3DES. It is a popular and recommended symmetric encryption algorithm. It offers nine combinations of key and block length by using a variable key length of 128-, 192-, or 256-bit key to encrypt data blocks that are 128, 192, or 256 bits long.
Software-Optimized Encryption Algorithm (SEAL)	SEAL is a faster alternative symmetric encryption algorithm to DES, 3DES, and AES. It uses a 160-bit encryption key and has a lower impact on the CPU compared to other software-based algorithms.
Rivest ciphers (RC) series algorithms	This algorithm was developed by Ron Rivest. Several variations have been developed, but RC4 is the most prevalent in use. RC4 is a stream cipher and is used to secure web traffic in SSL and TLS.

Refer to
Online Course
for Illustration

Asymmetric Encryption - 3.10.8

Asymmetric algorithms, also called public-key algorithms, are designed so that the key that is used for encryption is different from the key that is used for decryption, as shown in the figure. The decryption key cannot, in any reasonable amount of time, be calculated from the encryption key and vice versa.

Asymmetric algorithms use a public key and a private key. Both keys are capable of the encryption process, but the complementary paired key is required for decryption. The process is also reversible. Data encrypted with the public key requires the private key to decrypt. Asymmetric algorithms achieve confidentiality, authentication, and integrity by using this process.

Because neither party has a shared secret, very long key lengths must be used. Asymmetric encryption can use key lengths between 512 to 4,096 bits. Key lengths greater than or equal to 1,024 bits can be trusted while shorter key lengths are considered unreliable.

Examples of protocols that use asymmetric key algorithms include:

- **Internet Key Exchange (IKE)** - This is a fundamental component of IPsec VPNs.

- **Secure Socket Layer (SSL)** - This is now implemented as IETF standard Transport Layer Security (TLS).

- **Secure Shell (SSH)** - This protocol provides a secure remote access connection to network devices.

- **Pretty Good Privacy (PGP)** - This computer program provides cryptographic privacy and authentication. It is often used to increase the security of email communications.

Asymmetric algorithms are substantially slower than symmetric algorithms. Their design is based on computational problems, such as factoring extremely large numbers or computing discrete logarithms of extremely large numbers.

Because they are slow, asymmetric algorithms are typically used in low-volume cryptographic mechanisms, such as digital signatures and key exchange. However, the key management of asymmetric algorithms tends to be simpler than symmetric algorithms, because usually one of the two encryption or decryption keys can be made public.

Common examples of asymmetric encryption algorithms are described in the table.

Asymmetric Encryption Algorithm	Key Length	Description
Diffie-Hellman (DH)	512, 1024, 2048, 3072, 4096	The Diffie-Hellman algorithm allows two parties to agree on a key that they can use to encrypt messages they want to send to each other. The security of this algorithm depends on the assumption that it is easy to raise a number to a certain power, but difficult to compute which power was used given the number and the outcome.
Digital Signature Standard (DSS) and Digital Signature Algorithm (DSA)	512 - 1024	DSS specifies DSA as the algorithm for digital signatures. DSA is a public key algorithm based on the ElGamal signature scheme. Signature creation speed is similar to RSA, but is 10 to 40 times slower for verification.
Rivest, Shamir, and Adleman encryption algorithms (RSA)	512 to 2048	RSA is for public-key cryptography that is based on the current difficulty of factoring very large numbers. It is the first algorithm known to be suitable for signing as well as encryption. It is widely used in electronic commerce protocols and is believed to be secure given sufficiently long keys and the use of up-to-date implementations.
ElGamal	512 - 1024	An asymmetric key encryption algorithm for public-key cryptography which is based on the Diffie-Hellman key agreement. A disadvantage of the ElGamal system is that the encrypted message becomes very big, about twice the size of the original message and for this reason it is only used for small messages such as secret keys.
Elliptical curve techniques	160	Elliptic curve cryptography can be used to adapt many cryptographic algorithms, such as Diffie-Hellman or ElGamal. The main advantage of elliptic curve cryptography is that the keys can be much smaller.

Diffie-Hellman - 3.10.9

Refer to **Online Course** for Illustration

Diffie-Hellman (DH) is an asymmetric mathematical algorithm where two computers generate an identical shared secret key without having communicated before. The new shared key is never actually exchanged between the sender and receiver. However, because both parties know it, the key can be used by an encryption algorithm to encrypt traffic between the two systems.

Here are three examples of instances when DH is commonly used:

- Data is exchanged using an IPsec VPN.

- Data is encrypted on the internet using either SSL or TLS.

- SSH data is exchanged.

To help illustrate how DH operates, refer to the figure.

The colors in the figure will be used instead of complex long numbers to simplify the DH key agreement process. The DH key exchange begins with Alice and Bob agreeing on an arbitrary common color that does not need to be kept secret. The agreed on color in our example is yellow.

Next, Alice and Bob will each select a secret color. Alice chose red while Bob chose blue. These secret colors will never be shared with anyone. The secret color represents the chosen secret private key of each party.

Alice and Bob now mix the shared common color (yellow) with their respective secret color to produce a private color. Therefore, Alice will mix the yellow with her red color to produce a private color of orange. Bob will mix the yellow and the blue to produce a private color of green.

Alice sends her private color (orange) to Bob and Bob sends his private color (green) to Alice.

Alice and Bob each mix the color they received with their own, original secret color (Red for Alice and blue for Bob.). The result is a final brown color mixture that is identical to the other's final color mixture. The brown color represents the resulting shared secret key between Bob and Alice.

DH security uses unbelievably large numbers in its calculations. For example, a DH 1024-bit number is roughly equal to a decimal number of 309 digits. Considering that a billion is 10 decimal digits (1,000,000,000), one can easily imagine the complexity of working with not one, but many 309-digit decimal numbers.

Unfortunately, asymmetric key systems are extremely slow for any sort of bulk encryption. Therefore, it is common to encrypt the bulk of the traffic using a symmetric algorithm, such as 3DES or AES and then use the DH algorithm to create keys that will be used by the encryption algorithm.

Go to the online course to take the quiz and exam.

Check Your Understanding - Cryptography - 3.10.10

Module Practice and Quiz - 3.11

What did I learn in this module? - 3.11.1

Network security breaches can disrupt e-commerce, cause the loss of business data, threaten people's privacy, and compromise the integrity of information. Assets must be identified and protected. Vulnerabilities must be addressed before they become a threat and are exploited. Mitigation techniques are required before, during, and after an attack. An attack vector is a path by which a threat actor can gain access to a server, host, or network. Attack vectors originate from inside or outside the corporate network.

The term 'threat actor' includes hackers and any device, person, group, or nation state that is, intentionally or unintentionally, the source of an attack. There are "White Hat", "Gray Hat", and "Black Hat" hackers. Cyber criminals operate in an underground economy where they buy, sell, and trade attack toolkits, zero day exploit code, botnet services, banking

Trojans, keyloggers, and more. Hacktivists tend to rely on fairly basic, freely available tools. State-sponsored hackers create advanced, customized attack code, often using previously undiscovered software vulnerabilities called zero-day vulnerabilities.

Attack tools have become more sophisticated and highly automated. These new tools require less technical knowledge to implement. Ethical hacking involves many different types of tools used to test the network and keep its data secure. To validate the security of a network and its systems, many network penetration testing tools have been developed. Common types of attacks are: eavesdropping, data modification, IP address spoofing, password-based, denial-of-service, man-in-the-middle, compromised-key, and sniffer.

The three most common types of malware are worms, viruses, and Trojan horses. A worm executes arbitrary code and installs copies of itself in the memory of the infected computer. A virus executes a specific unwanted, and often harmful, function on a computer. A Trojan horse is non-self-replicating. When an infected application or file is downloaded and opened, the Trojan horse can attack the end device from within. Other types of malware are: adware, ransomware, rootkit, and spyware.

Networks are susceptible to the following types of attacks: reconnaissance, access, and DoS. Threat actors use reconnaissance (or recon) attacks to do unauthorized discovery and mapping of systems, services, or vulnerabilities. Access attacks exploit known vulnerabilities in authentication services, FTP services, and web services. Types of access attacks are: password, spoofing, trust exploitations, port redirections, man-in-the-middle, and buffer overflow. Social engineering is an access attack that attempts to manipulate individuals into performing actions or divulging confidential information. DoS and DDoS are attacks that create some sort of interruption of network services to users, devices, or applications.

Threat actors can send packets using a spoofed source IP address. Threat actors can also tamper with the other fields in the IP header to carry out their attacks. IP attack techniques include: ICMP, amplification and reflection, address spoofing, MITM, and session hijacking. Threat actors use ICMP for reconnaissance and scanning attacks. They launch information-gathering attacks to map out a network topology, discover which hosts are active (reachable), identify the host operating system (OS fingerprinting), and determine the state of a firewall. Threat actors often use amplification and reflection techniques to create DoS attacks.

TCP segment information appears immediately after the IP header. TCP provides reliable delivery, flow control, and stateful communication. TCP attacks include: TCPSYN Flood attack, TCP reset attack, and TCP Session hijacking. UDP is commonly used by DNS, TFTP, NFS, and SNMP. It is also used with real-time applications such as media streaming or VoIP. UDP is not protected by encryption. UDP Flood attacks send a flood of UDP packets, often from a spoofed host, to a server on the subnet. The result is very similar to a DoS attack.

Any client can send an unsolicited ARP Reply called a "gratuitous ARP." This mean that any host can claim to be the owner of any IP or MAC. A threat actor can poison the ARP cache of devices on the local network, creating an MITM attack to redirect traffic. ARP cache poisoning can be used to launch various man-in-the-middle attacks. DNS attacks include: open resolver attacks, stealth attacks, domain shadowing attacks, and tunneling attacks. To stop DNS tunneling, the network administrator must use a filter that inspects DNS traffic. A DHCP spoofing attack occurs when a rogue DHCP server is connected to the network and provides false IP configuration parameters to legitimate clients.

Most organizations follow the CIA information security triad: confidentiality, integrity, and availability. To ensure secure communications across both public and private networks, you must secure devices including routers, switches, servers, and hosts. This is known as defense-in-depth. A firewall is a system, or group of systems, that enforces an access control policy between networks. To defend against fast-moving and evolving attacks, you may need an intrusion detection systems (IDS), or the more scalable intrusion prevention systems (IPS).

The four elements of secure communications are data integrity, origin authentication, data confidentiality, and data non-repudiation. Hash functions guarantee that message data has not changed accidentally or intentionally. Three well-known hash functions are MD5 with 128-bit digest, SHA hashing algorithm, and SHA-2. To add authentication to integrity assurance, use a keyed-hash message authentication code (HMAC). HMAC is calculated using any cryptographic algorithm that combines a cryptographic hash function with a secret key. Symmetric encryption algorithms using DES, 3DES, AES, SEAL, and RC are based on the premise that each communicating party knows the pre-shared key. Data confidentiality can also be ensured using asymmetric algorithms, including Rivest, Shamir, and Adleman (RSA) and the public key infrastructure (PKI). Diffie-Hellman (DH) is an asymmetric mathematical algorithm where two computers generate an identical shared secret key without having communicated before.

Go to the online course to take the quiz and exam.

Chapter Quiz - Network Security Concepts

Your Chapter Notes

CHAPTER 4

ACL Concepts

Introduction - 4.0

Why should I take this module? - 4.0.1

Welcome to ACL Concepts!

You have arrived at your grandparents' residence. It is a beautiful gated community with walking paths and gardens. For the residents safety, no one is permitted to get into the community without stopping at the gate and presenting the guard with identification. You provide your ID and the guard verifies that you are expected as a visitor. He documents your information and lifts the gate. Imagine if the guard had to do this for the many staff members that entered each day. They have simplified this process by assigning a badge for each employee to automatically raise the gate once the badge is scanned. You greet your grandparents who are anxiously awaiting you at the front desk. You all get back into the car to go down the street for dinner. As you exit the parking lot, you must again stop and show your identification so that the guard will lift the gate. Rules have been put in place for all incoming and outgoing traffic.

Much like the guard in the gated community, network traffic passing through an interface configured with an access control list (ACL) has permitted and denied traffic. The router compares the information within the packet against each ACE, in sequential order, to determine if the packet matches one of the ACEs. This process is called packet filtering. Let's learn more!

What will I learn to do in this module? - 4.0.2

Module Title: ACL Concepts

Module Objective: Explain how ACLs are used as part of a network security policy.

Topic Title	Topic Objective
Purpose of ACLs	Explain how ACLs filter traffic.
Wildcard Masks in ACLs	Explain how ACLs use wildcard masks.
Guidelines for ACL Creation	Explain how to create ACLs.
Types of IPv4 ACLs	Compare standard and extended IPv4 ACLs.

Purpose of ACLs - 4.1

What is an ACL? - 4.1.1

Routers make routing decisions based on information in the packet header. Traffic entering a router interface is routed solely based on information within the routing table. The router compares the destination IP address with routes in the routing table to find the best match and then forwards the packet based on the best match route. That same process can be used to filter traffic using an access control list (ACL).

An ACL is a series of IOS commands that are used to filter packets based on information found in the packet header. By default, a router does not have any ACLs configured. However, when an ACL is applied to an interface, the router performs the additional task of evaluating all network packets as they pass through the interface to determine if the packet can be forwarded.

An ACL uses a sequential list of permit or deny statements, known as access control entries (ACEs).

Note: ACEs are also commonly called ACL statements.

When network traffic passes through an interface configured with an ACL, the router compares the information within the packet against each ACE, in sequential order, to determine if the packet matches one of the ACEs. This process is called packet filtering.

Several tasks performed by routers require the use of ACLs to identify traffic. The table lists some of these tasks with examples.

Task	Example
Limit network traffic to increase network performance	• A corporate policy prohibits video traffic on the network to reduce the network load. • A policy can be enforced using ACLs to block video traffic.
Provide traffic flow control	• A corporate policy requires that routing protocol traffic be limited to certain links only. • A policy can be implemented using ACLs to restrict the delivery of routing updates to only those that come from a known source.
Provide a basic level of security for network access	• Corporate policy demands that access to the Human Resources network be restricted to authorized users only. • A policy can be enforced using ACLs to limit access to specified networks.
Filter traffic based on traffic type	• Corporate policy requires that email traffic be permitted into a network, but that Telnet access be denied. • A policy can be implemented using ACLs to filter traffic by type.
Screen hosts to permit or deny access to network services	• Corporate policy requires that access to some file types (e.g., FTP or HTTP) be limited to user groups. • A policy can be implemented using ACLs to filter user access to services.
Provide priority to certain classes of network traffic	• Corporate traffic specifies that voice traffic be forwarded as fast as possible to avoid any interruption. • A policy can be implemented using ACLs and QoS services to identify voice traffic and process it immediately.

Refer to
Online Course
for Illustration

Packet Filtering - 4.1.2

Packet filtering controls access to a network by analyzing the incoming and/or outgoing packets and forwarding them or discarding them based on given criteria. Packet filtering can occur at Layer 3 or Layer 4, as shown in the figure.

Cisco routers support two types of ACLs:

- **Standard ACLs** - ACLs only filter at Layer 3 using the source IPv4 address only.

- **Extended ACLs** - ACLs filter at Layer 3 using the source and / or destination IPv4 address. They can also filter at Layer 4 using TCP, UDP ports, and optional protocol type information for finer control.

Refer to
Online Course
for Illustration

ACL Operation - 4.1.3

ACLs define the set of rules that give added control for packets that enter inbound interfaces, packets that relay through the router, and packets that exit outbound interfaces of the router.

ACLs can be configured to apply to inbound traffic and outbound traffic, as shown in the figure.

Note: ACLs do not act on packets that originate from the router itself.

An inbound ACL filters packets before they are routed to the outbound interface. An inbound ACL is efficient because it saves the overhead of routing lookups if the packet is discarded. If the packet is permitted by the ACL, it is then processed for routing. Inbound ACLs are best used to filter packets when the network attached to an inbound interface is the only source of packets that need to be examined.

An outbound ACL filters packets after being routed, regardless of the inbound interface. Incoming packets are routed to the outbound interface and then they are processed through the outbound ACL. Outbound ACLs are best used when the same filter will be applied to packets coming from multiple inbound interfaces before exiting the same outbound interface.

When an ACL is applied to an interface, it follows a specific operating procedure. For example, here are the operational steps used when traffic has entered a router interface with an inbound standard IPv4 ACL configured.

1. The router extracts the source IPv4 address from the packet header.

2. The router starts at the top of the ACL and compares the source IPv4 address to each ACE in a sequential order.

3. When a match is made, the router carries out the instruction, either permitting or denying the packet, and the remaining ACEs in the ACL, if any, are not analyzed.

4. If the source IPv4 address does not match any ACEs in the ACL, the packet is discarded because there is an implicit deny ACE automatically applied to all ACLs.

The last ACE statement of an ACL is always an implicit deny that blocks all traffic. By default, this statement is automatically implied at the end of an ACL even though it is hidden and not displayed in the configuration.

Note: An ACL must have at least one permit statement otherwise all traffic will be denied due to the implicit deny ACE statement.

Refer to **Packet Tracer Activity** for this chapter

Packet Tracer - ACL Demonstration - 4.1.4

In this activity, you will observe how an access control list (ACL) can be used to prevent a ping from reaching hosts on remote networks. After removing the ACL from the configuration, the pings will be successful.

Go to the online course to take the quiz and exam.

Check Your Understanding - Purpose of ACLs - 4.1.5

Wildcard Masks in ACLs - 4.2

Wildcard Mask Overview - 4.2.1

In the previous topic, you learned about the purpose of ACL. This topic explains how ACL uses wildcard masks. An IPv4 ACE uses a 32-bit wildcard mask to determine which bits of the address to examine for a match. Wildcard masks are also used by the Open Shortest Path First (OSPF) routing protocol.

A wildcard mask is similar to a subnet mask in that it uses the ANDing process to identify which bits in an IPv4 address to match. However, they differ in the way they match binary 1s and 0s. Unlike a subnet mask, in which binary 1 is equal to a match and binary 0 is not a match, in a wildcard mask, the reverse is true.

Wildcard masks use the following rules to match binary 1s and 0s:

- **Wildcard mask bit 0** - Match the corresponding bit value in the address
- **Wildcard mask bit 1** - Ignore the corresponding bit value in the address

The table lists some examples of wildcard masks and what they would identify.

Wildcard Mask	Last Octet (in Binary)	Meaning (0 - match, 1 - ignore)
0.0.0.0	00000000	• Match all octets.
0.0.0.63	00111111	• Match the first three octets • Match the two left most bits of the last octet • Ignore the last 6 bits
0.0.0.15	00001111	• Match the first three octets • Match the four left most bits of the last octet • Ignore the last 4 bits of the last octet
0.0.0.248	11111100	• Match the first three octets • Ignore the six left most bits of the last octet • Match the last two bits
0.0.0.255	11111111	• Match the first three octet • Ignore the last octet

Refer to
Interactive Graphic
in online course

Refer to
Online Course
for Illustration

Wildcard Mask Types - 4.2.2

Using wildcard masks will take some practice. Refer to the examples to learn how the wildcard mask is used to filter traffic for one host, one subnet, and a range IPv4 addresses.

Click each button to see how the wildcard mask is used in ACLs.

Wildcard to Match a Host

In this example, the wildcard mask is used to match a specific host IPv4 address. Assume ACL 10 needs an ACE that only permits the host with IPv4 address 192.168.1.1. Recall that "0" equals a match and "1" equals ignore. To match a specific host IPv4 address, a wildcard mask consisting of all zeroes (i.e., 0.0.0.0) is required.

The table lists in binary, the host IPv4 address, the wildcard mask, and the permitted IPv4 address.

The 0.0.0.0 wildcard mask stipulates that every bit must match exactly. Therefore, when the ACE is processed, the wildcard mask will permit only the 192.168.1.1 address. The resulting ACE in ACL 10 would be **access-list 10 permit 192.168.1.1 0.0.0.0**.

Wildcard Mask to Match an IPv4 Subnet

In this example, ACL 10 needs an ACE that permits all hosts in the 192.168.1.0/24 network. The wildcard mask 0.0.0.255 stipulates that the very first three octets must match exactly but the fourth octet does not.

The table lists in binary, the host IPv4 address, the wildcard mask, and the permitted IPv4 addresses.

When processed, the wildcard mask 0.0.0.255 permits all hosts in the 192.168.1.0/24 network. The resulting ACE in ACL 10 would be **access-list 10 permit 192.168.1.0 0.0.0.255**.

Wildcard Mask to Match an IPv4 Address Range

In this example, ACL 10 needs an ACE that permits all hosts in the 192.168.16.0/24, 192.168.17.0/24, ..., 192.168.31.0/24 networks. The wildcard mask 0.0.15.255 would correctly filter that range of addresses.

The table lists in binary the host IPv4 address, the wildcard mask, and the permitted IPv4 addresses.

The highlighted wildcard mask bits identify which bits of the IPv4 address must match. When processed, the wildcard mask 0.0.15.255 permits all hosts in the 192.168.16.0/24 to 192.168.31.0/24 networks. The resulting ACE in ACL 10 would be **access-list 10 permit 192.168.16.0 0.0.15.255**.

Refer to
Interactive Graphic
in online course

Refer to
Online Course
for Illustration

Wildcard Mask Calculation - 4.2.3

Calculating wildcard masks can be challenging. One shortcut method is to subtract the subnet mask from 255.255.255.255. Refer to the examples to learn how to calculate the wildcard mask using the subnet mask.

Click each button to see how to calculate each wildcard mask.

Example 1

Assume you wanted an ACE in ACL 10 to permit access to all users in the 192.168.3.0/24 network. To calculate the wildcard mask, subtract the subnet mask (i.e., 255.255.255.0) from 255.255.255.255, as shown in the table.

The solution produces the wildcard mask 0.0.0.255. Therefore, the ACE would be **access-list 10 permit 192.168.3.0 0.0.0.255**.

Example 2

In this example, assume you wanted an ACE in ACL 10 to permit network access for the 14 users in the subnet 192.168.3.32/28. Subtract the subnet (i.e., 255.255.255.240) from 255.255.255.255, as shown in the table.

This solution produces the wildcard mask 0.0.0.15. Therefore, the ACE would be **access-list 10 permit 192.168.3.32 0.0.0.15**.

Example 3

In this example, assume you needed an ACE in ACL 10 to permit only networks 192.168.10.0 and 192.168.11.0. These two networks could be summarized as 192.168.10.0/23 which is a subnet mask of 255.255.254.0. Again, you subtract 255.255.254.0 subnet mask from 255.255.255.255, as shown in the table.

This solution produces the wildcard mask 0.0.1.255. Therefore, the ACE would be **access-list 10 permit 192.168.10.0 0.0.1.255**.

Example 4

Consider an example in which you need an ACL number 10 to match networks in the range between 192.168.16.0/24 to 192.168.31.0/24. This network range could be summarized as 192.168.16.0/20 which is a subnet mask of 255.255.240.0. Therefore, subtract 255.255.240.0 subnet mask from 255.255.255.255, as shown in the table.

This solution produces the wildcard mask 0.0.15.255. Therefore, the ACE would be **access-list 10 permit 192.168.16.0 0.0.15.255**.

Wildcard Mask Keywords - 4.2.4

Working with decimal representations of binary wildcard mask bits can be tedious. To simplify this task, the Cisco IOS provides two keywords to identify the most common uses of wildcard masking. Keywords reduce ACL keystrokes and make it easier to read the ACE.

The two keywords are:

- **host** - This keyword substitutes for the 0.0.0.0 mask. This mask states that all IPv4 address bits must match to filter just one host address.

- **any** - This keyword substitutes for the 255.255.255.255 mask. This mask says to ignore the entire IPv4 address or to accept any addresses.

For example, in the command output, two ACLs are configured. The ACL 10 ACE permits only the 192.168.10.10 host and the ACL 11 ACE permits all hosts.

```
R1(config)# access-list 10 permit 192.168.10.10 0.0.0.0
R1(config)# access-list 11 permit 0.0.0.0 255.255.255.255
R1(config)#
```

Alternatively, the keywords **host** and **any** could have been used to replace the highlighted output.

The following commands accomplishes the same task as the previous commands.

```
R1(config)# access-list 10 permit host 192.168.10.10

R1(config)# access-list 11 permit any

R1(config)#
```

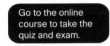

Check Your Understanding - Wildcard Masks in ACLs - 4.2.5

Guidelines for ACL Creation - 4.3

Refer to **Online Course** for Illustration

Limited Number of ACLs per Interface - 4.3.1

In a previous topic, you learned about how wildcard masks are used in ACLs. This topic will focus on the guidelines for ACL creation. There is a limit on the number of ACLs that can be applied on a router interface. For example, a dual-stacked (i.e, IPv4 and IPv6) router interface can have up to four ACLs applied, as shown in the figure.

Specifically, a router interface can have:

- one outbound IPv4 ACL
- one inbound IPv4 ACL
- one inbound IPv6 ACL
- one outbound IPv6 ACL

Assume R1 has two dual-stacked interfaces that require inbound and outbound IPv4 and IPv6 ACLs applied. As shown in the figure, R1 could have up to 8 ACLs configured and applied to interfaces. Each interface would have four ACLs; two ACLs for IPv4 and two ACLs for IPv6. For each protocol, one ACL is for inbound traffic and one for outbound traffic.

Note: ACLs do not have to be configured in both directions. The number of ACLs and their direction applied to the interface will depend on the security policy of the organization.

ACL Best Practices - 4.3.2

Using ACLs requires attention to detail and great care. Mistakes can be costly in terms of downtime, troubleshooting efforts, and poor network service. Basic planning is required before configuring an ACL.

The table presents guidelines that form the basis of an ACL best practices list.

Guideline	Benefit
Base ACLs on the organizational security policies.	This will ensure you implement organizational security guidelines.
Write out what you want the ACL to do.	This will help you avoid inadvertently creating potential access problems.
Use a text editor to create, edit, and save all of your ACLs.	This will help you create a library of reusable ACLs.
Document the ACLs using the **remark** command.	This will help you (and others) understand the purpose of an ACE.
Test the ACLs on a development network before implementing them on a production network.	This will help you avoid costly errors.

Go to the online course to take the quiz and exam.

Check Your Understanding - Guidelines for ACL Creation - 4.3.3

Types of IPv4 ACLs - 4.4

Standard and Extended ACLs - 4.4.1

The previous topics covered the purpose of ACL and the guidelines for ACL creation. This topic will cover standard and extended ACLs, named and numbered ACLs, and the examples of placement of these ACLs.

There are two types of IPv4 ACLs:

- **Standard ACLs** - These permit or deny packets based only on the source IPv4 address.

- **Extended ACLs** - These permit or deny packets based on the source IPv4 address and destination IPv4 address, protocol type, source and destination TCP or UDP ports and more.

For example, refer to the following standard ALC command.

```
R1(config)# access-list 10 permit 192.168.10.0 0.0.0.255
R1(config)#
```

ACL 10 permits hosts on the source network 192.168.10.0/24. Because of the implied "deny any" at the end, all traffic except for traffic coming from the 192.168.10.0/24 network is blocked with this ACL.

In the next example, an extended ACL 100 permits traffic originating from any host on the 192.168.10.0/24 network to any IPv4 network if the destination host port is 80 (HTTP).

```
R1(config)# access-list 100 permit tcp 192.168.10.0 0.0.0.255 any eq www
R1(config)#
```

Notice how the standard ACL 10 is only capable of filtering by source address while the extended ACL 100 is filtering on the source, and destination Layer 3, and Layer 4 protocol (i.e., TCP) information.

Note: Full ACL configuration is discussed in another module.

Numbered and Named ACLs - 4.4.2

Numbered ACLs

ACLs number 1 to 99, or 1300 to 1999 are standard ACLs while ACLs number 100 to 199, or 2000 to 2699 are extended ACLs, as shown in the output.

```
R1(config)# access-list ?

  <1-99>        IP standard access list

  <100-199>     IP extended access list

  <1100-1199>   Extended 48-bit MAC address access list

  <1300-1999>   IP standard access list (expanded range)

  <200-299>     Protocol type-code access list

  <2000-2699>   IP extended access list (expanded range)

  <700-799>     48-bit MAC address access list

  rate-limit    Simple rate-limit specific access list

  template      Enable IP template acls

Router(config)# access-list
```

Named ACLs

Named ACLs is the preferred method to use when configuring ACLs. Specifically, standard and extended ACLs can be named to provide information about the purpose of the ACL. For example, naming an extended ACL FTP-FILTER is far better than having a numbered ACL 100.

The **ip access-list** global configuration command is used to create a named ACL, as shown in the following example.

```
R1(config)# ip access-list extended FTP-FILTER

R1(config-ext-nacl)# permit tcp 192.168.10.0 0.0.0.255 any eq ftp

R1(config-ext-nacl)# permit tcp 192.168.10.0 0.0.0.255 any eq ftp-data

R1(config-ext-nacl)#
```

The following summarizes the rules to follow for named ACLs.

■ Assign a name to identify the purpose of the ACL.

■ Names can contain alphanumeric characters.

- Names cannot contain spaces or punctuation.
- It is suggested that the name be written in CAPITAL LETTERS.
- Entries can be added or deleted within the ACL.

Refer to
Online Course
for Illustration

Where to Place ACLs - 4.4.3

Every ACL should be placed where it has the greatest impact on efficiency.

The figure illustrates where standard and extended ACLs should be located in an enterprise network. Assume the objective to prevent traffic originating in the 192.168.10.0/24 network from reaching the 192.168.30.0/24 network.

Extended ACLs should be located as close as possible to the source of the traffic to be filtered. This way, undesirable traffic is denied close to the source network without crossing the network infrastructure.

Standard ACLs should be located as close to the destination as possible. If a standard ACL was placed at the source of the traffic, the "permit" or "deny" will occur based on the given source address no matter where the traffic is destined.

Placement of the ACL and therefore, the type of ACL used, may also depend on a variety of factors as listed in the table.

Factors Influencing ACL Placement	Explanation
The extent of organizational control	Placement of the ACL can depend on whether or not the organization has control of both the source and destination networks.
Bandwidth of the networks involved	It may be desirable to filter unwanted traffic at the source to prevent transmission of bandwidth-consuming traffic.
Ease of configuration	• It may be easier to implement an ACL at the destination, but traffic will use bandwidth unnecessarily. • An extended ACL could be used on each router where the traffic originated. This would save bandwidth by filtering the traffic at the source, but it would require creating extended ACLs on multiple routers.

Refer to
Online Course
for Illustration

Standard ACL Placement Example - 4.4.4

Following the guidelines for ACL placement, standard ACLs should be located as close to the destination as possible.

In the figure, the administrator wants to prevent traffic originating in the 192.168.10.0/24 network from reaching the 192.168.30.0/24 network.

Following the basic placement guidelines, the administrator would place a standard ACL on router R3. There are two possible interfaces on R3 to apply the standard ACL:

- **R3 S0/1/1 interface (inbound)** - The standard ACL can be applied inbound on the R3 S0/1/1 interface to deny traffic from .10 network. However, it would also filter .10 traffic to the 192.168.31.0/24 (.31 in this example) network. Therefore, the standard ACL should not be applied to this interface.

- R3 G0/0 interface (outbound) - The standard ACL can be applied outbound on the R3 G0/0/0 interface. This will not affect other networks that are reachable by R3. Packets from .10 network will still be able to reach the .31 network. This is the best interface to place the standard ACL to meet the traffic requirements.

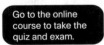

Extended ACL Placement Example - 4.4.5

Extended ACL should be located as close to the source as possible. This prevents unwanted traffic from being sent across multiple networks only to be denied when it reaches its destination.

However, the organization can only place ACLs on devices that they control. Therefore, the extended ACL placement must be determined in the context of where organizational control extends.

In the figure, for example, Company A wants to deny Telnet and FTP traffic to Company B's 192.168.30.0/24 network from their 192.168.11.0/24 network while permitting all other traffic.

There are several ways to accomplish these goals. An extended ACL on R3 would accomplish the task, but the administrator does not control R3. In addition, this solution allows unwanted traffic to cross the entire network, only to be blocked at the destination. This affects overall network efficiency.

The solution is to place an extended ACL on R1 that specifies both source and destination addresses.

There are two possible interfaces on R1 to apply the extended ACL:

- R1 S0/1/0 interface (outbound) - The extended ACL can be applied outbound on the S0/1/0 interface. However, this solution will process all packets leaving R1 including packets from 192.168.10.0/24.

- R1 G0/0/1 interface (inbound) - The extended ACL can be applied inbound on the G0/0/1 and only packets from the 192.168.11.0/24 network are subject to ACL processing on R1. Because the filter is to be limited to only those packets leaving the 192.168.11.0/24 network, applying the extended ACL to G0/1 is the best solution.

Go to the online course to take the quiz and exam.

Check Your Understanding - Guidelines for ACL Placement - 4.4.6

Module Practice and Quiz - 4.5

What did I learn in this module? - 4.5.1

Purpose of ACLs

Several tasks performed by routers require the use of ACLs to identify traffic. An ACL is a series of IOS commands that are used to filter packets based on information found in the packet header. A router does not have any ACLs configured by default. However, when an ACL is applied to an interface, the router performs the additional task of evaluating all network packets as they pass through the interface to determine if the packet can be forwarded. An ACL uses a sequential list of permit or deny statements, known as ACEs. Cisco routers support two types of ACLs: standard ACLs and extended ACLs. An inbound ACL filters packets before they are routed to the outbound interface. If the packet is permitted by the ACL, it is then processed for routing. An outbound ACL filters packets after being routed, regardless of the inbound interface. When an ACL is applied to an interface, it follows a specific operating procedure:

1. The router extracts the source IPv4 address from the packet header.

2. The router starts at the top of the ACL and compares the source IPv4 address to each ACE in a sequential order.

3. When a match is made, the router carries out the instruction, either permitting or denying the packet, and the remaining ACEs in the ACL, if any, are not analyzed.

4. If the source IPv4 address does not match any ACEs in the ACL, the packet is discarded because there is an implicit deny ACE automatically applied to all ACLs.

Wildcard Masks

An IPv4 ACE uses a 32-bit wildcard mask to determine which bits of the address to examine for a match. Wildcard masks are also used by the Open Shortest Path First (OSPF) routing protocol. A wildcard mask is similar to a subnet mask in that it uses the ANDing process to identify which bits in an IPv4 address to match. However, they differ in the way they match binary 1s and 0s. **Wildcard mask bit 0** matches the corresponding bit value in the address. **Wildcard mask bit 1** ignores the corresponding bit value in the address. A wildcard mask is used to filter traffic for one host, one subnet, and a range IPv4 addresses. A shortcut to calculating a wildcard mask is to subtract the subnet mask from 255.255.255.255. Working with decimal representations of binary wildcard mask bits can be simplified by using the Cisco IOS keywords **host** and **any** to identify the most common uses of wildcard masking. Keywords reduce ACL keystrokes and make it easier to read the ACE.

Guidelines for ACL creation

There is a limit on the number of ACLs that can be applied on a router interface. For example, a dual-stacked (i.e, IPv4 and IPv6) router interface can have up to four ACLs applied. Specifically, a router interface can have one outbound IPv4 ACL, one inbound IPv4 ACL, one inbound IPv6 ACL , and one outbound IPv6 ACL. ACLs do not have to be configured in both directions. The number of ACLs and their direction applied to the interface will depend on the security policy of the organization. Basic planning is required before configuring an ACL and includes the following best practices:

■ Base ACLs on the organizational security policies.

■ Write out what you want the ACL to do.

■ Use a text editor to create, edit, and save all of your ACLs.

■ Document the ACLs using the **remark** command.

■ Test the ACLs on a development network before implementing them on a production network.

Types of IPv4 ACLs

There are two types of IPv4 ACLs: standard ACLs and Extended ACLs. Standard ACLs permit or deny packets based only on the source IPv4 address. Extended ACLs permit or deny packets based on the source IPv4 address and destination IPv4 address, protocol type, source and destination TCP or UDP ports and more. ACLs number 1 -to 99, or 1300 to 1999, are standard ACLs. ACLs number 100-199, or 2000 to 2699, are extended ACLs. Named ACLs is the preferred method to use when configuring ACLs. Specifically, standard and extended ACLs can be named to provide information about the purpose of the ACL.

The following summarizes the rules to follow for named ACLs:

- Assign a name to identify the purpose of the ACL.

- Names can contain alphanumeric characters.

- Names cannot contain spaces or punctuation.

- It is suggested that the name be written in CAPITAL LETTERS.

- Entries can be added or deleted within the ACL.

Every ACL should be placed where it has the greatest impact on efficiency. Extended ACLs should be located as close as possible to the source of the traffic to be filtered. This way, undesirable traffic is denied close to the source network without crossing the network infrastructure. Standard ACLs should be located as close to the destination as possible. If a standard ACL was placed at the source of the traffic, the "permit" or "deny" will occur based on the given source address no matter where the traffic is destined. Placement of the ACL may depend on the extent of organizational control, bandwidth of the networks, and ease of configuration.

Go to the online
course to take the
quiz and exam.

Chapter Quiz - ACL Concepts

Your Chapter Notes

CHAPTER 5

ACLs for IPv4 Configuration

Introduction - 5.0

Why should I take this module? - 5.0.1

Welcome to ACLs for IPv4 Configuration!

In the gated community where your grandparents live, there are rules for who can enter and leave the premises. The guard will not raise the gate to let you in to the community until someone confirms that you are on an approved visitor list. Much like the guard in the gated community, network traffic passing through an interface configured with an access control list (ACL) has permitted and denied traffic. How do you configure these ACLs? How do you modify them if they are not working correctly or if they require other changes? How do ACLs provide secure remote administrative access? Get started with this module to learn more!

What will I learn to do in this module? - 5.0.2

Module Title: ACLs for IPv4 Configuration

Module Objective: Implement IPv4 ACLs to filter traffic and secure administrative access.

Topic Title	Topic Objective
Configure Standard IPv4 ACLs	Configure standard IPv4 ACLs to filter traffic to meet networking requirements.
Modify IPv4 ACLs	Use sequence numbers to edit existing standard IPv4 ACLs.
Secure VTY Ports with a Standard IPv4 ACL	Configure a standard ACL to secure VTY access.
Configure Extended IPv4 ACLs	Configure extended IPv4 ACLs to filter traffic according to networking requirements.

Configure Standard IPv4 ACLs - 5.1

Create an ACL - 5.1.1

In a previous module, you learned about what an ACL does and why it is important. In this topic, you will learn about creating ACLs.

All access control lists (ACLs) must be planned. However, this is especially true for ACLs requiring multiple access control entries (ACEs).

When configuring a complex ACL, it is suggested that you:

- Use a text editor and write out the specifics of the policy to be implemented.

- Add the IOS configuration commands to accomplish those tasks.

- Include remarks to document the ACL.

- Copy and paste the commands onto the device.

- Always thoroughly test an ACL to ensure that it correctly applies the desired policy.

These recommendations enable you to create the ACL thoughtfully without impacting the traffic on the network.

Numbered Standard IPv4 ACL Syntax - 5.1.2

To create a numbered standard ACL, use the following global configuration command:

```
Router(config)# access-list access-list-number {deny | permit | remark
text} source [source-wildcard] [log]
```

Use the **no access-list** *access-list-number* global configuration command to remove a numbered standard ACL.

The table provides a detailed explanation of the syntax for a standard ACL.

Parameter	Description
access-list-number	• This is the decimal number of the ACL. • Standard ACL number range is 1 to 99 or 1300 to 1999.
deny	This denies access if the condition is matched.
permit	This permits access if the condition is matched.
remark *text*	• (Optional) This adds a text entry for documentation purposes. • Each remark is limited to 100 characters.
source	• This identifies the source network or host address to filter. • Use the **any** keyword to specify all networks. • Use the **host** *ip-address* keyword or simply enter an *ip-address* (without the **host** keyword) to identify a specific IP address.
source-wildcard	(Optional) This is a 32-bit wildcard mask that is applied to the *source*. If omitted, a default 0.0.0.0 mask is assumed.
log	• (Optional) This keyword generates and sends an informational message whenever the ACE is matched. • Message includes ACL number, matched condition (i.e., permitted or denied), source address, and number of packets. • This message is generated for the first matched packet. • This keyword should only be implemented for troubleshooting or security reasons.

Named Standard IPv4 ACL Syntax - 5.1.3

Naming an ACL makes it easier to understand its function. To create a named standard ACL, use the following global configuration command:

```
Router(config)# ip access-list standard access-list-name
```

This command enters the named standard configuration mode where you configure the ACL ACEs.

ACL names are alphanumeric, case sensitive, and must be unique. Capitalizing ACL names is not required but makes them stand out when viewing the running-config output. It also makes it less likely that you will accidentally create two different ACLs with the same name but with different uses of capitalization.

Note: Use the **no ip access-list standard** *access-list-name* global configuration command to remove a named standard IPv4 ACL.

In the example, a named standard IPv4 ACL called NO-ACCESS is created. Notice that the prompt changes to named standard ACL configuration mode. ACE statements are entered in the named standard ACL sub configuration mode. Use the help facility to view all the named standard ACL ACE options.

The three highlighted options are configured similar to the numbered standard ACL. Unlike the numbered ACL method, there is no need to repeat the initial **ip access-list** command for each ACE.

```
R1(config)# ip access-list standard NO-ACCESS

R1(config-std-nacl)# ?

Standard Access List configuration commands:

  <1-2147483647>  Sequence Number

  default         Set a command to its defaults

  deny            Specify packets to reject

  exit            Exit from access-list configuration mode

  no              Negate a command or set its defaults

  permit          Specify packets to forward

  remark          Access list entry comment

R1(config-std-nacl)#
```

Apply a Standard IPv4 ACL - 5.1.4

After a standard IPv4 ACL is configured, it must be linked to an interface or feature. The following command can be used to bind a numbered or named standard IPv4 ACL to an interface:

```
Router(config-if) # ip access-group {access-list-number | access-list-name}
{in | out}
```

To remove an ACL from an interface, first enter the **no ip access-group** interface configuration command. However, the ACL will still be configured on the router. To remove the ACL from the router, use the **no access-list** global configuration command.

Refer to
Online Course
for Illustration

Numbered Standard IPv4 ACL Example - 5.1.5

The topology in the figure will be used to demonstrate configuring and applying numbered and named standard IPv4 ACLs to an interface. This first example shows a numbered standard IPv4 ACL implementation.

Assume only PC1 is allowed out to the internet. To enable this policy, a standard ACL ACE could be applied outbound on S0/1/0, as shown in the figure.

```
R1(config)# access-list 10 remark ACE permits ONLY host 192.168.10.10 to
the internet

R1(config)# access-list 10 permit host 192.168.10.10

R1(config)# do show access-lists

Standard IP access list 10

    10 permit 192.168.10.10

R1(config)#
```

Notice that the output of the **show access-lists** command does not display the **remark** statements. ACL remarks are displayed in the running configuration file. Although the **remark** command is not required to enable the ACL, it is strongly suggested for documentation purposes.

Now assume that a new network policy states that hosts in LAN 2 should also be permitted to the internet. To enable this policy, a second standard ACL ACE could be added to ACL 10, as shown in the output.

```
R1(config)# access-list 10 remark ACE permits all host in LAN 2

R1(config)# access-list 10 permit 192.168.20.0 0.0.0.255

R1(config)# do show access-lists

Standard IP access list 10

    10 permit 192.168.10.10

    20 permit 192.168.20.0, wildcard bits 0.0.0.255

R1(config)#
```

Apply ACL 10 outbound on the Serial 0/1/0 interface.

```
R1(config)# interface Serial 0/1/0

R1(config-if)# ip access-group 10 out

R1(config-if)# end

R1#
```

The resulting policy of ACL 10 will only permit host 192.168.10.10 and all host from LAN 2 to exit the Serial 0/1/0 interface. All other hosts in the 192.168.10.0 network will not be permitted to the internet.

Use the **show running-config** command to review the ACL in the configuration, as shown in the output.

```
R1# show run | section access-list
access-list 10 remark ACE permits host 192.168.10.10
access-list 10 permit 192.168.10.10
access-list 10 remark ACE permits all host in LAN 2
access-list 10 permit 192.168.20.0 0.0.0.255
R1#
```

Notice how the **remarks** statements are also displayed.

Finally, use the **show ip interface** command to verify if an interface has an ACL applied to it. In the example output, the output is specifically looking at the Serial 0/1/0 interface for lines that include "access list" text.

```
R1# show ip int Serial 0/1/0 | include access list
  Outgoing Common access list is not set
  Outgoing access list is 10
  Inbound Common access list is not set
  Inbound  access list is not set
R1#
```

Named Standard IPv4 ACL Example - 5.1.6

Refer to
Online Course
for Illustration

This second example shows a named standard IPv4 ACL implementation. The topology is repeated in the figure for your convenience.

Assume only PC1 is allowed out to the internet. To enable this policy, a named standard ACL called PERMIT-ACCESS could be applied outbound on S0/1/0.

Remove the previously configured named ACL 10 and create a named standard ACL called PERMIT-ACCESS, as shown here.

```
R1(config)# no access-list 10
R1(config)# ip access-list standard PERMIT-ACCESS
R1(config-std-nacl)# remark ACE permits host 192.168.10.10
R1(config-std-nacl)# permit host 192.168.10.10
R1(config-std-nacl)#
```

Now add an ACE permitting only host 192.168.10.10 and another ACE permitting all LAN 2 hosts to the internet.

```
R1(config-std-nacl)# remark ACE permits host 192.168.10.10
R1(config-std-nacl)# permit host 192.168.10.10
R1(config-std-nacl)# remark ACE permits all hosts in LAN 2
```

```
R1(config-std-nacl)# permit 192.168.20.0 0.0.0.255
R1(config-std-nacl)# exit
R1(config)#
```

Apply the new named ACL outbound to the Serial 0/1/0 interface.

```
R1(config)# interface Serial 0/1/0
R1(config-if)# ip access-group PERMIT-ACCESS out
R1(config-if)# end
R1#
```

Use the **show access-lists** and **show running-config** command to review the ACL in the configuration, as shown in the output.

```
R1# show access-lists
Standard IP access list PERMIT-ACCESS
    10 permit 192.168.10.10
    20 permit 192.168.20.0, wildcard bits 0.0.0.255
R1# show run | section ip access-list
ip access-list standard PERMIT-ACCESS
 remark ACE permits host 192.168.10.10
 permit 192.168.10.10
 remark ACE permits all hosts in LAN 2
 permit 192.168.20.0 0.0.0.255
R1#
```

Finally, use the **show ip interface** command to verify if an interface has an ACL applied to it. In the example output, the output is specifically looking at the Serial 0/1/0 interface for lines that include "access list" text.

```
R1# show ip int Serial 0/1/0 | include access list
  Outgoing Common access list is not set
  Outgoing access list is PERMIT-ACCESS
  Inbound Common access list is not set
  Inbound  access list is not set
R1#
```

> Refer to
> **Interactive Graphic**
> in online course

Syntax Check - Configure Standard IPv4 ACLs - 5.1.7

Configure a numbered and named ACLs on R1.

Refer to **Packet Tracer Activity** for this chapter

Packet Tracer - Configure Numbered Standard IPv4 ACLs - 5.1.8

Standard access control lists (ACLs) are router configuration scripts that control whether a router permits or denies packets based on the source address. This activity focuses on defining filtering criteria, configuring standard ACLs, applying ACLs to router interfaces, and verifying and testing the ACL implementation. The routers are already configured, including IPv4 addresses and EIGRP routing.

Refer to **Packet Tracer Activity** for this chapter

Packet Tracer - Configure Named Standard IPv4 ACLs - 5.1.9

The senior network administrator has asked you to create a named standard ACL to prevent access to a file server. All clients from one network and one specific workstation from a different network should be denied access.

Modify IPv4 ACLs - 5.2

Two Methods to Modify an ACL - 5.2.1

After an ACL is configured, it may need to be modified. ACLs with multiple ACEs can be complex to configure. Sometimes the configured ACE does not yield the expected behaviors. For these reasons, ACLs may initially require a bit of trial and error to achieve the desired filtering result.

This section will discuss two methods to use when modifying an ACL:

- Use a Text Editor
- Use Sequence Numbers

Text Editor Method - 5.2.2

ACLs with multiple ACEs should be created in a text editor. This allows you to plan the required ACEs, create the ACL, and then paste it into the router interface. It also simplifies the tasks to edit and fix an ACL.

For example, assume ACL 1 was entered incorrectly using **19** instead of **192** for the first octet, as shown in the running configuration.

```
R1# show run | section access-list
access-list 1 deny   19.168.10.10
access-list 1 permit 192.168.10.0 0.0.0.255
R1#
```

In the example, the first ACE should have been to deny the host at 192.168.10.10. However, the ACE was incorrectly entered.

To correct the error:

- Copy the ACL from the running configuration and paste it into the text editor.

- Make the necessary edits changes.

- Remove the previously configured ACL on the router otherwise, pasting the edited ACL commands will only append (i.e., add) to the existing ACL ACEs on the router.

- Copy and paste the edited ACL back to the router.

Assume that ACL 1 has now been corrected. Therefore, the incorrect ACL must be deleted, and the corrected ACL 1 statements must be pasted in global configuration mode, as shown in the output.

```
R1(config)# no access-list 1

R1(config)#

R1(config)# access-list 1 deny 192.168.10.10

R1(config)# access-list 1 permit 192.168.10.0 0.0.0.255

R1(config)#
```

Sequence Numbers Method - 5.2.3

An ACL ACE can also be deleted or added using the ACL sequence numbers. Sequence numbers are automatically assigned when an ACE is entered. These numbers are listed in the **show access-lists** command. The **show running-config** command does not display sequence numbers.

In the previous example, the incorrect ACE for ACL 1 is using sequence number 10, as shown in the example.

```
R1# show access-lists

Standard IP access list 1

    10 deny    19.168.10.10

    20 permit 192.168.10.0, wildcard bits 0.0.0.255

R1#
```

Use the **ip access-list standard** command to edit an ACL. Statements cannot be overwritten using the same sequence number as an existing statement. Therefore, the current statement must be deleted first with the **no 10** command. Then the correct ACE can be added using sequence number 10 is configured. Verify the changes using the **show access-lists** command, as shown in the example.

```
R1# conf t

R1(config)# ip access-list standard 1

R1(config-std-nacl)# no 10

R1(config-std-nacl)# 10 deny host 192.168.10.10

R1(config-std-nacl)# end

R1# show access-lists
```

```
Standard IP access list 1
    10 deny    192.168.10.10
    20 permit 192.168.10.0, wildcard bits 0.0.0.255
R1#
```

Modify a Named ACL Example - 5.2.4

Named ACLs can also use sequence numbers to delete and add ACEs. Refer to the example for ACL **NO-ACCESS**.

```
R1# show access-lists
Standard IP access list NO-ACCESS
    10 deny    192.168.10.10
    20 permit 192.168.10.0, wildcard bits 0.0.0.255
```

Assume that host 192.168.10.5 from the 192.168.10.0/24 network should also have been denied. If you entered a new ACE, it would be appended to the end of the ACL. Therefore, the host would never be denied because ACE 20 permits all hosts from that network.

The solution is to add an ACE denying host 192.168.10.5 in between ACE 10 and ACE 20, such as ACE 15, as shown in the example. Also notice that the new ACE was entered without using the **host** keyword. The keyword is optional when specifying a destination host.

Use the **show access-lists** command to verify the ACL now has a new ACE 15 inserted appropriately before the permit statement.

Notice that sequence number 15 is displayed prior to sequence number 10. We might expect the order of the statements in the output to reflect the order in which they were entered. However, the IOS puts host statements in an order using a special hashing function. The resulting order optimizes the search for a host ACL entry and then it searches for range statements.

Note: The hashing function is only applied to host statements in an IPv4 standard access list. The details of the hashing function are beyond the scope of this course.

```
R1# configure terminal
R1(config)# ip access-list standard NO-ACCESS
R1(config-std-nacl)# 15 deny 192.168.10.5
R1(config-std-nacl)# end
R1#
R1# show access-lists
Standard IP access list NO-ACCESS
    15 deny    192.168.10.5
    10 deny    192.168.10.10
    20 permit 192.168.10.0, wildcard bits 0.0.0.255
R1#
```

ACL Statistics - 5.2.5

Notice that the **show access-lists** command in the example shows statistics for each statement that has been matched. The deny ACE in the NO-ACCESS ACL has been matched 20 times and the permit ACE has been matched 64 times.

Note that the implied deny any the last statement does not display any statistics. To track how many implicit denied packets have been matched, you must manually configure the **deny any** command at the end of the ACL.

Use the **clear access-list counters** command to clear the ACL statistics. This command can be used alone or with the number or name of a specific ACL.

```
R1# show access-lists
Standard IP access list NO-ACCESS
    10 deny   192.168.10.10  (20 matches)
    20 permit 192.168.10.0, wildcard bits 0.0.0.255  (64 matches)
R1# clear access-list counters NO-ACCESS
R1# show access-lists
Standard IP access list NO-ACCESS
    10 deny   192.168.10.10
    20 permit 192.168.10.0, wildcard bits 0.0.0.255
R1#
```

Refer to **Interactive Graphic** in online course

Syntax Checker - Modify IPv4 ACLs - 5.2.6

Modify an ACL using sequence numbers.

Refer to **Packet Tracer Activity** for this chapter

Packet Tracer - Configure and Modify Standard IPv4 ACLs - 5.2.7

In this Packet Tracer activity, you will complete the following objectives:

- Part 1: Configure Devices and Verify Connectivity
- Part 2: Configure and Verify Standard Numbered and Named ACLs
- Part 3: Modify a Standard ACL

Secure VTY Ports with a Standard IPv4 ACL - 5.3

The access-class Command - 5.3.1

ACLs typically filter incoming or outgoing traffic on an interface. However, an ACL can also be used to secure remote administrative access to a device using the vty lines.

Use the following two steps to secure remote administrative access to the vty lines:

- Create an ACL to identify which administrative hosts should be allowed remote access.
- Apply the ACL to incoming traffic on the vty lines.

Use the following command to apply an ACL to the vty lines:

```
R1(config-line)# access-class {access-list-number | access-list-name} {
in | out }
```

The **in** keyword is the most commonly used option to filter incoming vty traffic. The **out** parameter filters outgoing vty traffic and is rarely applied.

The following should be considered when configuring access lists on vty lines:

- Both named and numbered access lists can be applied to vty lines.
- Identical restrictions should be set on all the vty lines, because a user can attempt to connect to any of them.

Refer to **Online Course** for Illustration

Secure VTY Access Example - 5.3.2

The topology in the figure is used to demonstrate how to configure an ACL to filter vty traffic. In this example, only PC1 will be allowed to Telnet in to R1.

Note: Telnet is used here for demonstration purposes only. SSH should be used in a production environment.

To increase secure access, a username and password will be created, and the **login local** authentication method will be used on the vty lines. The command in the example creates a local database entry for a user **ADMIN** and password **class.**

A named standard ACL called ADMIN-HOST is created and identifies PC1. Notice that the **deny any** has been configured to track the number of times access has been denied.

The vty lines are configured to use the local database for authentication, permit Telnet traffic, and use the ADMIN-HOST ACL to restrict traffic.

```
R1(config)# username ADMIN secret class

R1(config)# ip access-list standard ADMIN-HOST

R1(config-std-nacl)# remark This ACL secures incoming vty lines

R1(config-std-nacl)# permit 192.168.10.10

R1(config-std-nacl)# deny any

R1(config-std-nacl)# exit

R1(config)# line vty 0 4

R1(config-line)# login local

R1(config-line)# transport input telnet

R1(config-line)# access-class ADMIN-HOST in

R1(config-line)# end

R1#
```

In a production environment, you would set the vty lines to only allow SSH, as shown in the example.

```
R1(config)# line vty 0 4

R1(config-line)# login local

R1(config-line)# transport input ssh

R1(config-line)# access-class ADMIN-HOST in

R1(config-line)# end

R1#
```

Refer to
Online Course
for Illustration

Verify the VTY Port is Secured - 5.3.3

After the ACL to restrict access to the vty lines is configured, it is important to verify that it is working as expected.

As shown in the figure, when PC1 Telnets to R1, the host will be prompted for a username and password before successfully accessing the command prompt.

This verifies that PC1 can access R1 for administrative purposes.

Next, we test the connection from PC2. As shown in this figure, when PC2 attempts to Telnet, the connection is refused.

To verify the ACL statistics, issue the **show access-lists** command. Notice the informational message displayed on the console regarding the admin user. An informational console message is also generated when a user exits the vty line.

The match in the permit line of the output is a result of a successful Telnet connection by PC1. The match in the deny statement is due to the failed attempt to create a Telnet connection by PC2, a device on the 192.168.11.0/24 network.

```
R1#

Oct  9 15:11:19.544: %SEC_LOGIN-5-LOGIN_SUCCESS: Login Success [user:
admin] [Source: 192.168.10.10] [localport: 23] at 15:11:19 UTC Wed Oct 9
2019

R1# show access-lists

Standard IP access list ADMIN-HOST

    10 permit 192.168.10.10  (2 matches)

    20 deny   any  (2 matches)

R1#
```

Refer to
Interactive Graphic
in online course

Syntax Checker - Secure the VTY Ports - 5.3.4

Secure the vty lines for remote administrative access.

Configure Extended IPv4 ACLs - 5.4

Extended ACLs - 5.4.1

In the previous topics, you learned about how to configure and modify standard ACLs, and how to secure VTY ports with a standard IPv4 ACL. Standard ACLs only filter on source address. When more precise traffic-filtering control is required, extended IPv4 ACLs can be created.

Extended ACLs are used more often than standard ACLs because they provide a greater degree of control. They can filter on source address, destination address, protocol (i.e., IP, TCP, UDP, ICMP), and port number. This provides a greater range of criteria on which to base the ACL. For example, one extended ACL can allow email traffic from a network to a specific destination while denying file transfers and web browsing.

Like standard ACLs, extended ACLs can be created as:

- **Numbered Extended ACL** - Created using the **access-list** *access-list-number* global configuration command.

- **Named Extended ACL** - Created using the **ip access-list extended** *access-list-name*.

Numbered Extended IPv4 ACL Syntax - 5.4.2

The procedural steps for configuring extended ACLs are the same as for standard ACLs. The extended ACL is first configured, and then it is activated on an interface. However, the command syntax and parameters are more complex to support the additional features provided by extended ACLs.

To create a numbered extended ACL, use the following global configuration command:

```
Router(config)# access-list access-list-number {deny | permit | remark
text} protocol source source-wildcard [operator [port]] destination
destination-wildcard [operator [port]] [established] [log]
```

Use the **no ip access-list extended** *access-list-name* global configuration command to remove an extended ACL.

Although there are many keywords and parameters for extended ACLs, it is not necessary to use all of them when configuring an extended ACL. The table provides a detailed explanation of the syntax for an extended ACL.

Parameter	Description
access-list-number	• This is the decimal number of the ACL. • Extended ACL number range is 100 to 199 and 2000 to 2699.
deny	This denies access if the condition is matched.
permit	This permits access if the condition is matched.
remark text	• (Optional) Adds a text entry for documentation purposes. • Each remark is limited to 100 characters.
protocol	• Name or number of an internet protocol. • Common keywords include **ip**, **tcp**, **udp**, and **icmp**. • The **ip** keyword matches all IP protocols.

Parameter	Description
source	• This identifies the source network or host address to filter. • Use the **any** keyword to specify all networks. • Use the **host** *ip-address* keyword or simply enter an *ip-address* (without the **host** keyword) to identify a specific IP address.
source-wildcard	(Optional) A 32-bit wildcard mask that is applied to the source.
destination	• This identifies the destination network or host address to filter. • Use the **any** keyword to specify all networks. • Use the **host** *ip-address* keyword or *ip-address*.
destination-wildcard	(Optional) This is a 32-bit wildcard mask that is applied to the destination.
operator	• (Optional) This compares source or destination ports. • Possible operands include **lt** (less than), **gt** (greater than), **eq** (equal), **neq** (not equal), and **range** (inclusive range).
port	(Optional) The decimal number or name of a TCP or UDP port.
established	• (Optional) For the TCP protocol only. • This is a 1st generation firewall feature.
log	• (Optional) This keyword generates and sends an informational message whenever the ACE is matched. • This message includes ACL number, matched condition (i.e., permitted or denied), source address, and number of packets. • This message is generated for the first matched packet. • This keyword should only be implemented for troubleshooting or security reasons.

The command to apply an extended IPv4 ACL to an interface is the same as the command used for standard IPv4 ACLs.

```
Router(config-if) # ip access-group {access-list-number | access-list-
name} {in | out}
```

To remove an ACL from an interface, first enter the **no ip access-group** interface configuration command. To remove the ACL from the router, use the **no access-list** global configuration command.

Note: The internal logic applied to the ordering of standard ACL statements does not apply to extended ACLs. The order in which the statements are entered during configuration is the order they are displayed and processed.

Refer to
Interactive Graphic
in online course

Protocols and Ports - 5.4.3

Extended ACLs can filter on many different types of internet protocols and ports. Click each button for more information about the internet protocols and ports on which extended ACLs can filter.

Protocol Options

The four highlighted protocols are the most popular options.

Note: Use the **?** to get help when entering a complex ACE.

Note: If an internet protocol is not listed, then the IP protocol number could be specified. For instance, the ICMP protocol number 1, TCP is 6, and UDP is 17.

```
R1(config)# access-list 100 permit ?
  <0-255>        An IP protocol number
  ahp            Authentication Header Protocol
  dvmrp          dvmrp
  eigrp          Cisco's EIGRP routing protocol
  esp            Encapsulation Security Payload
  gre            Cisco's GRE tunneling
  icmp           Internet Control Message Protocol
  igmp           Internet Gateway Message Protocol
  ip             Any Internet Protocol
  ipinip         IP in IP tunneling
  nos            KA9Q NOS compatible IP over IP tunneling
  object-group   Service object group
  ospf           OSPF routing protocol
  pcp            Payload Compression Protocol
  pim            Protocol Independent Multicast
  tcp            Transmission Control Protocol
  udp            User Datagram Protocol
R1(config)# access-list 100 permit
```

Port Keyword Options

Selecting a *protocol* influences *port* options. For instance, selecting the:

- **tcp** protocol would provide TCP related ports options
- **udp** protocol would provide UDP specific ports options
- **icmp** protocol would provide ICMP related ports (i.e., message) options

Again, notice how many TCP port options are available. The highlighted ports are popular options.

```
R1(config)# access-list 100 permit tcp any any eq ?
  <0-65535>      Port number
  bgp            Border Gateway Protocol (179)
```

chargen	Character generator (19)
cmd	Remote commands (rcmd, 514)
daytime	Daytime (13)
discard	Discard (9)
domain	Domain Name Service (53)
echo	Echo (7)
exec	Exec (rsh, 512)
finger	Finger (79)
ftp	File Transfer Protocol (21)
ftp-data	FTP data connections (20)
gopher	Gopher (70)
hostname	NIC hostname server (101)
ident	Ident Protocol (113)
irc	Internet Relay Chat (194)
klogin	Kerberos login (543)
kshell	Kerberos shell (544)
login	Login (rlogin, 513)
lpd	Printer service (515)
msrpc	MS Remote Procedure Call (135)
nntp	Network News Transport Protocol (119)
onep-plain	Onep Cleartext (15001)
onep-tls	Onep TLS (15002)
pim-auto-rp	PIM Auto-RP (496)
pop2	Post Office Protocol v2 (109)
pop3	Post Office Protocol v3 (110)
smtp	Simple Mail Transport Protocol (25)
sunrpc	Sun Remote Procedure Call (111)
syslog	Syslog (514)
tacacs	TAC Access Control System (49)
talk	Talk (517)
telnet	Telnet (23)
time	Time (37)
uucp	Unix-to-Unix Copy Program (540)
whois	Nicname (43)
www	World Wide Web (HTTP, 80)

Protocols and Port Numbers Configuration Examples - 5.4.4

Extended ACLs can filter on different port number and port name options. This example configures an extended ACL 100 to filter HTTP traffic. The first ACE uses the **www** port name. The second ACE uses the port number **80**. Both ACEs achieve exactly the same result.

```
R1(config)# access-list 100 permit tcp any any eq www

!or...

R1(config)# access-list 100 permit tcp any any eq 80
```

Configuring the port number is required when there is not a specific protocol name listed such as SSH (port number 22) or an HTTPS (port number 443), as shown in the next example.

```
R1(config)# access-list 100 permit tcp any any eq 22

R1(config)# access-list 100 permit tcp any any eq 443

R1(config)#
```

Refer to
Online Course
for Illustration

Apply a Numbered Extended IPv4 ACL - 5.4.5

The topology in the figure will be used to demonstrate configuring and applying numbered and named extended IPv4 ACLs to an interface. This first example shows a numbered extended IPv4 ACL implementation.

In this example, the ACL permits both HTTP and HTTPS traffic from the 192.168.10.0 network to go to any destination.

Extended ACLs can be applied in various locations. However, they are commonly applied close to the source. Therefore, ACL 110 was applied inbound on the R1 G0/0/0 interface.

```
R1(config)# access-list 110 permit tcp 192.168.10.0 0.0.0.255 any eq www

R1(config)# access-list 110 permit tcp 192.168.10.0 0.0.0.255 any eq 443

R1(config)# interface g0/0/0

R1(config-if)# ip access-group 110 in

R1(config-if)# exit

R1(config)#
```

Refer to
Online Course
for Illustration

TCP Established Extended ACL - 5.4.6

TCP can also perform basic stateful firewall services using the TCP **established** keyword. The keyword enables inside traffic to exit the inside private network and permits the returning reply traffic to enter the inside private network, as shown in the figure.

However, TCP traffic generated by an outside host and attempting to communicate with an inside host is denied.

The **established** keyword can be used to permit only the return HTTP traffic from requested websites, while denying all other traffic.

In the topology, the design for this example shows that ACL 110, which was previously configured, will filter traffic from the inside private network. ACL 120, using the **established** keyword, will filter traffic coming into the inside private network from the outside public network.

In the example, ACL 120 is configured to only permit returning web traffic to the inside hosts. The new ACL is then applied outbound on the R1 G0/0/0 interface. The **show access-lists** command displays both ACLs. Notice from the match statistics that inside hosts have been accessing the secure web resources from the internet.

```
R1(config)# access-list 120 permit tcp any 192.168.10.0 0.0.0.255
established

R1(config)# interface g0/0/0

R1(config-if)# ip access-group 120 out

R1(config-if)# end

R1# show access-lists

Extended IP access list 110

    10 permit tcp 192.168.10.0 0.0.0.255 any eq www

    20 permit tcp 192.168.10.0 0.0.0.255 any eq 443 (657 matches)

Extended IP access list 120

    10 permit tcp any 192.168.10.0 0.0.0.255 established (1166 matches)

R1#
```

Notice that the permit secure HTTPS counters (i.e., eq 443) in ACL 110 and the return established counters in ACL 120 have increased.

The **established** parameter allows only responses to traffic that originates from the 192.168.10.0/24 network to return to that network. Specifically, a match occurs if the returning TCP segment has the ACK or reset (RST) flag bits set. This indicates that the packet belongs to an existing connection. Without the **established** parameter in the ACL statement, clients could send traffic to a web server, but not receive traffic returning from the web server.

Named Extended IPv4 ACL Syntax - 5.4.7

Naming an ACL makes it easier to understand its function. To create a named extended ACL, use the following global configuration command:

```
Router(config)# ip access-list extended access-list-name
```

This command enters the named extended configuration mode. Recall that ACL names are alphanumeric, case sensitive, and must be unique.

In the example, a named extended ACL called NO-FTP-ACCESS is created and the prompt changed to named extended ACL configuration mode. ACE statements are entered in the named extended ACL sub configuration mode.

```
R1(config)# ip access-list extended NO-FTP-ACCESS
R1(config-ext-nacl)#
```

Named Extended IPv4 ACL Example - 5.4.8

Refer to
Online Course
for Illustration

Named extended ACLs are created in essentially the same way that named standard ACLs are created.

The topology in the figure is used to demonstrate configuring and applying two named extended IPv4 ACLs to an interface:

- **SURFING** - This will permit inside HTTP and HTTPS traffic to exit to the internet.

- **BROWSING** - This will only permit returning web traffic to the inside hosts while all other traffic exiting the R1 G0/0/0 interface is implicitly denied.

The example shows the configuration for the inbound SURFING ACL and the outbound BROWSING ACL.

The SURFING ACL permits HTTP and HTTPS traffic from inside users to exit the G0/0/1 interface connected to the internet. Web traffic returning from the internet is permitted back into the inside private network by the BROWSING ACL.

The SURFING ACL is applied inbound and the BROWSING ACL applied outbound on the R1 G0/0/0 interface, as shown in the output.

Inside hosts have been accessing the secure web resources from the internet. The **show access-lists** command is used to verify the ACL statistics. Notice that the permit secure HTTPS counters (i.e., eq 443) in the SURFING ACL and the return established counters in the BROWSING ACL have increased.

```
R1(config)# ip access-list extended SURFING
R1(config-ext-nacl)# Remark Permits inside HTTP and HTTPS traffic
R1(config-ext-nacl)# permit tcp 192.168.10.0 0.0.0.255 any eq 80
R1(config-ext-nacl)# permit tcp 192.168.10.0 0.0.0.255 any eq 443
R1(config-ext-nacl)# exit
R1(config)#
R1(config)# ip access-list extended BROWSING
R1(config-ext-nacl)# Remark Only permit returning HTTP and HTTPS traffic
R1(config-ext-nacl)# permit tcp any 192.168.10.0 0.0.0.255 established
R1(config-ext-nacl)# exit
R1(config)# interface g0/0/0
R1(config-if)# ip access-group SURFING in
R1(config-if)# ip access-group BROWSING out
R1(config-if)# end
R1# show access-lists
Extended IP access list SURFING
```

```
     10 permit tcp 192.168.10.0 0.0.0.255 any eq www

     20 permit tcp 192.168.10.0 0.0.0.255 any eq 443 (124 matches)
Extended IP access list BROWSING

     10 permit tcp any 192.168.10.0 0.0.0.255 established (369 matches)
R1#
```

Edit Extended ACLs - 5.4.9

Like standard ACLs, an extended ACL can be edited using a text editor when many changes are required. Otherwise, if the edit applies to one or two ACEs, then sequence numbers can be used.

For example, assume you have just entered the SURFING and BROWSING ACLs and wish to verify their configuration using the **show access-lists** command.

```
R1# show access-lists

Extended IP access list BROWSING

     10 permit tcp any 192.168.10.0 0.0.0.255 established
Extended IP access list SURFING

     10 permit tcp 19.168.10.0 0.0.0.255 any eq www

     20 permit tcp 192.168.10.0 0.0.0.255 any eq 443
R1#
```

You notice that the ACE sequence number 10 in the SURFING ACL has an incorrect source IP networks address.

To correct this error using sequence numbers, the original statement is removed with the **no** *sequence_#* command and the corrected statement is added replacing the original statement.

```
R1# configure terminal

R1(config)# ip access-list extended SURFING

R1(config-ext-nacl)# no 10

R1(config-ext-nacl)# 10 permit tcp 192.168.10.0 0.0.0.255 any eq www

R1(config-ext-nacl)# end
```

The output verifies the configuration change using the **show access-lists** command.

```
R1# show access-lists

Extended IP access list BROWSING

     10 permit tcp any 192.168.10.0 0.0.0.255 established
Extended IP access list SURFING

     10 permit tcp 192.168.10.0 0.0.0.255 any eq www

     20 permit tcp 192.168.10.0 0.0.0.255 any eq 443
R1#
```

Refer to
Online Course
for Illustration

Another Named Extended IPv4 ACL Example - 5.4.10

The figure shows another scenario for implementing a named extended IPv4 ACL. Assume that PC1 in the inside private network is permitted FTP, SSH, Telnet, DNS, HTTP, and HTTPS traffic. However, all other users in the inside private network should be denied access.

Two named extended ACLs will be created:

- **PERMIT-PC1** - This will only permit PC1 TCP access to the internet and deny all other hosts in the private network.

- **REPLY-PC1** - This will only permit specified returning TCP traffic to PC1 implicitly deny all other traffic.

The example shows the configuration for the inbound PERMIT-PC1 ACL and the outbound REPLY-PC1.

The **PERMIT-PC1** ACL permits PC1 (i.e., 192.168.10.10) TCP access to the FTP (i.e., ports 20 and 21), SSH (22), Telnet (23), DNS (53), HTTP (80), and HTTPS (443) traffic.

The **REPLY-PC1** ACL will permit return traffic to PC1.

There are many factors to consider when applying an ACL including:

- The device to apply it on

- The interface to apply it on

- The direction to apply it

Careful consideration must be taken to avoid undesired filtering results. The PERMIT-PC1 ACL is applied inbound and the REPLY-PC1 ACL applied outbound on the R1 G0/0/0 interface.

```
R1(config)# ip access-list extended PERMIT-PC1

R1(config-ext-nacl)# Remark Permit PC1 TCP access to internet

R1(config-ext-nacl)# permit tcp host 192.168.10.10 any eq 20

R1(config-ext-nacl)# permit tcp host 192.168.10.10 any eq 21

R1(config-ext-nacl)# permit tcp host 192.168.10.10 any eq 22

R1(config-ext-nacl)# permit tcp host 192.168.10.10 any eq 23

R1(config-ext-nacl)# permit tcp host 192.168.10.10 any eq 53

R1(config-ext-nacl)# permit tcp host 192.168.10.10 any eq 80

R1(config-ext-nacl)# permit tcp host 192.168.10.10 any eq 443

R1(config-ext-nacl)# deny ip 192.168.10.0 0.0.0.255 any

R1(config-ext-nacl)# exit

R1(config)#

R1(config)# ip access-list extended REPLY-PC1

R1(config-ext-nacl)# Remark Only permit returning traffic to PC1

R1(config-ext-nacl)# permit tcp any host 192.168.10.10 established
```

```
R1(config-ext-nacl)# exit
R1(config)# interface g0/0/0
R1(config-if)# ip access-group PERMIT-PC1 in
R1(config-if)# ip access-group REPLY-PC1 out
R1(config-if)# end
R1#
```

Refer to
Interactive Graphic
in online course

Verify Extended ACLs - 5.4.11

After an ACL has been configured and applied to an interface, use Cisco IOS **show** commands to verify the configuration.

Click each button for more information about verifying the configuration of an ACL.

show ip interface

The **show ip interface** command is used to verify the ACL on the interface and the direction in which it was applied, as shown in the output.

The command generates quite a bit of output but notice how the capitalized ACL names stand out in the output.

To reduce the command output, use filtering techniques, as shown in the second command.

```
R1# show ip interface g0/0/0
GigabitEthernet0/0/0 is up, line protocol is up (connected)
    Internet address is 192.168.10.1/24
    Broadcast address is 255.255.255.255
    Address determined by setup command
    MTU is 1500 bytes
    Helper address is not set
    Directed broadcast forwarding is disabled
    Outgoing access list is REPLY-PC1
    Inbound  access list is PERMIT-PC1
    Proxy ARP is enabled
    Security level is default
    Split horizon is enabled
    ICMP redirects are always sent
    ICMP unreachables are always sent
    ICMP mask replies are never sent
    IP fast switching is disabled
    IP fast switching on the same interface is disabled
```

```
    IP Flow switching is disabled

    IP Fast switching turbo vector

    IP multicast fast switching is disabled

    IP multicast distributed fast switching is disabled

    Router Discovery is disabled
R1#

R1# show ip interface g0/0/0 | include access list
Outgoing access list is REPLY-PC1
Inbound access list is PERMIT-PC1
R1#
```

show access-lists

The **show access-lists** command can be used to confirm that the ACLs work as expected. The command displays statistic counters that increase whenever an ACE is matched.

Note: Traffic must be generated to verify the operation of the ACL.

In the top example, the Cisco IOS command is used to display the contents of all ACLs.

Notice how the IOS is displaying the keyword even though port numbers were configured.

Also, notice that extended ACLs do not implement the same internal logic and hashing function as standard ACLs. The output and sequence numbers displayed in the **show access-lists** command output is the order in which the statements were entered. Host entries are not automatically listed prior to range entries.

```
R1# show access-lists
Extended IP access list PERMIT-PC1
10 permit tcp host 192.168.10.10 any eq 20

20 permit tcp host 192.168.10.10 any eq ftp

30 permit tcp host 192.168.10.10 any eq 22

40 permit tcp host 192.168.10.10 any eq telnet

50 permit tcp host 192.168.10.10 any eq domain

60 permit tcp host 192.168.10.10 any eq www

70 permit tcp host 192.168.10.10 any eq 443

80 deny ip 192.168.10.0 0.0.0.255 any
Extended IP access list REPLY-PC1
10 permit tcp any host 192.168.10.10 established
R1#
```

show running-config

The **show running-config** command can be used to validate what was configured. The command also displays configured remarks.

The command can be filtered to display only pertinent information, as shown in the following.

```
R1# show running-config | begin ip access-list
ip access-list extended PERMIT-PC1
remark Permit PC1 TCP access to internet
permit tcp host 192.168.10.10 any eq 20
permit tcp host 192.168.10.10 any eq ftp
permit tcp host 192.168.10.10 any eq 22
permit tcp host 192.168.10.10 any eq telnet
permit tcp host 192.168.10.10 any eq domain
permit tcp host 192.168.10.10 any eq www
permit tcp host 192.168.10.10 any eq 443
deny ip 192.168.10.0 0.0.0.255 any
ip access-list extended REPLY-PC1
remark Only permit returning traffic to PC1
permit tcp any host 192.168.10.10 established
!
```

*Refer to **Packet Tracer Activity** for this chapter*

Packet Tracer - Configure Extended IPv4 ACLs - Scenario 1 - 5.4.12

In this Packet Tracer activity, you will complete the following objectives:

- Part 1: Configure, Apply, and Verify an Extended Numbered IPv4 ACL
- Part 2: Configure, Apply, and Verify an Extended Named IPv4 ACL

*Refer to **Packet Tracer Activity** for this chapter*

Packet Tracer - Configure Extended IPv4 ACLs - Scenario 2 - 5.4.13

In this Packet Tracer activity, you will complete the following objectives:

- Part 1: Configure a Named Extended IPv4 ACL
- Part 2: Apply and Verify the Extended IPv4 ACL

Module Practice and Quiz - 5.5

*Refer to **Packet Tracer Activity** for this chapter*

Packet Tracer - IPv4 ACL Implementation Challenge - 5.5.1

In this Packet Tracer challenge, you will configure extended, standard named, and extended named IPv4 ACLs to meet specified communication requirements.

Refer to
Lab Activity
for this chapter

Lab - Configure and Verify Extended IPv4 ACLs - 5.5.2

In this lab, you will complete the following objectives:

- Part 1: Build the Network and Configure Basic Device Settings
- Part 2: Configure and Verify Extended IPv4 ACLs

What did I learn in this module? - 5.5.3

Configure Standard IPv4 ACLs

All access control lists (ACLs) must be planned, especially for ACLs requiring multiple access control entries (ACEs). When configuring a complex ACL, it is suggested that you use a text editor and write out the specifics of the policy to be implemented, add the IOS configuration commands to accomplish those tasks, include remarks to document the ACL, copy and paste the commands on a lab device, and always thoroughly test an ACL to ensure that it correctly applies the desired policy. To create a numbered standard ACL, use the use the **ip access-list standard** *access-list-name* global configuration command. Use the **no access-list** *access-list-number* global configuration command to remove a numbered standard ACL. Use the **show ip interface** command to verify if an interface has an ACL applied to it. In addition to standard numbered ACLs, there are named standard ACLs. ACL names are alphanumeric, case sensitive, and must be unique. Capitalizing ACL names is not required but makes them stand out when viewing the running-config output. To create a named standard ACL, use the **ip access-list standard** *access-list-name* global configuration command. Use the **no ip access-list standard** *access-list-name* global configuration command to remove a named standard IPv4 ACL. After a standard IPv4 ACL is configured, it must be linked to an interface or feature. To bind a numbered or named standard IPv4 ACL to an interface, use the **ip access-group** {*access-list-number* | *access-list-name*} { **in** | **out** } global configuration command. To remove an ACL from an interface, first enter the **no ip access-group** interface configuration command. To remove the ACL from the router, use the **no access-list** global configuration command.

Modify IPv4 ACLs

To modify an ACL, use a text editor or use sequence numbers. ACLs with multiple ACEs should be created in a text editor. This allows you to plan the required ACEs, create the ACL, and then paste it into the router interface. An ACL ACE can also be deleted or added using the ACL sequence numbers. Sequence numbers are automatically assigned when an ACE is entered. These numbers are listed in the **show access-lists** command. The **show running-config** command does not display sequence numbers. Named ACLs can also use sequence numbers to delete and add ACEs. The **show access-lists** command shows statistics for each statement that has been matched. The **clear access-list counters** command to clear the ACL statistics.

Secure VTY Ports with a Standard IPv4 ACL

ACLs typically filter incoming or outgoing traffic on an interface. However, a standard ACL can also be used to secure remote administrative access to a device using the vty lines. The two steps to secure remote administrative access to the vty lines are to create an ACL to identify which administrative hosts should be allowed remote access and to apply the ACL to incoming traffic on the vty lines. The **in** keyword is the most commonly used option to filter incoming vty traffic. The **out** parameter filters outgoing vty traffic and is rarely applied. Both named and numbered access lists can be applied to vty lines. Identical restrictions should be set on all the vty lines, because a user can attempt to connect to any of them. After the ACL to restrict access to the vty lines is configured, it is important to verify that it is working as expected. Use the **show ip interface** command to verify if an interface has an ACL applied to it. To verify the ACL statistics, issue the **show access-lists** command.

Configure Extended IPv4 ACLs

Extended ACLs are used more often than standard ACLs because they provide a greater degree of control. They can filter on source address, destination address, protocol (i.e., IP, TCP, UDP, ICMP), and port number. This provides a greater range of criteria on which to base the ACL. Like standard ACLs, extended ACLs can be created as numbered extended ACL and named extended ACL. Numbered Extended ACLs are created using the same global configuration commands that are used for standard ACLs. The procedural steps for configuring extended ACLs are the same as for standard ACLs. However, the command syntax and parameters are more complex to support the additional features provided by extended ACLs. To create a numbered extended ACL, use the Router(config)# **access-list** *access-list-number* {**deny** | **permit** | **remark** *text*} *protocol source source-wildcard* [*operator* [*port*]] *destination destination-wildcard* [*operator* [*port*]] [**established**] [**log**] global configuration command. Extended ACLs can filter on many different types of internet protocols and ports. Selecting a *protocol* influences *port* options. For instance, selecting the **tcp** protocol would provide TCP related ports options. Configuring the port number is required when there is not a specific protocol name listed such as SSH (port number 22) or HTTPS (port number 443). TCP can also perform basic stateful firewall services using the TCP **established** keyword. The keyword enables inside traffic to exit the inside private network and permits the returning reply traffic to enter the inside private network. After an ACL has been configured and applied to an interface, use Cisco IOS **show** commands to verify the configuration. The **show ip interface** command is used to verify the ACL on the interface and the direction in which it was applied.

Go to the online course to take the quiz and exam.

Chapter Quiz - ACLs for IPv4 Implementation

Your Chapter Notes

CHAPTER 6

NAT for IPv4

Introduction - 6.0

Why should I take this module? - 6.0.1

Welcome to NAT for IPv4!

IPv4 addresses are 32-bit numbers. Mathematically, this means that there can be just over 4 billion unique IPv4 addresses. In the 1980s, this seemed like more than enough IPv4 addresses. Then came the development of affordable desktop and laptop computers, smart phones and tablets, many other digital technologies, and of course, the internet. Rather quickly it became apparent that 4 billion IPv4 addresses would not be nearly enough to handle the growing demand. This is why IPv6 was developed. Even with IPv6, most networks today are IPv4-only, or a combination of IPv4 and IPv6. The transition to IPv6-only networks is still ongoing, that is why Network Address Translation (NAT) was developed. NAT is designed to help manage those 4 billion addresses so that we can all use our many devices to access the internet. As you can see, it is important that you understand the purpose of (NAT) and how it works. As a bonus, this module contains multiple Packet Tracer activities where you get to configure different types of NAT. Get going!

What will I learn to do in this module? - 6.0.2

Module Title: NAT for IPv4

Module Objective: Configure NAT services on the edge router to provide IPv4 address scalability.

Topic Title	Topic Objective
NAT Characteristics	Explain the purpose and function of NAT.
Types of NAT	Explain the operation of different types of NAT.
NAT Advantages and Disadvantages	Describe the advantages and disadvantages of NAT.
Static NAT	Configure static NAT using the CLI.
Dynamic NAT	Configure dynamic NAT using the CLI.
PAT	Configure PAT using the CLI.
NAT64	Describe NAT for IPv6.

NAT Characteristics - 6.1

IPv4 Private Address Space - 6.1.1

Refer to
Online Course
for Illustration

As you know, there are not enough public IPv4 addresses to assign a unique address to each device connected to the internet. Networks are commonly implemented using private IPv4 addresses, as defined in RFC 1918. The range of addresses included in RFC 1918 are included in the following table. It is very likely that the computer that you use to view this course is assigned a private address.

Private Internet Addresses are Defined in RFC 1918

Class	RFC 1918 Internal Address Range	Prefix
A	10.0.0.0 - 10.255.255.255	10.0.0.0/8
B	172.16.0.0 - 172.31.255.255	172.16.0.0/12
C	192.168.0.0 - 192.168.255.255	192.168.0.0/16

These private addresses are used within an organization or site to allow devices to communicate locally. However, because these addresses do not identify any single company or organization, private IPv4 addresses cannot be routed over the internet. To allow a device with a private IPv4 address to access devices and resources outside of the local network, the private address must first be translated to a public address.

NAT provides the translation of private addresses to public addresses, as shown in the figure. This allows a device with a private IPv4 address to access resources outside of their private network, such as those found on the internet. NAT, combined with private IPv4 addresses, has been the primary method of preserving public IPv4 addresses. A single, public IPv4 address can be shared by hundreds, even thousands of devices, each configured with a unique private IPv4 address.

Without NAT, the exhaustion of the IPv4 address space would have occurred well before the year 2000. However, NAT has limitations and disadvantages, which will be explored later in this module. The solution to the exhaustion of IPv4 address space and the limitations of NAT is the eventual transition to IPv6.

Refer to
Online Course
for Illustration

What is NAT - 6.1.2

NAT has many uses, but its primary use is to conserve public IPv4 addresses. It does this by allowing networks to use private IPv4 addresses internally and providing translation to a public address only when needed. NAT has a perceived benefit of adding a degree of privacy and security to a network, because it hides internal IPv4 addresses from outside networks.

NAT-enabled routers can be configured with one or more valid public IPv4 addresses. These public addresses are known as the NAT pool. When an internal device sends traffic out of the network, the NAT-enabled router translates the internal IPv4 address of the device to a public address from the NAT pool. To outside devices, all traffic entering and exiting the network appears to have a public IPv4 address from the provided pool of addresses.

A NAT router typically operates at the border of a stub network. A stub network is one or more networks with a single connection to its neighboring network, one way in and one way out of the network. In the example in the figure, R2 is a border router. As seen from the ISP, R2 forms a stub network.

When a device inside the stub network wants to communicate with a device outside of its network, the packet is forwarded to the border router. The border router performs the NAT process, translating the internal private address of the device to a public, outside, routable address.

Note: The connection to the ISP may use a private address or a public address that is shared among customers. For the purposes of this module, a public address is shown.

Refer to
Interactive Graphic
in online course

How NAT Works - 6.1.3

In this example, PC1 with private address 192.168.10.10 wants to communicate with an outside web server with public address 209.165.201.1.

Click the Play button in the figure to start the animation.

Refer to
Online Course
for Illustration

NAT Terminology - 6.1.4

In NAT terminology, the inside network is the set of networks that is subject to translation. The outside network refers to all other networks.

Refer to
Interactive Graphic
in online course

When using NAT, IPv4 addresses have different designations based on whether they are on the private network, or on the public network (internet), and whether the traffic is incoming or outgoing.

NAT includes four types of addresses:

- Inside local address
- Inside global address
- Outside local address
- Outside global address

When determining which type of address is used, it is important to remember that NAT terminology is always applied from the perspective of the device with the translated address:

- **Inside address** - The address of the device which is being translated by NAT.
- **Outside address** - The address of the destination device.

NAT also uses the concept of local or global with respect to addresses:

- **Local address** - A local address is any address that appears on the inside portion of the network.
- **Global address** - A global address is any address that appears on the outside portion of the network.

The terms, inside and outside, are combined with the terms local and global to refer to specific addresses. The NAT router, R2 in the figure, is the demarcation point between the inside and outside networks. R2 is configured with a pool of public addresses to assign to inside hosts. Refer to the network and NAT table in the figure for the following discussion of each of the NAT address types.

Click each button for more information about the different address types.

Inside Local

Inside local address

The address of the source as seen from inside the network. This is typically a private IPv4 address. In the figure, the IPv4 address 192.168.10.10 is assigned to PC1. This is the inside local address of PC1.

Inside Global

Inside global address

The address of source as seen from the outside network. This is typically a globally routable IPv4 address. In the figure, when traffic from PC1 is sent to the web server at 209.165.201.1, R2 translates the inside local address to an inside global address. In this case, R2 changes the IPv4 source address from 192.168.10.10 to 209.165.200.226. In NAT terminology, the inside local address of 192.168.10.10 is translated to the inside global address of 209.165.200.226.

Outside Global

Outside global address

The address of the destination as seen from the outside network. It is a globally routable IPv4 address assigned to a host on the internet. For example, the web server is reachable at IPv4 address 209.165.201.1. Most often the outside local and outside global addresses are the same.

Outside Local

Outside local address

The address of the destination as seen from the inside network. In this example, PC1 sends traffic to the web server at the IPv4 address 209.165.201.1. While uncommon, this address could be different than the globally routable address of the destination.

PC1 has an inside local address of 192.168.10.10. From the perspective of PC1, the web server has an outside address of 209.165.201.1. When packets are sent from PC1 to the global address of the web server, the inside local address of PC1 is translated to 209.165.200.226 (inside global address). The address of the outside device is not typically translated because that address is usually a public IPv4 address.

Notice that PC1 has different local and global addresses, whereas the web server has the same public IPv4 address for both. From the perspective of the web server, traffic originating from PC1 appears to have come from 209.165.200.226, the inside global address.

Go to the online course to take the quiz and exam.

Check Your Understanding - NAT Characteristics - 6.1.5

Types of NAT - 6.2

Refer to **Online Course** for Illustration

Static NAT - 6.2.1

Now that you have learned about NAT and how it works, this topic will discuss the many versions of NAT that are available to you.

Static NAT uses a one-to-one mapping of local and global addresses. These mappings are configured by the network administrator and remain constant.

In the figure, R2 is configured with static mappings for the inside local addresses of Svr1, PC2, and PC3. When these devices send traffic to the internet, their inside local addresses are translated to the configured inside global addresses. To outside networks, these devices appear to have public IPv4 addresses.

Static NAT is particularly useful for web servers or devices that must have a consistent address that is accessible from the internet, such as a company web server. It is also useful for devices that must be accessible by authorized personnel when offsite, but not by the general public on the internet. For example, a network administrator from PC4 can use SSH to gain access to the inside global address of Svr1 (209.165.200.226). R2 translates this inside global address to the inside local address 192.168.10.10 and connects the session to Svr1.

Static NAT requires that enough public addresses are available to satisfy the total number of simultaneous user sessions.

Refer to **Online Course** for Illustration

Dynamic NAT - 6.2.2

Dynamic NAT uses a pool of public addresses and assigns them on a first-come, first-served basis. When an inside device requests access to an outside network, dynamic NAT assigns an available public IPv4 address from the pool.

In the figure, PC3 has accessed the internet using the first available address in the dynamic NAT pool. The other addresses are still available for use. Similar to static NAT, dynamic NAT requires that enough public addresses are available to satisfy the total number of simultaneous user sessions.

Refer to **Interactive Graphic** in online course

Port Address Translation - 6.2.3

Port Address Translation (PAT), also known as NAT overload, maps multiple private IPv4 addresses to a single public IPv4 address or a few addresses. This is what most home routers do. The ISP assigns one address to the router, yet several members of the household can simultaneously access the internet. This is the most common form of NAT for both the home and the enterprise.

With PAT, multiple addresses can be mapped to one or to a few addresses, because each private address is also tracked by a port number. When a device initiates a TCP/IP session, it generates a TCP or UDP source port value, or a specially assigned query ID for ICMP, to uniquely identify the session. When the NAT router receives a packet from the client, it uses its source port number to uniquely identify the specific NAT translation.

PAT ensures that devices use a different TCP port number for each session with a server on the internet. When a response comes back from the server, the source port number, which becomes the destination port number on the return trip, determines to which device the router forwards the packets. The PAT process also validates that the incoming packets were requested, thus adding a degree of security to the session.

Click Play in the figure to view an animation of the PAT process. PAT adds unique source port numbers to the inside global address to distinguish between translations.

As R2 processes each packet, it uses a port number (1331 and 1555, in this example) to identify the device from which the packet originated. The source address (SA) is the inside local address with the TCP/UDP assigned port number added. The destination address (DA) is the outside global address with the service port number added. In this example, the service port is 80, which is HTTP.

For the source address, R2 translates the inside local address to an inside global address with the port number added. The destination address is not changed but is now referred to as the outside global IPv4 address. When the web server replies, the path is reversed.

Refer to
Interactive Graphic
in online course

Next Available Port - 6.2.4

In the previous example, the client port numbers, 1331 and 1555, did not change at the NAT-enabled router. This is not a very likely scenario, because there is a good chance that these port numbers may have already been attached to other active sessions.

PAT attempts to preserve the original source port. However, if the original source port is already used, PAT assigns the first available port number starting from the beginning of the appropriate port group 0-511, 512-1,023, or 1,024-65,535. When there are no more ports available and there is more than one external address in the address pool, PAT moves to the next address to try to allocate the original source port. This process continues until there are no more available ports or external IPv4 addresses.

Click Play in the figure to view an animation of PAT operation. In this example, PAT has assigned the next available port (1445) to the second host address.

In the animation, the hosts have chosen the same port number of 1444. This is acceptable for the inside address, because the hosts have unique private IPv4 addresses. However, at the NAT router, the port numbers must be changed; otherwise, packets from two different hosts would exit R2 with the same source address. This example assumes that the first 420 ports in the range 1,024 - 65,535 are already in use, so the next available port number, 1445, is used.

When packets are returned from outside the network, if the source port number was previously modified by the NAT-enable router, the destination port number will now be changed back to the original port number by the NAT-enabled router.

Refer to
Online Course
for Illustration

NAT and PAT Comparison - 6.2.5

The table provides a summary of the differences between NAT and PAT.

Refer to
Interactive Graphic
in online course

NAT	PAT
One-to-one mapping between Inside Local and Inside Global addresses.	One Inside Global address can be mapped to many Inside Local addresses.
Uses only IPv4 addresses in translation process.	Uses IPv4 addresses and TCP or UDP source port numbers in translation process.
A unique Inside Global address is required for each inside host accessing the outside network.	A single unique Inside Global address can be shared by many inside hosts accessing the outside network.

Click each button for an example and explanation of the differences between NAT and PAT.

NAT

The figure shows a simple example of a NAT table. In this example, four hosts on the internal network are communicating to the outside network. The left column lists the addresses in the global address pool that are used by NAT to translate the Inside Local address of each host. Note the one-to-one relationship of Inside Global addresses to Inside Local addresses for each of the four hosts accessing the outside network. With NAT, an Inside Global address is needed for each host that needs to connect to the outside network.

Note: NAT forwards the incoming return packets to the original inside host by referring to the table and translating the Inside Global address back to the corresponding Inside Local address of the host.

PAT

While NAT only modifies the IPv4 addresses, PAT modifies both the IPv4 address and the port number. With PAT, there is generally only one, or very few, publicly exposed IPv4 addresses. The example NAT table shows one Inside Global address being used to translate the Inside Local addresses of the four inside hosts. PAT uses the Layer 4 port number to track the conversations of the four hosts.

Packets without a Layer 4 Segment - 6.2.6

What about IPv4 packets carrying data other than a TCP or UDP segment? These packets do not contain a Layer 4 port number. PAT translates most common protocols carried by IPv4 that do not use TCP or UDP as a transport layer protocol. The most common of these is ICMPv4. Each of these types of protocols is handled differently by PAT. For example, ICMPv4 query messages, echo requests, and echo replies include a Query ID. ICMPv4 uses the Query ID to identify an echo request with its corresponding echo reply. The Query ID is incremented with each echo request sent. PAT uses the Query ID instead of a Layer 4 port number.

Note: Other ICMPv4 messages do not use the Query ID. These messages and other protocols that do not use TCP or UDP port numbers vary and are beyond the scope of this curriculum.

Refer to **Packet Tracer Activity** for this chapter

Packet Tracer - Investigate NAT Operations - 6.2.7

You know that as a frame travels across a network, the MAC addresses change. But IPv4 addresses can also change when a packet is forwarded by a device configure with NAT. In this activity we will see what happens to IPv4 addresses during the NAT process.

In this Packet Tracer activity, you will:

■ Investigate NAT operation across the intranet

■ Investigate NAT operation across the internet

■ Conduct further investigations

NAT Advantages and Disadvantages - 6.3

Advantages of NAT - 6.3.1

NAT solves our problem of not having enough IPv4 addresses, but it can also create other problems. This topic addresses the advantages and disadvantage of NAT.

NAT provides many benefits, including the following:

- NAT conserves the legally registered addressing scheme by allowing the privatization of intranets. NAT conserves addresses through application port-level multiplexing. With NAT overload (PAT), internal hosts can share a single public IPv4 address for all external communications. In this type of configuration, very few external addresses are required to support many internal hosts.

- NAT increases the flexibility of connections to the public network. Multiple pools, backup pools, and load-balancing pools can be implemented to ensure reliable public network connections.

- NAT provides consistency for internal network addressing schemes. On a network not using private IPv4 addresses and NAT, changing the public IPv4 address scheme requires the readdressing of all hosts on the existing network. The costs of readdressing hosts can be significant. NAT allows the existing private IPv4 address scheme to remain while allowing for easy change to a new public addressing scheme. This means an organization could change ISPs and not need to change any of its inside clients.

- Using RFC 1918 IPv4 addresses, NAT hides the IPv4 addresses of users and other devices. Some people consider this a security feature; however, most experts agree that NAT does not provide security. A stateful firewall is what provides security on the edge of the network.

Disadvantages of NAT - 6.3.2

NAT does have drawbacks. The fact that hosts on the internet appear to communicate directly with the NAT-enabled device, rather than with the actual host inside the private network, creates a number of issues.

One disadvantage of using NAT is related to network performance, particularly for real time protocols such as VoIP. NAT increases forwarding delays because the translation of each IPv4 address within the packet headers takes time. The first packet is always process-switched going through the slower path. The router must look at every packet to decide whether it needs translation. The router must alter the IPv4 header, and possibly alter the TCP or UDP header. The IPv4 header checksum, along with the TCP or UDP checksum must be recalculated each time a translation is made. Remaining packets go through the fast-switched path if a cache entry exists; otherwise, they too are delayed.

This becomes more of an issue as the pools of public IPv4 addresses for ISPs become depleted. Many ISPs are having to assign customers a private IPv4 address instead of a public IPv4 address. This means the customer's router translates the packet from its private IPv4 address to the private IPv4 address of the ISP. Before forwarding the packet to another provider, the ISP will then perform NAT again, translating its private IPv4 addresses to one of its limited number of public IPv4 addresses. This process of two layers of NAT translation is known as Carrier Grade NAT (CGN).

Another disadvantage of using NAT is that end-to-end addressing is lost. This is known as the end-to-end principle. Many internet protocols and applications depend on end-to-end addressing from the source to the destination. Some applications do not work with NAT. For example, some security applications, such as digital signatures, fail because the source IPv4 address changes before reaching the destination. Applications that use physical addresses, instead of a qualified domain name, do not reach destinations that are translated across the NAT router. Sometimes, this problem can be avoided by implementing static NAT mappings.

End-to-end IPv4 traceability is also lost. It becomes much more difficult to trace packets that undergo numerous packet address changes over multiple NAT hops, making trouble-shooting challenging.

Using NAT also complicates the use of tunneling protocols, such as IPsec, because NAT modifies values in the headers, causing integrity checks to fail.

Services that require the initiation of TCP connections from the outside network, or state-less protocols, such as those using UDP, can be disrupted. Unless the NAT router has been configured to support such protocols, incoming packets cannot reach their destination. Some protocols can accommodate one instance of NAT between participating hosts (passive mode FTP, for example), but fail when both systems are separated from the internet by NAT.

Go to the online course to take the quiz and exam.

Check Your Understanding - NAT Advantages and Disadvantages - 6.3.3

Static NAT - 6.4

Refer to **Online Course** for Illustration

Static NAT Scenario - 6.4.1

In this topic, you will learn how to configure and verify static NAT. It includes a Packet Tracer activity to test your skills and knowledge. Static NAT is a one-to-one mapping between an inside address and an outside address. Static NAT allows external devices to initiate connections to internal devices using the statically assigned public address. For instance, an internal web server may be mapped to a specific inside global address so that it is accessible from outside networks.

The figure shows an inside network containing a web server with a private IPv4 address. Router R2 is configured with static NAT to allow devices on the outside network (internet) to access the web server. The client on the outside network accesses the web server using a public IPv4 address. Static NAT translates the public IPv4 address to the private IPv4 address.

Configure Static NAT - 6.4.2

There are two basic tasks when configuring static NAT translations:

Step 1. The first task is to create a mapping between the inside local address and the inside global addresses. For example, the 192.168.10.254 inside local address and the 209.165.201.5 inside global address in the figure are configured as a static NAT translation.

```
R2(config)# ip nat inside source static 192.168.10.254 209.165.201.5
```

Step 2. After the mapping is configured, the interfaces participating in the translation are configured as inside or outside relative to NAT. In the example, the R2 Serial 0/1/0 interface is an inside interface and Serial 0/1/1 is an outside interface.

```
R2(config)# interface serial 0/1/0

R2(config-if)# ip address 192.168.1.2 255.255.255.252

R2(config-if)# ip nat inside

R2(config-if)# exit

R2(config)# interface serial 0/1/1

R2(config-if)# ip address 209.165.200.1 255.255.255.252

R2(config-if)# ip nat outside
```

With this configuration in place, packets arriving on the inside interface of R2 (Serial 0/1/0) from the configured inside local IPv4 address (192.168.10.254) are translated and then forwarded towards the outside network. Packets arriving on the outside interface of R2 (Serial 0/1/1), that are addressed to the configured inside global IPv4 address (209.165.201.5), are translated to the inside local address (192.168.10.254) and then forwarded to the inside network.

> Refer to
> **Online Course**
> for Illustration

Analyze Static NAT - 6.4.3

Using the previous configuration, the figure illustrates the static NAT translation process between the client and the web server. Usually static translations are used when clients on the outside network (internet) need to reach servers on the inside (internal) network.

Verify Static NAT - 6.4.4

To verify NAT operation, issue the **show ip nat translations** command. This command shows active NAT translations. Because the example is a static NAT configuration, the translation is always present in the NAT table regardless of any active communications.

```
R2# show ip nat translations

Pro   Inside global     Inside local      Outside local     Outside global

---   209.165.201.5     192.168.10.254    ---               ---

Total number of translations: 1
```

If the command is issued during an active session, the output also indicates the address of the outside device as shown in the following example.

```
R2# show ip nat translations

Pro   Inside global     Inside local      Outside local     Outside global

tcp   209.165.201.5     192.168.10.254    209.165.200.254   209.165.200.254

---   209.165.201.5     192.168.10.254    ---               ---

Total number of translations: 2
```

Another useful command is **show ip nat statistics**, which displays information about the total number of active translations, NAT configuration parameters, the number of addresses in the pool, and the number of addresses that have been allocated.

To verify that the NAT translation is working, it is best to clear statistics from any past translations using the **clear ip nat statistics** command before testing.

```
R2# show ip nat statistics
Total active translations: 1 (1 static, 0 dynamic; 0 extended)
Outside interfaces:
  Serial0/1/1
Inside interfaces:
  Serial0/1/0
Hits: 0  Misses: 0
```
(output omitted)

After the client establishes a session with the web server, the **show ip nat statistics** displays an increase to four hits on the inside (Serial0/1/0) interface. This verifies that the static NAT translation is taking place on R2.

```
R2# show ip nat statistics
Total active translations: 1 (1 static, 0 dynamic; 0 extended)
Outside interfaces:
  Serial0/1/1
Inside interfaces:
  Serial0/1/0
Hits: 4  Misses: 1
```
(output omitted)

Refer to **Packet Tracer Activity** for this chapter

Packet Tracer - Configure Static NAT - 6.4.5

In IPv4 configured networks, clients and servers use private addressing. Before packets with private addressing can cross the internet, they need to be translated to public addressing. Servers that are accessed from outside the organization are usually assigned both a public and a private static IPv4 address. In this activity, you will configure static NAT so that outside devices can access an inside server at its public address.

In this Packet Tracer activity, you will:

- Test Access without NAT
- Configure Static NAT
- Test Access with NAT

Dynamic NAT - 6.5

Refer to
Online Course
for Illustration

Dynamic NAT Scenario - 6.5.1

In this topic, you will learn how to configure and verify dynamic NAT. It includes a Packet Tracer activity to test your skills and knowledge. Although static NAT provides a permanent mapping between an inside local address and an inside global address, dynamic NAT automatically maps inside local addresses to inside global addresses. These inside global addresses are typically public IPv4 addresses. Dynamic NAT, like static NAT, requires the configuration of the inside and outside interfaces participating in NAT with the **ip nat inside** and **ip nat outside** interface configuration commands. However, where static NAT creates a permanent mapping to a single address, dynamic NAT uses a pool of addresses.

The example topology shown in the figure has an inside network using addresses from the RFC 1918 private address space. Attached to router R1 are two LANs, 192.168.10.0/24 and 192.168.11.0/24. Router R2, the border router, is configured for dynamic NAT using a pool of public IPv4 addresses 209.165.200.226 through 209.165.200.240.

The pool of public IPv4 addresses (inside global address pool) is available to any device on the inside network on a first-come first-served basis. With dynamic NAT, a single inside address is translated to a single outside address. With this type of translation there must be enough addresses in the pool to accommodate all the inside devices needing concurrent access to the outside network. If all addresses in the pool are in use, a device must wait for an available address before it can access the outside network.

Note: Translating between public and private IPv4 addresses is by far the most common use of NAT. However, NAT translations can occur between pair of IPv4 addresses.

Refer to
Online Course
for Illustration

Refer to
Interactive Graphic
in online course

Configure Dynamic NAT - 6.5.2

The figure shows an example topology where the NAT configuration allows translation for all hosts on the 192.168.0.0/16 network. This includes the 192.168.10.0 and 192.168.11.0 LANs when the hosts generate traffic that enters interface S0/1/0 and exits S0/1/1. The host inside local addresses are translated to an available pool address in the range of 209.165.200.226 to209.165.200.240.

Click each button for a description and example of each step to configure static NAT.

Step 1

Define the pool of addresses that will be used for translation using the **ip nat pool** command. This pool of addresses is typically a group of public addresses. The addresses are defined by indicating the starting IPv4 address and the ending IPv4 address of the pool. The **netmask** or **prefix-length** keyword indicates which address bits belong to the network and which bits belong to the host for that range of addresses.

In the scenario, define a pool of public IPv4 addresses under the pool name NAT-POOL1.

```
R2(config)# ip nat pool NAT-POOL1 209.165.200.226 209.165.200.240 netmask
255.255.255.224
```

Step 2

Configure a standard ACL to identify (permit) only those addresses that are to be trans-lated. An ACL that is too permissive can lead to unpredictable results. Remember there is an implicit **deny all** statement at the end of each ACL.

In the scenario, define which addresses are eligible to be translated.

```
R2(config)# access-list 1 permit 192.168.0.0 0.0.255.255
```

Step 3

Bind the ACL to the pool, using the following command syntax:

Router(config)# **ip nat inside source list** {*access-list-number* | *access-list-name*} **pool** *pool-name*

This configuration is used by the router to identify which devices (**list**) receive which addresses (**pool**). In the scenario, bind NAT-POOL1 with ACL 1.

```
R2(config-if)# ip nat inside source list 1 pool NAT-POOL1
```

Step 4

Identify which interfaces are inside, in relation to NAT; this will be any interface that connects to the inside network.

In the scenario, identify interface serial 0/1/0 as an inside NAT interface.

```
R2(config)# interface serial 0/1/0
R2(config-if)# ip nat inside
```

Step 5

Identify which interfaces are outside, in relation to NAT; this will be any interface that connects to the outside network.

In the scenario, identify interface serial 0/1/1 as the outside NAT interface.

```
R2(config)# interface serial 0/1/1
R2(config-if)# ip nat outside
```

Analyze Dynamic NAT - Inside to Outside - 6.5.3

Refer to Online Course for Illustration

Using the previous configuration, the next two figures illustrate the dynamic NAT transla-tion process between two clients and the web server.

The figure below is used to illustrate the traffic flow from the inside network to the outside.

Analyze Dynamic NAT - Outside to Inside - 6.5.4

Refer to Online Course for Illustration

The figure below illustrates the remainder of the traffic flow between the clients and the server from the outside to the inside direction.

Verify Dynamic NAT - 6.5.5

The output of the **show ip nat translations** command displays all static translations that have been configured and any dynamic translations that have been created by traffic.

```
R2# show ip nat translations

Pro Inside global     Inside local     Outside local     Outside global

--- 209.165.200.228   192.168.10.10    ---               ---

--- 209.165.200.229   192.168.11.10    ---               ---

R2#
```

Adding the **verbose** keyword displays additional information about each translation, including how long ago the entry was created and used.

```
R2# show ip nat translation verbose

Pro Inside global     Inside local     Outside local     Outside global

tcp 209.165.200.228   192.168.10.10    ---               ---

    create 00:02:11, use 00:02:11 timeout:86400000, left 23:57:48, Map-Id(In): 1,

    flags:

none, use_count: 0, entry-id: 10, lc_entries: 0

tcp 209.165.200.229   192.168.11.10    ---               ---

    create 00:02:10, use 00:02:10 timeout:86400000, left 23:57:49, Map-Id(In): 1,

    flags:

none, use_count: 0, entry-id: 12, lc_entries: 0

R2#
```

By default, translation entries time out after 24 hours, unless the timers have been reconfigured with the **ip nat translation timeout** *timeout-seconds* command in global configuration mode.

To clear dynamic entries before the timeout has expired, use the **clear ip nat translation** privileged EXEC mode command as shown.

```
R2# clear ip nat translation *

R2# show ip nat translation
```

It is useful to clear the dynamic entries when testing the NAT configuration. The **clear ip nat translation** command can be used with keywords and variables to control which entries are cleared, as shown in the table. Specific entries can be cleared to avoid disrupting active sessions. Use the **clear ip nat translation *** privileged EXEC command to clear all translations from the table.

Command	Description
clear ip nat translation *	Clears all dynamic address translation entries from the NAT translation table.
clear ip nat translation inside *global-ip local-ip* [outside *local-ip global-ip*]	Clears a simple dynamic translation entry containing an inside translation or both inside and outside translation.
clear ip nat translation *protocol* inside *global-ip global-port local-ip local-port* [outside *local-ip local-port global-ip global-port*]	Clears an extended dynamic translation entry.

Note: Only the dynamic translations are cleared from the table. Static translations cannot be cleared from the translation table.

Another useful command, **show ip nat statistics**, displays information about the total number of active translations, NAT configuration parameters, the number of addresses in the pool, and how many of the addresses have been allocated.

```
R2# show ip nat statistics

Total active translations: 4 (0 static, 4 dynamic; 0 extended)

Peak translations: 4, occurred 00:31:43 ago

Outside interfaces:

  Serial0/1/1

Inside interfaces:

  Serial0/1/0

Hits: 47  Misses: 0

CEF Translated packets: 47, CEF Punted packets: 0

Expired translations: 5

Dynamic mappings:

-- Inside Source

[Id: 1] access-list 1 pool NAT-POOL1 refcount 4

 pool NAT-POOL1: netmask 255.255.255.224

      start 209.165.200.226 end 209.165.200.240

      type generic, total addresses 15, allocated 2 (13%), misses 0

(output omitted)

R2#
```

Alternatively, you can use the **show running-config** command and look for NAT, ACL, interface, or pool commands with the required values. Examine these carefully and correct any errors discovered. The example shows the NAT pool configuration.

```
R2# show running-config | include NAT

ip nat pool NAT-POOL1 209.165.200.226 209.165.200.240 netmask 255.255.255.224

ip nat inside source list 1 pool NAT-POOL1
```

Refer to **Packet Tracer Activity** for this chapter

Packet Tracer - Configure Dynamic NAT - 6.5.6

In this Packet Tracer, you will complete the following objectives:

- Configure Dynamic NAT
- Verify NAT Implementation

PAT - 6.6

Refer to **Online Course** for Illustration

PAT Scenario - 6.6.1

In this topic, you will learn how to configure and verify PAT. It includes a Packet Tracer activity to test your skills and knowledge. There are two ways to configure PAT, depending on how the ISP allocates public IPv4 addresses. In the first instance, the ISP allocates a single public IPv4 address that is required for the organization to connect to the ISP and in the other, it allocates more than one public IPv4 address to the organization.

Both methods will be demonstrated using the scenario shown in the figure.

Configure PAT to Use a Single IPv4 Address - 6.6.2

To configure PAT to use a single IPv4 address, simply add the keyword **overload** to the **ip nat inside source** command. The rest of the configuration is the similar to static and dynamic NAT configuration except that with PAT, multiple hosts can use the same public IPv4 address to access the internet.

In the example, all hosts from network 192.168.0.0/16 (matching ACL 1) that send traffic through router R2 to the internet will be translated to IPv4 address 209.165.200.225 (IPv4 address of interface S0/1/1). The traffic flows will be identified by port numbers in the NAT table because the **overload** keyword is configured.

```
R2(config)# ip nat inside source list 1 interface serial 0/1/0 overload

R2(config)# access-list 1 permit 192.168.0.0 0.0.255.255

R2(config)# interface serial0/1/0

R2(config-if)# ip nat inside

R2(config-if)# exit

R2(config)# interface Serial0/1/1

R2(config-if)# ip nat outside
```

Refer to
Online Course
for Illustration

Configure PAT to Use an Address Pool - 6.6.3

An ISP may allocate more than one public IPv4 address to an organization. In this scenario the organization can configure PAT to use a pool of IPv4 public addresses for translation.

If a site has been issued more than one public IPv4 address, these addresses can be part of a pool that is used by PAT. The small pool of addresses is shared among a larger number of devices, with multiple hosts using the same public IPv4 address to access the internet. To configure PAT for a dynamic NAT address pool, simply add the keyword **overload** to the **ip nat inside source** command.

The topology for this scenario is repeated in the figure for your convenience.

In the example, NAT-POOL2 is bound to an ACL to permit 192.168.0.0/16 to be translated. These hosts can share an IPv4 address from the pool because PAT is enabled with the keyword **overload**.

```
R2(config)# ip nat pool NAT-POOL2 209.165.200.226 209.165.200.240 netmask
255.255.255.224

R2(config)# access-list 1 permit 192.168.0.0 0.0.255.255

R2(config)# ip nat inside source list 1 pool NAT-POOL2 overload

R2(config)# interface serial0/1/0

R2(config-if)# ip nat inside

R2(config)# interface serial0/1/0

R2(config-if)# ip nat outside
```

Refer to
Online Course
for Illustration

Analyze PAT - PC to Server - 6.6.4

The process of NAT overload is the same whether a pool of addresses is used, or a single address is used. In this figure, PAT is configured to use a single public IPv4 address, instead of a pool of addresses. PC1 wants to communicate with the web server, Svr1. At the same time another client, PC2, wants to establish a similar session with the web server Svr2. Both PC1 and PC2 are configured with private IPv4 addresses, with R2 enabled for PAT.

Refer to
Online Course
for Illustration

Analyze PAT - Server to PC - 6.6.5

Although PC1 and PC2 are using the same translated address, the inside global address of 209.165.200.225, and the same source port number of 1444; the modified port number for PC2 (1445) makes each entry in the NAT table unique. This will become evident with the packets sent from the servers back to the clients, as shown in the figure.

Verify PAT - 6.6.6

Router R2 has been configured to provide PAT to the 192.168.0.0/16 clients. When the internal hosts exit router R2 to the internet, they are translated to an IPv4 address from the PAT pool with a unique source port number.

The same commands used to verify static and dynamic NAT are used to verify PAT, as shown in the example output. The **show ip nat translations** command displays the translations from two different hosts to different web servers. Notice that two different inside hosts are allocated the same IPv4 address of 209.165.200.226 (inside global address). The source port numbers in the NAT table differentiate the two transactions.

```
R2# show ip nat translations
Pro Inside global          Inside local  Outside local   Outside global
tcp 209.165.200.225:1444   192.168.10.10:1444  209.165.201.1:80   209.165.201.1:80
tcp 209.165.200.225:1445   192.168.11.10:1444  209.165.202.129:80 209.165.202.129:80
R2#
```

In the next example, the **show ip nat statistics** command verifies that NAT-POOL2 has allocated a single address for both translations. Included in the output is information about the number and type of active translations, NAT configuration parameters, the number of addresses in the pool, and how many have been allocated.

```
R2# show ip nat statistics
Total active translations: 4 (0 static, 2 dynamic; 2 extended)
Peak translations: 2, occurred 00:31:43 ago
Outside interfaces:
  Serial0/1/1
Inside interfaces:
  Serial0/1/0
Hits: 4  Misses: 0
CEF Translated packets: 47, CEF Punted packets: 0
Expired translations: 0
Dynamic mappings:
-- Inside Source
[Id: 3] access-list 1 pool NAT-POOL2 refcount 2
 pool NAT-POOL2: netmask 255.255.255.224
       start 209.165.200.225 end 209.165.200.240
       type generic, total addresses 15, allocated 1 (6%), misses 0
(output omitted)
R2#
```

Refer to Packet Tracer Activity for this chapter

Packet Tracer - Configure PAT - 6.6.7

In this Packet Tracer, you will complete the following objectives:

- Part 1: Configure Dynamic NAT with Overload
- Part 2: Verify Dynamic NAT with Overload Implementation
- Part 3: Configure PAT using an Interface
- Part 4: Verify PAT Interface Implementation

NAT64 - 6.7

NAT for IPv6? - 6.7.1

Because many networks use both IPv4 and IPv6, there needs to be a way to use IPv6 with NAT. This topic discusses how IPv6 can be integrated with NAT. IPv6, with a 128-bit address, provides 340 undecillion addresses. Therefore, address space is not an issue. IPv6 was developed with the intention of making NAT for IPv4 with translation between public and private IPv4 addresses unnecessary. However, IPv6 does include its own IPv6 private address space, unique local addresses (ULAs).

IPv6 unique local addresses (ULA) are similar to RFC 1918 private addresses in IPv4 but have a different purpose. ULA addresses are meant for only local communications within a site. ULA addresses are not meant to provide additional IPv6 address space, nor to provide a level of security.

IPv6 does provide for protocol translation between IPv4 and IPv6 known as NAT64.

Refer to
Online Course
for Illustration

NAT64 - 6.7.2

NAT for IPv6 is used in a much different context than NAT for IPv4. The varieties of NAT for IPv6 are used to transparently provide access between IPv6-only and IPv4-only networks, as shown in the figure. It is not used as a form of private IPv6 to global IPv6 translation.

Ideally, IPv6 should be run natively wherever possible. This means IPv6 devices communicating with each other over IPv6 networks. However, to aid in the move from IPv4 to IPv6, the IETF has developed several transition techniques to accommodate a variety of IPv4-to-IPv6 scenarios, including dual-stack, tunneling, and translation.

Dual-stack is when the devices are running protocols associated with both IPv4 and IPv6. Tunneling for IPV6 is the process of encapsulating an IPv6 packet inside an IPv4 packet. This allows the IPv6 packet to be transmitted over an IPv4-only network.

NAT for IPv6 should not be used as a long-term strategy, but as a temporary mechanism to assist in the migration from IPv4 to IPv6. Over the years, there have been several types of NAT for IPv6 including Network Address Translation-Protocol Translation (NAT-PT). NAT-PT has been deprecated by IETF in favor of its replacement, NAT64. NAT64 is beyond the scope of this curriculum.

Module Practice and Quiz - 6.8

Packet Tracer - Configure NAT for IPv4 - 6.8.1

In this Packet Tracer, you will complete the following objectives:

■ Configure Dynamic NAT with PAT

■ Configure Static NAT

Refer to **Lab Activity** for this chapter

Lab - Configure NAT for IPv4 - 6.8.2

In this lab, you will complete the following objectives:

- Part 1: Build the Network and Configure Basic Device Settings
- Part 2: Configure and verify NAT for IPv4
- Part 3: Configure and verify PAT for IPv4
- Part 4: Configure and verify Static NAT for IPv4

What did I learn in this module? - 6.8.3

NAT Characteristics

There are not enough public IPv4 addresses to assign a unique address to each device connected to the internet. Private IPv4 addresses cannot be routed over the internet. To allow a device with a private IPv4 address to access devices and resources outside of the local network, the private address must first be translated to a public address. NAT provides the translation of private addresses to public addresses. The primary use of NAT is to conserve public IPv4 addresses. It allows networks to use private IPv4 addresses internally and provides translation to a public address only when needed. When an internal device sends traffic out of the network, the NAT-enabled router translates the internal IPv4 address of the device to a public address from the NAT pool. In NAT terminology, the inside network is the set of networks that is subject to translation. The outside network refers to all other networks. When determining which type of address is used, it is important to remember that NAT terminology is always applied from the perspective of the device with the translated address:

- **Inside address** - The address of the device which is being translated by NAT.
- **Outside address** - The address of the destination device.

NAT also uses the concept of local or global with respect to addresses:

- **Local address** - A local address is any address that appears on the inside portion of the network.
- **Global address** - A global address is any address that appears on the outside portion of the network.

Types of NAT

Static NAT uses a one-to-one mapping of local and global addresses. These mappings are configured by the network administrator and remain constant. Static NAT is particularly useful for web servers or devices that must have a consistent address that is accessible from the internet, such as a company web server. Static NAT requires that enough public addresses are available to satisfy the total number of simultaneous user sessions. Dynamic NAT uses a pool of public addresses and assigns them on a first-come, first-served basis. When an inside device requests access to an outside network, dynamic NAT assigns an available public IPv4 address from the pool. Similar to static NAT, dynamic NAT requires that enough public addresses are available to satisfy the total number of simultaneous user sessions. Port Address Translation (PAT), also known as NAT overload, maps multiple private IPv4 addresses to a single public IPv4 address or a few addresses. This is the most common form of NAT for both the home and the enterprise. PAT ensures that devices use a different TCP port number for each session with a server on the internet. PAT attempts to preserve the original source port. However, if the original source port is already used, PAT assigns the first available port number starting from the beginning of the appropriate port group. PAT translates most common protocols carried by IPv4 that do not use TCP or UDP as a transport layer protocol. The most common of these is ICMPv4.

NAT	PAT
One-to-one mapping between Inside Local and Inside Global addresses.	One Inside Global address can be mapped to many Inside Local address.
Uses only IPv4 addresses in translation process.	Uses IPv4 addresses and TCP or UDP source port numbers in translation process.
A unique Inside Global address is required for each inside host accessing the outside network.	A single unique Inside Global address can be shared by many inside hosts accessing the outside network.

NAT Advantages and Disadvantages

Advantages: NAT conserves the legally registered addressing scheme by allowing the privatization of intranets. NAT increases the flexibility of connections to the public network. NAT provides consistency for internal network addressing schemes. NAT hides user IPv4 addresses.

Disadvantages: NAT increases forwarding delays because the translation of each IPv4 address within the packet headers takes time. The process of two layers of NAT translation is known as Carrier Grade NAT (CGN). End-to-end addressing is lost. Many internet protocols and applications depend on end-to-end addressing from the source to the destination. End-to-end IPv4 traceability is also lost. Using NAT also complicates the use of tunneling protocols, such as IPsec, because NAT modifies values in the headers, causing integrity checks to fail.

Static NAT

Static NAT is a one-to-one mapping between an inside address and an outside address. Static NAT allows external devices to initiate connections to internal devices using the statically assigned public address. The first task is to create a mapping between the inside local address and the inside global addresses using the **ip nat inside source static** command. After the mapping is configured, the interfaces participating in the translation are configured as inside or outside relative to NAT using the **ip nat inside** and **ip nat outside** commands. To verify NAT operation use the **show ip nat translations** command. To verify that the NAT translation is working, it is best to clear statistics from any past translations using the **clear ip nat statistics** command before testing.

Dynamic NAT

Dynamic NAT automatically maps the inside local addresses to inside global addresses. Dynamic NAT, like static NAT, requires the configuration of the inside and outside interfaces participating in NAT. Dynamic NAT uses a pool of addresses translating a single inside address to a single outside address. The pool of public IPv4 addresses (inside global address pool) is available to any device on the inside network on a first-come first-served basis. With this type of translation there must be enough addresses in the pool to accommodate all the inside devices needing concurrent access to the outside network. If all addresses in the pool are in use, a device must wait for an available address before it can access the outside network.

To configure dynamic NAT, first define the pool of addresses that will be used for translation using the **ip nat pool** command. The addresses are defined by indicating the starting IPv4 address and the ending IPv4 address of the pool. The **netmask** or **prefix-length** keyword indicates which address bits belong to the network and which bits belong to the host for the range of addresses. Configure a standard ACL to identify (permit) only those addresses that are to be translated. Bind the ACL to the pool, using the following command syntax: Router(config)# **ip nat inside source list** {*access-list-number* | *access-list-name*} **pool** *pool-name*. Identify which interfaces are inside, in relation to NAT. Identify which interfaces are outside, in relation to NAT.

To verify dynamic NAT configurations, The output of the **show ip nat translations** command shown displays all static translations that have been configured and any dynamic translations that have been created by traffic. Adding the **verbose** keyword displays additional information about each translation, including how long ago the entry was created and used. By default, translation entries time out after 24 hours, unless the timers have been reconfigured with the **ip nat translation timeout** *timeout-seconds* command in global configuration mode. To clear dynamic entries before the timeout has expired, use the **clear ip nat translation** privileged EXEC mode command.

PAT

There are two ways to configure PAT, depending on how the ISP allocates public IPv4 addresses. In the first instance, the ISP allocates a single public IPv4 address that is required for the organization to connect to the ISP and in the other, it allocates more than one public IPv4 address to the organization. To configure PAT to use a single IPv4 address, simply add the keyword **overload** to the **ip nat inside source** command. The rest of the configuration is the similar to static and dynamic NAT configuration except that with PAT, multiple hosts can use the same public IPv4 address to access the internet. To configure PAT for a dynamic NAT address pool, simply add the keyword **overload** to the **ip nat inside source** command. Multiple hosts can share an IPv4 address from the pool because PAT is enabled with the keyword **overload**.

To verify PAT configurations us the **show ip nat translations** command. The source port numbers in the NAT table differentiate the transactions. The **show ip nat statistics** command verifies that the NAT-POOL has allocated a single address for multiple translations. Included in the output is information about the number and type of active translations, NAT configuration parameters, the number of addresses in the pool, and how many have been allocated.

NAT64

IPv6 was developed with the intention of making NAT for IPv4 with translation between public and private IPv4 addresses unnecessary. However, IPv6 does include its own IPv6 private address space, unique local addresses (ULAs). IPv6 unique local addresses (ULA) are similar to RFC 1918 private addresses in IPv4 but have a different purpose. ULA addresses are meant for only local communications within a site. ULA addresses are not meant to provide additional IPv6 address space, nor to provide a level of security; however, IPv6 does provide for protocol translation between IPv4 and IPv6 known as NAT64. NAT for IPv6 is used in a much different context than NAT for IPv4. The varieties of NAT for IPv6 are used to transparently provide access between IPv6-only and IPv4-only networks. To aid in the move from IPv4 to IPv6, the IETF has developed several transition techniques to accommodate a variety of IPv4-to-IPv6 scenarios, including dual-stack, tunneling, and translation. Dual-stack is when the devices are running protocols associated with both the IPv4 and IPv6. Tunneling for IPV6 is the process of encapsulating an IPv6 packet inside an IPv4 packet. This allows the IPv6 packet to be transmitted over an IPv4-only network. NAT for IPv6 should not be used as a long-term strategy, but as a temporary mechanism to assist in the migration from IPv4 to IPv6.

Go to the online
course to take the
quiz and exam.

Chapter Quiz - NAT for IPv4

Your Chapter Notes

CHAPTER 7

WAN Concepts

Introduction - 7.0

Why should I take this module? - 7.0.1

Welcome to WAN Concepts!

As you know, local area networks are called LANs. The name implies that your LAN is local to you and your small home or office business. But what if your network is for a larger business, perhaps even a global enterprise? You cannot operate a large business with multiple sites without a wide area network, which is called a WAN. This module explains what WANs are and how they connect to the internet and also back to your LAN. Understanding the purpose and functions of WANs is foundational to your understanding of modern networks. So let's jump in to WAN Concepts!

What will I learn to do in this module? - 7.0.2

Module Title: WAN Concepts

Module Objective: Explain how WAN access technologies can be used to satisfy business requirements.

Topic Title	Topic Objective
Purpose of WANs	Explain the purpose of a WAN.
WAN Operations	Explain how WANs operate.
Traditional WAN Connectivity	Compare traditional WAN connectivity options.
Modern WAN Connectivity	Compare modern WAN connectivity options.
Internet-Based Connectivity	Compare internet-based connectivity options.

Purpose of WANs - 7.1

LANs and WANs - 7.1.1

Refer to Online Course for Illustration

Whether at work or at home, we all use Local Area Networks (LANs). However, LANs are limited to a small geographical area.

A Wide Area Network (WAN) is required to connect beyond the boundary of the LAN. A WAN is a telecommunications network that spans over a relatively large geographical area. A WAN operates beyond the geographic scope of a LAN.

In the figure, WAN services are required to interconnect an enterprise campus network to remote LANs at branch sites, telecommuter sites, and remote users.

The table highlights differences between LANs and WANs.

Local Area Networks (LANs)	Wide Area Networks (WANs)
LANs provide networking services within a small geographic area (i.e., home network, office network, building network, or campus network).	WANs provide networking services over large geographical areas (i.e., in and between cities, countries, and continents).
LANs are used to interconnect local computers, peripherals, and other devices.	WANs are used to interconnect remote users, networks, and sites.
A LAN is owned and managed by an organization or home user.	WANs are owned and managed by internet service, telephone, cable, and satellite providers.
Other than the network infrastructure costs, there is no fee to use a LAN.	WAN services are provided for a fee.
LANs provide high bandwidth speeds using wired Ethernet and Wi-Fi services.	WANs providers offer low to high bandwidth speeds, over long distances using complex physical networks.

Private and Public WANs - 7.1.2

WANs may be built by a variety of different types of organizations, as follows:

- An organization that wants to connect users in different locations

- An ISP that wants to connect customers to the internet

- An ISP or telecommunications that wants to interconnect ISPs

A private WAN is a connection that is dedicated to a single customer. This provides for the following:

- Guaranteed service level

- Consistent bandwidth

- Security

A public WAN connection is typically provided by an ISP or telecommunications service provider using the internet. In this case, the service levels and bandwidth may vary, and the shared connections do not guarantee security.

Refer to
Interactive Graphic
in online course

Refer to
Online Course
for Illustration

WAN Topologies - 7.1.3

Physical topologies describe the physical network infrastructure used by data when it is travelling from a source to a destination. The physical WAN topology used in WANs is complex and for the most part, unknown to users. Consider a user in New York establishing a video conference call with a user in Tokyo, Japan. Other than the user's internet connection in New York, it would not be feasible to identify the all of the actual physical connections that are needed to support the video call.

Instead, WAN topologies are described using a logical topology. Logical topologies describe the virtual connection between the source and destination. For example, the video conference call between the user in New York and Japan would be a logical point-to-point connection.

WANs are implemented using the following logical topology designs:

- Point-to-Point Topology
- Hub-and-Spoke Topology
- Dual-homed Topology
- Fully Meshed Topology
- Partially Meshed Topology

Note: Large networks usually deploy a combination of these topologies.

Click each button for an illustration and explanation of each WAN logical topology.

Point-to-Point Topology

A point-to-point topology, as shown in the figure, employs a point-to-point circuit between two endpoints.

Point-to-point links often involve dedicated, leased-line connections from the corporate edge point to the provider networks. A point-to-point connection involves a Layer 2 transport service through the service provider network. Packets sent from one site are delivered to the other site and vice versa. A point-to-point connection is transparent to the customer network. It seems as if there is a direct physical link between two endpoints.

It can become expensive if many point-to-point connections are required.

Hub-and-Spoke Topology

A hub-and-spoke topology enables a single interface on the hub router to be shared by all spoke circuits. Spoke routers can be interconnected through the hub router using virtual circuits and routed subinterfaces. The figure displays a sample hub-and-spoke topology consisting of three spoke routers connecting to a hub router across a WAN cloud.

A hub-and-spoke topology is a single-homed topology. There is only one hub router and all communication must go through it. Therefore, spoke routers can only communicate with each other through the hub router. Consequently, the hub router represents a single point of failure. If it fails, inter-spoke communication also fails.

Dual-homed Topology

A dual-homed topology provides redundancy. As shown in in the figure, two hub routers are dual-homed and redundantly attached to three spoke routers across a WAN cloud.

The advantage of dual-homed topologies is that they offer enhanced network redundancy, load balancing, distributed computing and processing, and the ability to implement backup service provider connections.

The disadvantage is that they are more expensive to implement than single-homed topologies. This is because they require additional networking hardware, such as additional routers and switches. Dual-homed topologies are also more difficult to implement because they require additional, and more complex, configurations.

Fully Meshed Topology

A fully meshed topology uses multiple virtual circuits to connect all sites, as shown in the figure.

This is the most fault-tolerant topology of the five shown. For instance, if site B lost connectivity to site A, it could send the data through either site C or site D.

Partially Meshed Topology

A partially meshed topology connects many but not all sites. For example, in the figure sites A, B, C are still fully meshed. Site D must connect to site A to reach sites B and C.

Refer to
Interactive Graphic
in online course

Refer to
Online Course
for Illustration

Carrier Connections - 7.1.4

Another aspect of WAN design is how an organization connects to the internet. An organization usually signs a service level agreement (SLA) with a service provider. The SLA outlines the expected services relating to the reliability and availability of the connection. The service provider may or may not be the actual carrier. A carrier owns and maintains the physical connection and equipment between the provider and the customer. Typically, an organization will choose either a single-carrier or dual-carrier WAN connection.

Click each button for an illustration and explanation of each carrier connection type.

Single-Carrier WAN Connection

A single-carrier connection is when an organization connects to only one service provider, as shown in the figure. An SLA is negotiated between the organization and the service provider. The disadvantage of this design is the carrier connection and service provider are both single points of failure. Connectivity to the internet would be lost if the carrier link or the provider router failed.

Dual-Carrier WAN Connection

A dual-carrier connection provides redundancy and increases network availability, as shown in the figure. The organization negotiates separate SLAs with two different service providers. The organization should ensure that the two providers each use a different carrier. Although more expensive to implement, the second connection can be used for redundancy as a backup link. It could also be used to improve network performance and load balance internet traffic.

Refer to
Interactive Graphic
in online course

Refer to
Online Course
for Illustration

Evolving Networks - 7.1.5

Network requirements of a company can change dramatically as the company grows over time. Distributing employees saves costs in many ways, but it puts increased demands on the network. Not only must a network meet the day-to-day operational needs of the business, but it must be able to adapt and grow as the company changes. Network designers and administrators meet these challenges by carefully choosing network technologies, protocols, and service providers. They must also optimize their networks by using a variety of network design techniques and architectures.

To illustrate differences between network size, we will use a fictitious company called SPAN Engineering as it grows from a small, local, business into a global enterprise. SPAN Engineering, an environmental consulting firm, has developed a special process for converting household waste into electricity and is developing a small pilot project for a municipal government in its local area.

Click each button for an illustration and description of the SPAN network as it evolves from a small network to a global enterprise.

Small Network

The company initially consisted of 15 employees working in a small office, as shown in the figure.

They used a single LAN that was connected to a wireless router for sharing data and peripherals. The connection to the internet is through a common broadband service called Digital Subscriber Line (DSL), which is supplied by their local telephone service provider. To support their IT requirements, they contracted services from the DSL provider.

Campus Network

Within a few years, the company grew and required several floors of a building, as shown in the figure.

The company now required a Campus Area Network (CAN). A CAN interconnects several LANs within a limited geographical area. Multiple LANs are required to segment the various departments that are connecting to multiple switches in a campus network environment.

The network includes dedicated servers for email, data transfer, and file storage, and web-based productivity tools and applications. A firewall secures internet access to corporate users. The business now requires in-house IT staff to support and maintain the network.

Branch Network

A few years later, the company expanded and added a branch site in the city, and remote and regional sites in other cities, as shown in the figure.

The company now required a metropolitan area network (MAN) to interconnect sites within the city. A MAN is larger than a LAN, but smaller than a WAN.

To connect to the central office, branch offices in nearby cities used private dedicated lines through their local service provider. Offices in other cities and countries require the services of a WAN or may use internet services to connect distant locations. However, the internet introduces security and privacy issues that the IT team must address.

Distributed Network

SPAN Engineering has now been in business for 20 years and has grown to thousands of employees distributed in offices worldwide, as shown in the figure.

To reduce network costs, SPAN encouraged teleworking and virtual teams using web-based applications, including web-conferencing, e-learning, and online collaboration tools to increase productivity and reduce costs. Site-to-site and remote access Virtual Private Networks (VPNs) enable the company to use the internet to connect easily and securely with employees and facilities around the world.

Go to the online course to take the quiz and exam.

Check Your Understanding - Purpose of WANs - 7.1.6

WAN Operations - 7.2

WAN Standards - 7.2.1

Now that you understand how critical WANs are to large networks, this topic discusses how they work. The concept of a WAN has been around for many years. Consider that the telegraph system was the first large-scale WAN, followed by radio, telephone system, television, and now data networks. Many of the technologies and standards developed for these WANs were used as the basis for network WANs.

Modern WAN standards are defined and managed by a number of recognized authorities including the following:

- **TIA/EIA** - Telecommunications Industry Association and Electronic Industries Alliance

- **ISO** - International Organization for Standardization

- **IEEE** - Institute of Electrical and Electronics Engineers

Refer to **Online Course** for Illustration

WANs in the OSI Model - 7.2.2

Most WAN standards focus on the physical layer (OSI Layer 1) and the data link layer (OSI Layer 2), as shown in the figure.

Layer 1 Protocols

Layer 1 protocols describe the electrical, mechanical, and operational components needed to transmit bits over a WAN. For example, service providers commonly use high-bandwidth optical fiber media to span long distances (i.e., long haul) using the following Layer 1 optical fiber protocol standards:

- Synchronous Digital Hierarchy (SDH)

- Synchronous Optical Networking (SONET)

- Dense Wavelength Division Multiplexing (DWDM)

SDH and SONET essentially provide the same services and their transmission capacity can be increased by using DWDM technology.

Layer 2 Protocols

Layer 2 protocols define how data will be encapsulated into a frame.

Several Layer 2 protocols have evolved over the years including the following:

- Broadband (i.e., DSL and Cable)
- Wireless
- Ethernet WAN (Metro Ethernet)
- Multiprotocol Label Switching (MPLS)
- Point-to-Point Protocol (PPP) (less used)
- High-Level Data Link Control (HDLC) (less used)
- Frame Relay (legacy)
- Asynchronous Transfer Mode (ATM) (legacy)

Refer to
Online Course
for Illustration

Common WAN Terminology - 7.2.3

The WAN physical layer describes the physical connections between the company network and the service provider network.

There are specific terms used to describe WAN connections between the subscriber (i.e., the company / client) and the WAN service provider, as shown in the figure.

Refer to the table for an explanation of the term shown in the figure, as well as some additional WAN-related terms.

WAN Term	Description
Data Terminal Equipment (DTE)	• This is the device that connects the subscriber LANs to the WAN communication device (i.e., DCE). • Inside hosts send their traffic to the DTE device. • The DTE connects to the local loop through the DCE. • The DTE device is usually a router but could be a host or server.
Data Communications Equipment (DCE)	• Also called data circuit-terminating equipment, this is the device used to communicate with the provider. • The DCE primarily provides an interface to connect subscribers to a communication link on the WAN cloud.
Customer Premises Equipment (CPE)	• This is the DTE and DCE devices (i.e., router, modem, optical converter) located on the enterprise edge. • The subscriber either owns the CPE or leases the CPE from the service provider.
Point-of-Presence (POP)	• This is the point where the subscriber connects to the service provider network.

WAN Term	Description
Demarcation Point	• This is a physical location in a building or complex that officially separates the CPE from service provider equipment.
	• The demarcation point is typically a cabling junction box, located on the customer premises, that connects the CPE wiring to the local loop.
	• It identifies the location where the network operation responsibility changes from the subscriber to the service provider.
	• When problems arise, it is necessary to determine whether the user or the service provider is responsible for troubleshooting or repair.
Local Loop (or last mile)	• This is the actual copper or fiber cable that connects the CPE to the CO of the service provider.
Central Office (CO)	• This is the local service provider facility or building that connects the CPE to the provider network.
Toll network	• This includes backhaul, long-haul, all-digital, fiber-optic communications lines, switches, routers, and other equipment inside the WAN provider network.
Backhaul network	• (Not shown) Backhaul networks connect multiple access nodes of the service provider network.
	• Backhaul networks can span over municipalities, countries and regions.
	• Backhaul networks are also connected to internet service providers and to the backbone network.
Backbone network	• (Not shown) These are large, high-capacity networks used to interconnect service provider networks and to create a redundant network.
	• Other service providers can connect to the backbone directly or through another service provider.
	• Backbone network service providers are also called Tier-1 providers.

Refer to
Online Course
for Illustration

WAN Devices - 7.2.4

There are many types of devices that are specific to WAN environments. However, the end-to-end data path over a WAN is usually from source DTE to the DCE, then to the WAN cloud, then to the DCE to and finally to the destination DTE, as shown in the figure.

Refer to table for an explanation of the WAN devices shown in the figure.

WAN Device	Description
Voiceband Modem	• Also known as dial-up modem.
	• Legacy device that converted (i.e., modulated) the digital signals produced by a computer into analog voice frequencies.
	• Uses telephone lines to transmit data.

WAN Device	Description
DSL Modem and **Cable Modem**	• Collectively known as broadband modems, these high-speed digital modems connect to the DTE router using Ethernet. • DSL modems connect to the WAN using telephone lines. • Cable modems connect to the WAN using coaxial lines. • Both operate in a similar manner to the voiceband modem but use higher broadband frequencies and transmission speeds.
CSU/DSU	• Digital-leased lines require a CSU and a DSU. • It connects a digital device to a digital line. • A CSU/DSU can be a separate device like a modem or it can be an interface on a router. • The CSU provides termination for the digital signal and ensures connection integrity through error correction and line monitoring. • The DSU converts the line frames into frames that the LAN can interpret and vice versa.
Optical Converter	• Also known as an optical fiber converter. • These devices connect fiber-optic media to copper media and convert optical signals to electronic pulses.
Wireless Router or **Access Point**	• Devices are used to wirelessly connect to a WAN provider. • Routers could also use cellular wireless connectivity.
WAN Core devices	• The WAN backbone consists of multiple high-speed routers and Layer 3 switches. • A router or multilayer switch must be able to support multiple telecommunications interfaces of the highest speed used in the WAN core. • It must also be able to forward IP packets at full speed on all those interfaces. • The router or multilayer switch must also support the routing protocols being used in the core.

Note: The preceding list is not exhaustive and other devices may be required, depending on the WAN access technology chosen.

Refer to **Interactive Graphic** in online course

Serial Communication - 7.2.5

Almost all network communications occur using a serial communication delivery. Serial communication transmits bits sequentially over a single channel. In contrast, parallel communications simultaneously transmit several bits using multiple wires.

Click Play to see an illustration of the difference between serial and parallel connections.

Although a parallel connection theoretically transfers data eight times faster than a serial connection, it is prone to synchronization problems. As the cable length increases, the synchronization timing between multiple channels becomes more sensitive to distance.

For this reason, parallel communication is limited to very short distances only (e.g., copper media is limited to less than 8 meters (i.e., 26 feet).

Therefore, parallel communication is not a viable WAN communication method because of its length restriction. It is however a viable solution in data centers where distances between servers and switches are relatively short.

For instance, the Cisco Nexus switches in data centers support parallel optics solutions to transfer more data signals and achieve higher speeds (i.e., 40 Gbps and 100 Gbps).

Refer to
Interactive Graphic
in online course

Circuit-Switched Communication - 7.2.6

Network communication can be implemented using circuit-switched communication. A circuit-switched network establishes a dedicated circuit (or channel) between endpoints before the users can communicate.

Specifically, circuit switching dynamically establishes a dedicated virtual connection through the service provider network before voice or data communication can start.

For example, when a user makes a telephone call using a landline, the number called is used by the provider equipment to create a dedicated circuit from the caller to the called party.

Note: A landline describes a telephone situated in a fixed location that is connected to the provider using copper or fiber-optic media.

During transmission over a circuit-switched network, all communication uses the same path. The entire fixed capacity allocated to the circuit is available for the duration of the connection, regardless of whether there is information to transmit or not. This can lead to inefficiencies in circuit usage. For this reason, circuit switching is generally not suited for data communication.

The two most common types of circuit-switched WAN technologies are the public switched telephone network (PSTN) and the legacy Integrated Services Digital Network (ISDN).

Click Play in the figure to see how circuit switching works.

Refer to
Interactive Graphic
in online course

Packet-Switched Communications - 7.2.7

Network communication is most commonly implemented using packet-switched communication. In contrast to circuit-switching, packet-switching segments traffic data into packets that are routed over a shared network. Packet-switched networks do not require a circuit to be established, and they allow many pairs of nodes to communicate over the same channel.

Packet switching is much less expensive and more flexible than circuit switching. Although susceptible to delays (latency) and variability of delay (jitter), modern technology allows satisfactory transport of voice and video communications on these networks.

Common types of packet-switched WAN technologies are Ethernet WAN (Metro Ethernet), Multiprotocol Label Switching (MPLS), as well as legacy Frame Relay and legacy Asynchronous Transfer Mode (ATM).

Click Play in the figure to see a packet-switching example.

SDH, SONET, and DWDM - 7.2.8

Refer to **Online Course** for Illustration

Service provider networks use fiber-optic infrastructures to transport user data between destinations. Fiber-optic cable is far superior to copper cable for long distance transmissions due to its much lower attenuation and interference.

There are two optical fiber OSI layer 1 standards available to service providers:

- **SDH** - Synchronous Digital Hierarchy (SDH) is a global standard for transporting data over fiber-optic cable.

- **SONET** - Synchronous Optical Networking (SONET) is the North American standard that provides the same services as SDH.

Both standards are essentially the same and therefore, they are often listed as SONET/SDH.

SDH/SONET define how to transfer multiple data, voice, and video communications over optical fiber using lasers or light-emitting diodes (LEDs) over great distances. Both standards are used on the ring network topology that contains the redundant fiber paths that allow traffic to flow in both directions.

Dense Wavelength Division Multiplexing (DWDM) is a newer technology that increases the data-carrying capacity of SDH and SONET by simultaneously sending multiple streams of data (multiplexing) using different wavelengths of light, as shown in the figure.

Go to the online course to take the quiz and exam.

Check Your Understanding - WAN Operations - 7.2.9

Traditional WAN Connectivity - 7.3

Traditional WAN Connectivity Options - 7.3.1

Refer to **Online Course** for Illustration

To understand the WANs of today, it helps to know where they started. This topic discusses WAN connectivity options from the beginning. When LANs appeared in the 1980s, organizations began to see the need to interconnect with other locations. To do so, they needed their networks to connect to the local loop of a service provider. This was accomplished by using dedicated lines, or by using switched services from a service provider.

The figure summarizes the traditional WAN connectivity options.

Note: There are several WAN access connection options that the enterprise edge can use to connect over the local loop to the provider. These WAN access options differ in technology, bandwidth, and cost. Each has distinct advantages and disadvantages. Familiarity with these technologies is an important part of network design.

Common WAN Terminology - 7.3.2

When permanent dedicated connections were required, a point-to-point link using copper media was used to provide a pre-established WAN communications path from the customer premises to the provider network. Point-to-point lines could be leased from a service provider and were called "leased lines". The term refers to the fact that the organization pays a monthly lease fee to a service provider to use the line.

Leased lines have existed since the early 1950s and for this reason, are referred to by different names such as leased circuits, serial link, serial line, point-to-point link, and T1/E1 or T3/E3 lines.

Leased lines are available in different fixed capacities and are generally priced based on the bandwidth required and the distance between the two connected points.

There are two systems used to define the digital capacity of a copper media serial link:

- **T-carrier** - Used in North America, T-carrier provides T1 links supporting bandwidth up to 1.544 Mbps and T3 links supporting bandwidth up to 43.7 Mbps.

- **E-carrier** - Used in Europe, E-carrier provides E1 links supporting bandwidth up to 2.048 Mbps and E3 links supporting bandwidth up to 34.368 Mbps.

Note: The copper cable physical infrastructure has largely been replaced by optical fiber network. Transmission rates in optical fiber network are given in terms of Optical Carrier (OC) transmission rates, which define the digital transmitting capacity of a fiber-optic network.

The table summarizes the advantages and disadvantages of leased lines.

Advantages	
Simplicity	Point-to-point communication links require minimal expertise to install and maintain.
Quality	Point-to-point communication links usually offer high quality service, if they have adequate bandwidth. The dedicated capacity removes latency or jitter between the endpoints.
Availability	Constant availability is essential for some applications, such as e-commerce. Point-to-point communication links provide permanent, dedicated capacity which is required for VoIP or Video over IP.
Disadvantages	
Cost	Point-to-point links are generally the most expensive type of WAN access. The cost of leased line solutions can become significant when they are used to connect many sites over increasing distances. In addition, each endpoint requires an interface on the router, which increases equipment costs.
Limited flexibility	WAN traffic is often variable, and leased lines have a fixed capacity, so that the bandwidth of the line seldom matches the need exactly. Any change to the leased line generally requires a site visit by ISP personnel to adjust capacity.

Circuit-Switched Options - 7.3.3

Circuit-switched connections are provided by Public Service Telephone Network (PSTN) carriers. The local loop connecting the CPE to the CO is copper media. There are two traditional circuit-switched options.

Public Service Telephone Network (PSTN)

Dialup WAN access uses the PSTN as its WAN connection. Traditional local loops can transport binary computer data through the voice telephone network using a voiceband modem. The modem modulates the digital data into an analog signal at the source and demodulates the analog signal to digital data at the destination. The physical characteristics of the local loop and its connection to the PSTN limit the rate of the signal to less than 56 kbps.

Dialup access is considered a legacy WAN technology. However, it may still be viable solution when no other WAN technology is available.

Integrated Services Digital Network (ISDN)

ISDN is a circuit-switching technology that enables the PSTN local loop to carry digital signals. This provided higher capacity switched connections than dialup access. ISDN provides for data rates from 45 Kbps to 2.048 Mbps.

ISDN has declined greatly in popularity due to high-speed DSL and other broadband services. ISDN is considered a legacy technology with most major providers discontinuing this service.

Packet-Switched Options - 7.3.4

Packet switching segments data into packets that are routed over a shared network. Circuit-switched networks require a dedicated circuit to be established. In contrast, packet-switching networks allow many pairs of nodes to communicate over the same channel.

There are two traditional (legacy) packet-switched connectivity options.

Frame Relay

Frame Relay is a simple Layer 2 non-broadcast multi-access (NBMA) WAN technology that is used to interconnect enterprise LANs. A single router interface can be used to connect to multiple sites using different PVCs. PVCs are used to carry both voice and data traffic between a source and destination, and support data rates up to 4 Mbps, with some providers offering even higher rates.

Frame Relay creates PVCs which are uniquely identified by a data-link connection identifier (DLCI). The PVCs and DLCIs ensure bidirectional communication from one DTE device to another.

Frame Relay networks have been largely replaced by faster Metro Ethernet and internet-based solutions.

Asynchronous Transfer Mode (ATM)

Asynchronous Transfer Mode (ATM) technology is capable of transferring voice, video, and data through private and public networks. It is built on a cell-based architecture rather than on a frame-based architecture. ATM cells are always a fixed length of 53 bytes. The ATM cell contains a 5-byte ATM header followed by 48 bytes of ATM payload. Small, fixed-length cells are well-suited for carrying voice and video traffic because this traffic is intolerant of delay. Video and voice traffic do not have to wait for larger data packets to be transmitted.

The 53-byte ATM cell is less efficient than the bigger frames and packets of Frame Relay. Furthermore, the ATM cell has at least five bytes of overhead for each 48-byte payload. When the cell is carrying segmented network layer packets, the overhead is higher because the ATM switch must be able to reassemble the packets at the destination. A typical ATM line needs almost 20 percent greater bandwidth than Frame Relay to carry the same volume of network layer data.

ATM networks have been largely replaced by faster Metro Ethernet and internet-based solutions.

Go to the online course to take the quiz and exam.

Check Your Understanding - Traditional WAN Connectivity - 7.3.5

Modern WAN Connectivity - 7.4

Refer to **Online Course** for Illustration

Modern WANs - 7.4.1

Modern WANs have more connectivity options that traditional WANs. Enterprises now require faster and more flexible WAN connectivity options. Traditional WAN connectivity options have rapidly declined in use because they are either no longer available, too expensive, or have limited bandwidth.

The figure displays the local loop connections most likely encountered today.

Refer to **Online Course** for Illustration

Refer to **Interactive Graphic** in online course

Modern WAN Connectivity Options - 7.4.2

New technologies are continually emerging. The figure summarizes the modern WAN connectivity options.

Click each button for a detailed description of the three major types of modern WAN connectivity options.

Dedicated Broadband

In the late 1990s, many telecommunication companies built optical fiber networks with enough fiber to satisfy projected next generation needs. However, optical technologies such as wavelength division multiplexing (WDM) were developed and dramatically increased the transmitting ability of a single strand of optical fiber. Consequently, many fiber-optic cable runs are not in use. Fiber-optic cable that is not in use, and therefore, "un-lit" (i.e. dark) is referred to as dark fiber.

Fiber can be installed independently by an organization to connect remote locations directly together. However, dark fiber could also be leased or purchased from a supplier. Leasing dark fiber is typically more expensive than any other WAN option available today. However, it provides the greatest flexibility, control, speed, and security.

Packet-Switched

Two packet-switched WAN network options are available.

Advances in Ethernet LAN technology have enabled it to expand into the MAN and WAN areas. Metro Ethernet provides fast bandwidth links and has been responsible for replacing many traditional WAN connectivity options.

Multi-protocol Label Switching (MPLS) enables the WAN provider network to carry any protocol (e.g., IPv4 packets, IPv6 packets, Ethernet, DSL) as payload data. This enables different sites to connect to the provider network regardless of its access technologies.

Internet-based Broadband

Organizations are now commonly using the global internet infrastructure for WAN connectivity. To address the security concerns, the connectivity options are often combined with VPN technologies.

Valid WAN network options include Digital Subscriber Line (DSL), cable, wireless, and fiber.

Note: There are several WAN access connection options that the enterprise edge can use to connect over the local loop to the provider. These WAN access options differ in technology, bandwidth, and cost. Each has distinct advantages and disadvantages. Familiarity with these technologies is an important part of network design.

Refer to
Online Course
for Illustration

Ethernet WAN - 7.4.3

Ethernet was originally developed as a LAN access technology and was not suitable as a WAN access technology due primarily to the limited distance provided by copper media.

However, newer Ethernet standards using fiber-optic cables have made Ethernet a reasonable WAN access option. For instance, the IEEE 1000BASE-LX standard supports fiber-optic cable lengths of 5 km, while the IEEE 1000BASE-ZX standard supports cable lengths up to 70 km.

Service providers now offer Ethernet WAN service using fiber-optic cabling. The Ethernet WAN service can go by many names, including the following:

- Metropolitan Ethernet (Metro E)

- Ethernet over MPLS (EoMPLS)

- Virtual Private LAN Service (VPLS)

The figure displays a simple Metro Ethernet topology example.

The following are several benefits to an Ethernet WAN:

- **Reduced expenses and administration** - Ethernet WAN provides a switched, high-bandwidth Layer 2 network capable of managing data, voice, and video all on the same infrastructure. This increases bandwidth and eliminates expensive conversions to other WAN technologies. The technology enables businesses to inexpensively connect numerous sites in a metropolitan area, to each other, and to the internet.

- **Easy integration with existing networks** - Ethernet WAN connects easily to existing Ethernet LANs, reducing installation costs and time.

■ **Enhanced business productivity** - Ethernet WAN enables businesses to take advantage of productivity-enhancing IP applications that are difficult to implement on TDM or Frame Relay networks, such as hosted IP communications, VoIP, and streaming and broadcast video.

Note: Ethernet WANs have gained in popularity and are now commonly being used to replace the traditional serial point-to-point, Frame Relay and ATM WAN links.

Refer to **Online Course** for Illustration

MPLS - 7.4.4

Multiprotocol Label Switching (MPLS) is a high-performance service provider WAN routing technology to interconnect clients without regard to access method or payload. MPLS supports a variety of client access methods (e.g., Ethernet, DSL, Cable, Frame Relay). MPLS can encapsulate all types of protocols including IPv4 and IPv6 traffic.

Refer to the sample topology of a simple MPLS enabled network.

An MPLS router can be a customer edge (CE) router, a provider edge (PE) router, or an internal provider (P) router. Notice that MPLS supports a variety of client access connections.

MPLS routers are label switched routers (LSRs). This means that they attach labels to packets that are then used by other MPLS routers to forward traffic. When traffic is leaving the CE, the MPLS PE router adds a short fixed-length label in between the frame header and packet header. MPLS P routers use the label to determine the next hop of the packet. The label is removed by the egress PE router when the packet leaves the MPLS network.

MPLS also provides services for QoS support, traffic engineering, redundancy, and VPNs.

Go to the online course to take the quiz and exam.

Check Your Understanding - Modern WAN Connectivity - 7.4.5

Internet-Based Connectivity - 7.5

Refer to **Online Course** for Illustration

Internet-Based Connectivity Options - 7.5.1

Modern WAN connectivity options do not end with Ethernet WAN and MPLS. Today, there are a host of internet-based wired and wireless options from which to choose. Internet-based broadband connectivity is an alternative to using dedicated WAN options.

The figure lists the internet-based connectivity options.

Internet-based connectivity can be divided into wired and wireless options.

Wired options

Wired options use permanent cabling (e.g., copper or fiber) to provide consistent bandwidth, and reduce error rates and latency. Examples of wired broadband connectivity are Digital Subscriber Line (DSL), cable connections, and optical fiber networks.

Wireless options

Wireless options are less expensive to implement compared to other WAN connectivity options because they use radio waves instead of wired media to transmit data. However, wireless signals can be negatively affected by factors such as distance from radio towers, interference from other sources, weather, and number of users accessing the shared space. Examples of wireless broadband include cellular 3G/4G/5G or satellite internet services. Wireless carrier options vary depending on location.

DSL Technology - 7.5.2

Refer to **Online Course** for Illustration

A Digital Subscriber Line (DSL) is a high-speed, always-on, connection technology that uses existing twisted-pair telephone lines to provide IP services to users. DSL is a popular choice for home users and for enterprise IT departments to support teleworkers.

The figure shows a representation of bandwidth space allocation on a copper wire for Asymmetric DSL (ADSL).

The area labeled POTS (Plain Old Telephone System) identifies the frequency range used by the voice-grade telephone service. The area labeled ADSL represents the frequency space used by the upstream and downstream DSL signals. The area that encompasses both the POTS area and the ADSL area represents the entire frequency range supported by the copper wire pair.

There are several xDSL varieties offering different upload and download transmission rates. However, all forms of DSL are categorized as either Asymmetric DSL (ADSL) or Symmetric DSL (SDSL). ADSL and ADSL2+ provide higher downstream bandwidth to the user than upload bandwidth. SDSL provides the same capacity in both directions.

The transfer rates are also dependent on the actual length of the local loop, and the type and condition of the cabling. For example, an ADSL loop must be less than 5.46 km (3.39 miles) for guaranteed signal quality.

Security risks are incurred in this process but can be mediated with security measures such as VPNs.

DSL Connections - 7.5.3

Refer to **Online Course** for Illustration

Service providers deploy DSL connections in the local loop. As shown in the figure, the connection is set up between the DSL modem and the DSL access multiplexer (DSLAM).

The DSL modem converts the Ethernet signals from the teleworker device to a DSL signal, which is transmitted to a DSL access multiplexer (DSLAM) at the provider location.

A DSLAM is the device located at the Central Office (CO) of the provider and concentrates connections from multiple DSL subscribers. A DSLAM is often built into an aggregation router.

The advantage that DSL has over cable technology is that DSL is not a shared medium. Each user has a separate direct connection to the DSLAM. Adding users does not impede performance, unless the DSLAM internet connection to the ISP, or to the internet, becomes saturated.

Refer to
Interactive Graphic
in online course

Refer to
Online Course
for Illustration

DSL and PPP - 7.5.4

Point-to-Point protocol (PPP) is a Layer 2 protocol that was commonly used by telephone service providers to establish router-to-router and host-to-network connections over dial-up and ISDN access networks.

ISPs still use PPP as the Layer 2 protocol for broadband DSL connections because of the following factors:

- PPP can be used to authenticate the subscriber.
- PPP can assign a public IPv4 address to the subscriber.
- PPP provides link-quality management features.

A DSL modem has a DSL interface to connect to the DSL network, and an Ethernet interface to connect to the client device. However, Ethernet links do not natively support PPP.

Click each button for an illustration and explanation of two ways PPP over Ethernet (PPPoE) can be deployed.

Host with PPPoE Client

As shown in the figure, the host runs a PPPoE client to obtain a public IP address from a PPPoE server located at the provider site. The PPPoE client software communicates with the DSL modem using PPPoE and the modem communicates with the ISP using PPP. In this topology, only one client can use the connection. Also, notice that there is no router to protect the inside network.

Router PPPoE Client

Another solution is to configure a router to be a PPPoE client, as shown in the figure. The router is the PPPoE client and obtains its configuration from the provider. The client(s) communicate with the router using only Ethernet and are unaware of the DSL connection. In this topology, multiple clients can share the DSL connection.

Refer to
Online Course
for Illustration

Cable Technology - 7.5.5

Cable technology is a high-speed always-on connection technology that uses a coaxial cable from the cable company to provide IP services to users. Like DSL, cable technology is a popular choice for home users and for enterprise IT departments to support remote workers.

Modern cable systems offer customers advanced telecommunications services, including high-speed internet access, digital cable television, and residential telephone service.

The Data over Cable Service Interface Specification (DOCSIS) is the international standard for adding high-bandwidth data to an existing cable system.

Cable operators deploy hybrid fiber-coaxial (HFC) networks to enable high-speed transmission of data to cable modems. The cable system uses a coaxial cable to carry radio frequency (RF) signals to the end user.

HFC uses fiber-optic and coaxial cable in different portions of the network. For example, the connection between the cable modem and optical node is coaxial cable, as shown in the figure.

The optical node performs optical to RF signal conversion. Specifically, it converts RF signals to light pulses over fiber-optic cable. The fiber media enables the signals to travel over long distances to the provider headend where a Cable Modem Termination System (CMTS) is located.

The headend contains the databases needed to provide internet access while the CMTS is responsible for communicating with the cable modems.

All the local subscribers share the same cable bandwidth. As more users join the service, available bandwidth may drop below the expected rate.

Optical Fiber - 7.5.6

Many municipalities, cities, and providers install fiber-optic cable to the user location. This is commonly referred to as Fiber to the x (FTTx) and includes the following:

- **Fiber to the Home (FTTH)** - Fiber reaches the boundary of the residence. Passive optical networks and point-to-point Ethernet are architectures that can deliver cable TV, internet, and phone services over FTTH networks directly from an the service provider central office.

- **Fiber to the Building (FTTB)** - Fiber reaches the boundary of the building, such as the basement in a multi-dwelling unit, with the final connection to the individual living space being made via alternative means, like curb or pole technologies.

- **Fiber to the Node/Neighborhood (FTTN)** - Optical cabling reaches an optical node that converts optical signals to a format acceptable for twisted pair or coaxial cable to the premise.

FTTx can deliver the highest bandwidth of all broadband options.

Refer to **Interactive Graphic** in online course

Wireless Internet-Based Broadband - 7.5.7

Wireless technology uses the unlicensed radio spectrum to send and receive data. The unlicensed spectrum is accessible to anyone who has a wireless router and wireless technology in the device they are using.

Until recently, one limitation of wireless access has been the need to be within the local transmission range (typically less than 100 feet) of a wireless router or a wireless modem that had a wired connection to the internet.

Click each button for a description of new developments that are enabling broadband wireless technology.

Municipal Wi-Fi

Many cities have begun setting up municipal wireless networks. Some of these networks provide high-speed internet access for free, or for substantially less than the price of other broadband services. Others are for city use only, allowing police and fire departments and other city employees to do certain aspects of their jobs remotely. To connect to a municipal Wi-Fi, a subscriber typically needs a wireless modem, which provides a stronger radio and directional antenna than conventional wireless adapters. Most service providers provide the necessary equipment for free or for a fee, much like they do with DSL or cable modems.

Cellular

Increasingly, cellular service is another wireless WAN technology being used to connect users and remote locations where no other WAN access technology is available. Many users with smart phones and tablets can use cellular data to email, surf the web, download apps, and watch videos.

Phones, tablet computers, laptops, and even some routers can communicate through to the internet using cellular technology. These devices use radio waves to communicate through a nearby mobile phone tower. The device has a small radio antenna, and the provider has a much larger antenna sitting at the top of a tower somewhere within miles of the phone.

The following are two common cellular industry terms:

- **3G/4G/5G Wireless** - These are abbreviations for 3rd generation, 4th generation, and the emerging 5th generation mobile wireless technologies. These technologies support wireless internet access. The 4G standards supports bandwidths up to 450 Mbps download and 100 Mbps upload. The emerging 5G standard should support 100 Mbps to 10 Gbps and beyond.

- **Long-Term Evolution (LTE)** - This refers to a newer and faster technology and is part of fourth generation (4G) technology.

Satellite Internet

Typically used by rural users or in remote locations where cable and DSL are not available. To access satellite internet services, subscribers need a satellite dish, two modems (uplink and downlink), and coaxial cables between the dish and the modem.

Specifically, a router connects to a satellite dish which is pointed to a service provider satellite. This satellite is in geosynchronous orbit in space. The signals must travel approximately 35,786 kilometers (22,236 miles) to the satellite and back.

The primary installation requirement is for the antenna to have a clear view toward the equator, where most orbiting satellites are stationed. Trees and heavy rains can affect reception of the signals.

Satellite internet provides two-way (upload and download) data communications. Upload speeds are about one-tenth of the download speed. Download speeds range from 5 Mbps to 25 Mbps.

WiMAX

Worldwide Interoperability for Microwave Access (WiMAX) is a new technology that is just beginning to come into use. It is described in the IEEE standard 802.16. WiMAX provides high-speed broadband service with wireless access and provides broad coverage like a cell phone network rather than through small Wi-Fi hotspots.

WiMAX operates in a similar way to Wi-Fi, but at higher speeds, over greater distances, and for a greater number of users. It uses a network of WiMAX towers that are like cell phone towers. To access a WiMAX network, users must subscribe to an ISP with a WiMAX tower that is within 30 miles of their location. They also need some type of WiMAX receiver and a special encryption code to get access to the base station.

WiMAX has largely been replaced by LTE for mobile access and cable, or DSL for fixed access.

VPN Technology - 7.5.8

Security risks are incurred when a teleworker or a remote office worker uses a broadband service to access the corporate WAN over the internet.

To address security concerns, broadband services provide Virtual Private Networks (VPN) connections to a network device that accepts VPN connections. The network device is typically located at the corporate site.

A VPN is an encrypted connection between private networks over a public network, such as the internet. Instead of using a dedicated Layer 2 connection, such as a leased line, a VPN uses virtual connections called VPN tunnels. VPN tunnels are routed through the internet from the private network of the company to the remote site or employee host.

The following are several benefits to using VPN:

- **Cost savings** - VPNs enable organizations to use the global internet to connect remote offices, and to connect remote users to the main corporate site. This eliminates expensive, dedicated WAN links and modem banks.

- **Security** - VPNs provide the highest level of security by using advanced encryption and authentication protocols that protect data from unauthorized access.

- **Scalability** - Because VPNs use the internet infrastructure within ISPs and devices, it is easy to add new users. Corporations can add large amounts of capacity without adding significant infrastructure.

- **Compatibility with broadband technology** - VPN technology is supported by broadband service providers such as DSL and cable. VPNs allow mobile workers and telecommuters to take advantage of their home high-speed internet service to access their corporate networks. Business-grade, high-speed broadband connections can also provide a cost-effective solution for connecting remote offices.

VPNs are commonly implemented as the following:

- **Site-to-site VPN** - VPN settings are configured on routers. Clients are unaware that their data is being encrypted.

- **Remote Access** - The user is aware and initiates remote access connection. For example, using HTTPS in a browser to connect to your bank. Alternatively, the user can run VPN client software on their host to connect to and authenticate with the destination device.

Note: VPNs are discussed in more detail later in this course.

Refer to
Interactive Graphic
in online course

Refer to
Online Course
for Illustration

ISP Connectivity Options - 7.5.9

Click each button for an illustration and explanation about the different ways an organization can connect to an ISP. The choice depends on the needs and budget of the organization.

Single-homed

Single-homed ISP connectivity is used by organization when internet access is not crucial to the operation. As shown in the figure, the client connects to the ISP using one link. The topology provides no redundancy. This is the least expensive solution of the four shown.

Dual-homed

Dual-homed ISP connectivity is used by an organization when internet access is somewhat crucial to the operation. As shown in the figure, the client connects to the same ISP using two links. The topology provides both redundancy and load balancing. If one link fails, the other link can carry the traffic. If both links are operational, traffic can be load balanced over them. However, the organization loses internet connectivity if the ISP experiences an outage.

Multihomed

Multihomed ISP connectivity is used by an organization when internet access is crucial to the operation. The client connects to two different ISPs, as shown in the figure. This design provides increased redundancy and enables load-balancing, but it can be expensive.

Dual-multihomed

Dual-multihomed is the most resilient topology of the four shown. The client connects with redundant links to multiple ISPs, as shown in the figure. This topology provides the most redundancy possible. It is the most expensive option of the four.

Broadband Solution Comparison - 7.5.10

Each broadband solution has advantages and disadvantages. The ideal solution is to have a fiber-optic cable directly connected to the client network. Some locations have only one option, such as cable or DSL. Some locations only have broadband wireless options for internet connectivity.

If there are multiple broadband solutions available, a cost-versus-benefit analysis should be performed to determine the best solution.

Some factors to consider include the following:

- **Cable** - Bandwidth is shared by many users. Therefore, upstream data rates are often slow during high-usage hours in areas with over-subscription.
- **DSL** - Limited bandwidth that is distance sensitive (in relation to the ISP central office). Upload rate is proportionally lower compared to download rate.
- **Fiber-to-the-Home** - This option requires fiber installation directly to the home.
- **Cellular/Mobile** - With this option, coverage is often an issue, even within a small office or home office where bandwidth is relatively limited.
- **Municipal Wi-Fi** - Most municipalities do not have a mesh Wi-Fi network deployed. If is available and in range, then it is a viable option.
- **Satellite** - This option is expensive and provides limited capacity per subscriber. Typically used when no other option is available.

Lab - Research Broadband Internet Access Options - 7.5.11

In this lab, you will complete the following objectives:

- Part 1: Investigate Broadband Distribution
- Part 2: Research Broadband Access Options for Specific Scenarios

Module Practice and Quiz - 7.6

Packet Tracer - WAN Concepts - 7.6.1

In this Packet Tracer activity, you will explore various WAN technologies and implementations.

What did I learn in this module? - 7.6.2

Purpose of WANs

A Wide Area Network (WAN) is required to connect beyond the boundary of the LAN. A WAN is a telecommunications network that spans over a relatively large geographical area. A WAN operates beyond the geographic scope of a LAN. A private WAN is a connection that is dedicated to a single customer. A public WAN connection is typically provided by an ISP or telecommunications service provider using the internet. WAN topologies are described using a logical topology. WANs are implemented using the following logical topologies: Point-to-Point, Hub-and-Spoke, Dual-homed, Fully Meshed, and Partially Meshed. A single-carrier connection is when an organization connects to only one service provider. A dual-carrier connection provides redundancy and increases network availability. The organization negotiates separate SLAs with two different service providers. Network requirements of a company can change dramatically as the company grows over time. Distributing employees saves costs in many ways, but it puts increased demands on the network. Small companies may use a single LAN connected to a wireless router to share data and peripherals. Connection to the internet is through a broadband service provider. A slightly larger company may use a Campus Area Network (CAN). A CAN interconnects several LANs within a limited geographical area. An even larger company may require a metropolitan area network (MAN) to interconnect sites within the city. A MAN is larger than a LAN but smaller than a WAN. A global company may require teleworking and virtual teams using web-based applications, including web-conferencing, e-learning, and online collaboration tools. Site-to-site and remote access Virtual Private Networks (VPNs) enable the company to use the internet to securely connect with employees and facilities around the world.

WAN Operations

Modern WAN standards are defined and managed by a number of recognized authorities: TIA/EIA, ISO, and IEEE. Most WAN standards focus on the physical layer (OSI Layer 1) and the data link layer (OSI Layer 2). Layer 1 protocols describe the electrical, mechanical, and operational components needed to transmit bits over a WAN. Layer 1 optical fiber protocol standards include SDH, SONET, and DWDM. Layer 2 protocols define how data will be encapsulated into a frame. Layer 2 protocols include broadband, wireless, Ethernet WAN, MPLS, PPP, HDLC. The WAN physical layer describes the physical connections between the company network and the service provider network. There are specific terms used to describe WAN connections between the subscriber (i.e., the company / client) and the WAN service provider: DTE, DCE, CPE, POP, Demarcation Point, Local Loop, CO, Toll network, Backhaul network, and Backbone network. The end-to-end data path over a WAN is usually from source DTE to the DCE, then to the WAN cloud, then to the DCE to and finally to the destination DTE. Devices used in this path include voiceband modem, DSL and Cable modems, CSU/DSU, Optical converter, wireless router or access point, and other WAN core devices. Serial communication transmits bits sequentially over a single channel. In contrast, parallel communications simultaneously transmit several bits using multiple wires. A circuit-switched network establishes a dedicated circuit (or channel) between endpoints before the users can communicate. During transmission over a circuit-switched network, all communication uses the same path. The two most common types of circuit-switched WAN technologies are PSTN and ISDN. Packet-switching segments traffic data into packets that are routed over a shared network. Common types of packet-switched WAN technologies are Ethernet WAN and MPLS. There are two optical fiber OSI layer 1 standards. SDH/SONET define how to transfer multiple data, voice, and video communications over optical fiber using lasers or LEDs over great distances. Both standards are used on the ring network topology that contains redundant fiber paths allowing traffic to flow in both directions. DWDM is a newer technology that increases the data-carrying capacity SDH and SONET by simultaneously sending multiple streams of data (multiplexing) using different wavelengths of light.

Traditional WAN Connectivity

In the 1980s, organizations started to see the need to interconnect their LANs with other locations. They needed their networks to connect to the local loop of a service provider by using dedicated lines or by using switched services from a service provider. When permanent dedicated connections were required, a point-to-point link using copper media was used to provide a pre-established WAN communications path from the customer premises to the provider network. Dedicated leased lines were T1/E1 or T3/E3 lines. Circuit-switched connections were provided by PSTN carriers. The local loop connecting the CPE to the CO was copper media. ISDN is a circuit-switching technology that enables the PSTN local loop to carry digital signals. This provided higher capacity switched connections than dialup access. Packet switching segments data into packets that are routed over a shared network. Packet-switching networks allow many pairs of nodes to communicate over the same channel. Frame Relay is a simple Layer 2 NBMA WAN technology used to interconnect enterprise LANs. ATM technology is capable of transferring voice, video, and data through private and public networks. It is built on a cell-based architecture rather than on a frame-based architecture.

Modern WAN Connectivity

Modern WAN connectivity options include dedicated broadband, Ethernet WAN and MPLS (packet-switched), along with various wired and wireless version of internet-based broadband. Service providers now offer Ethernet WAN service using fiber-optic cabling. Ethernet WAN reduces expenses and administration, is easily integrated with existing networks, and enhances business productivity. MPLS is a high-performance service provider

WAN routing technology to interconnect clients. MPLS supports a variety of client access methods (e.g., Ethernet, DSL, Cable, Frame Relay). MPLS can encapsulate all types of protocols including IPv4 or IPv6 traffic.

Internet-Based Connectivity

Internet-based broadband connectivity is an alternative to using dedicated WAN options. There are wired and wireless versions of broadband VPN. Wired options use permanent cabling e.g., copper or fiber) to provide consistent bandwidth, and reduce error rates and latency. Examples of wired broadband connectivity are Digital Subscriber Line (DSL), cable connections, and optical fiber networks. Examples of wireless broadband include cellular 3G/4G/5G or satellite internet services. DSL is a high-speed, always-on, connection technology that uses existing twisted-pair telephone lines to provide IP services to users. All forms of DSL are categorized as either ADSL or SDSL. The DSL modem converts the Ethernet signals from the teleworker device to a DSL signal, which is transmitted to a DSLAM at the provider location. The advantage that DSL has over cable technology is that DSL is not a shared medium. ISPs still use PPP as the Layer 2 protocol for broadband DSL connections. A DSL modem has a DSL interface to connect to the DSL network and an Ethernet interface to connect to the client device. Ethernet links do not natively support PPP. Cable technology is a high-speed always-on connection technology that uses a cable company coaxial cable to provide IP services to users. Cable operators deploy hybrid fiber-coaxial (HFC) networks to enable high-speed transmission of data to cable modems. The cable system uses a coaxial cable to carry radio frequency (RF) signals to the end user. Many municipalities, cities, and providers install fiber optic cable to the user location. This is commonly referred to as Fiber to the x (FTTx) and versions are FTTH, FTTB, and FTTN.

Wireless technology uses the unlicensed radio spectrum to send and receive data. The unlicensed spectrum is accessible to anyone who has a wireless router and wireless technology in the device they are using. Until recently, one limitation of wireless access has been the need to be within the local transmission range (typically less than 100 feet) of a wireless router or a wireless modem that has a wired connection to the internet. Newer developments in wireless technology include Municipal Wi-Fi, Cellular, Satellite internet, and WiMAX. To address security concerns, broadband services provide capabilities for using Virtual Private Networks (VPN) connections to a network device that accepts VPN connections, which is typically located at the corporate site. A VPN is an encrypted connection between private networks over a public network, such as the internet. Instead of using a dedicated Layer 2 connection, such as a leased line, a VPN uses virtual connections called VPN tunnels. VPN tunnels are routed through the internet from the private network of the company to the remote site or employee host. Common VPN implementations include site-to-site and remote access. ISP connectivity options include single-homed, dual-homed, multihomed, and dual-multihomed. Cable, DSL, fiber-to-the-home, cellular/mobile, municipal Wi-Fi, and satellite internet all have advantages and disadvantages. Perform a cost-versus-benefit analysis before choosing an internet-based connectivity solution.

Go to the online
course to take the
quiz and exam.

Chapter Quiz - WAN Concepts

Your Chapter Notes

CHAPTER 8

VPN and IPsec Concepts

Introduction - 8.0

Why should I take this module? - 8.0.1

Welcome to VPN and IPsec Concepts!

Have you, or someone you know, ever been hacked while using public WiFi? It's surprisingly easy to do. But there is a solution to this problem: Virtual Private Networks (VPNs) and the additional protection of IP Security (IPsec). VPNs are commonly used by remote workers around the globe. There are also personal VPNs that you can use when you are on public WiFi. In fact, there are many different kinds of VPNs using IPsec to protect and authenticate IP packets between their source and destination. Want to know more? Click Next!

What will I learn in this module? - 8.0.2

Module Title: VPN and IPsec Concepts

Module Objective: Explain how VPNs and IPsec are used to secure site-to-site and remote access connectivity.

Topic Title	Topic Objective
VPN Technology	Describe benefits of VPN technology.
Types of VPNs	Describe different types of VPNs.
IPsec	Explain how the IPsec framework is used to secure network traffic.

VPN Technology - 8.1

Virtual Private Networks - 8.1.1

Refer to
Online Course
for Illustration

To secure network traffic between sites and users, organizations use virtual private networks (VPNs) to create end-to-end private network connections. A VPN is virtual in that it carries information within a private network, but that information is actually transported over a public network. A VPN is private in that the traffic is encrypted to keep the data confidential while it is transported across the public network.

The figure shows a collection of various types of VPNs managed by an enterprise's main site. The tunnel enables remote sites and users to access main site's network resources securely.

The first types of VPNs were strictly IP tunnels that did not include authentication or encryption of the data. For example, Generic Routing Encapsulation (GRE) is a tunneling protocol developed by Cisco and which does not include encryption services. It is used to encapsulate IPv4 and IPv6 traffic inside an IP tunnel to create a virtual point-to-point link.

VPN Benefits - 8.1.2

Modern VPNs now support encryption features, such as Internet Protocol Security (IPsec) and Secure Sockets Layer (SSL) VPNs to secure network traffic between sites.

Major benefits of VPNs are shown in the table.

Benefit	Description
Cost Savings	With the advent of cost-effective, high-bandwidth technologies, organizations can use VPNs to reduce their connectivity costs while simultaneously increasing remote connection bandwidth.
Security	VPNs provide the highest level of security available, by using advanced encryption and authentication protocols that protect data from unauthorized access.
Scalability	VPNs allow organizations to use the internet, making it easy to add new users without adding significant infrastructure.
Compatibility	VPNs can be implemented across a wide variety of WAN link options including all the popular broadband technologies. Remote workers can take advantage of these high-speed connections to gain secure access to their corporate networks.

Refer to
Interactive Graphic
in online course

Refer to
Online Course
for Illustration

Site-to-Site and Remote-Access VPNs - 8.1.3

VPNs are commonly deployed in one of the following configurations: site-to-site or remote-access.

Click each for VPN type for more information.

Site-to-Site VPN

A site-to-site VPN is created when VPN terminating devices, also called VPN gateways, are preconfigured with information to establish a secure tunnel. VPN traffic is only encrypted between these devices. Internal hosts have no knowledge that a VPN is being used.

Remote-Access VPN

A remote-access VPN is dynamically created to establish a secure connection between a client and a VPN terminating device. For example, a remote access SSL VPN is used when you check your banking information online.

Refer to
Online Course
for Illustration

Enterprise and Service Provider VPNs - 8.1.4

There are many options available to secure enterprise traffic. These solutions vary depending on who is managing the VPN.

VPNs can be managed and deployed as:

- **Enterprise VPNs** - Enterprise-managed VPNs are a common solution for securing enterprise traffic across the internet. Site-to-site and remote access VPNs are created and managed by the enterprise using both IPsec and SSL VPNs.

- **Service Provider VPNs** - Service provider-managed VPNs are created and managed over the provider network. The provider uses Multiprotocol Label Switching (MPLS) at Layer 2 or Layer 3 to create secure channels between an enterprise's sites. MPLS is a routing technology the provider uses to create virtual paths between sites. This effectively segregates the traffic from other customer traffic. Other legacy solutions include Frame Relay and Asynchronous Transfer Mode (ATM) VPNs.

The figure lists the different types of enterprise-managed and service provider-managed VPN deployments that will be discussed in more detail in this module.

Go to the online course to take the quiz and exam.

Check Your Understanding - VPN Technology - 8.1.5

Types of VPNs - 8.2

Refer to Online Course for Illustration

Remote-Access VPNs - 8.2.1

In the previous topic you learned about the basics of a VPN. Here you will learn about the types of VPNs.

VPNs have become the logical solution for remote-access connectivity for many reasons. As shown in the figure, remote-access VPNs let remote and mobile users securely connect to the enterprise by creating an encrypted tunnel. Remote users can securely replicate their enterprise security access including email and network applications. Remote-access VPNs also allow contractors and partners to have limited access to the specific servers, web pages, or files as required. This means that these users can contribute to business productivity without compromising network security.

Remote-access VPNs are typically enabled dynamically by the user when required. Remote access VPNs can be created using either IPsec or SSL. As shown in the figure, a remote user must initiate a remote access VPN connection.

The figure displays two ways that a remote user can initiate a remote access VPN connection: clientless VPN and client-based VPN.

SSL VPNs - 8.2.2

When a client negotiates an SSL VPN connection with the VPN gateway, it actually connects using Transport Layer Security (TLS). TLS is the newer version of SSL and is sometimes expressed as SSL/TLS. However, both terms are often used interchangeably.

SSL uses the public key infrastructure and digital certificates to authenticate peers. Both IPsec and SSL VPN technologies offer access to virtually any network application or resource. However, when security is an issue, IPsec is the superior choice. If support and ease of deployment are the primary issues, consider SSL. The type of VPN method implemented is based on the access requirements of the users and the organization's IT processes. The table compares IPsec and SSL remote access deployments.

Feature	IPsec	SSL
Applications supported	Extensive - All IP-based applications are supported.	Limited - Only web-based applications and file sharing are supported.
Authentication strength	Strong - Uses two-way authentication with shared keys or digital certificates.	Moderate - Using one-way or two-way authentication.
Encryption strength	Strong - Uses key lengths from 56 bits to 256 bits.	Moderate to strong - With key lengths from 40 bits to 256 bits.
Connection complexity	Medium - Because it requires a VPN client pre-installed on a host.	Low - It only requires a web browser on a host.
Connection option	Limited - Only specific devices with specific configurations can connect.	Extensive - Any device with a web browser can connect.

It is important to understand that IPsec and SSL VPNs are not mutually exclusive. Instead, they are complementary; both technologies solve different problems, and an organization may implement IPsec, SSL, or both, depending on the needs of its telecommuters.

Refer to **Online Course** for Illustration

Site-to-Site IPsec VPNs - 8.2.3

Site-to-site VPNs are used to connect networks across another untrusted network such as the internet. In a site-to-site VPN, end hosts send and receive normal unencrypted TCP/IP traffic through a VPN terminating device. The VPN terminating is typically called a VPN gateway. A VPN gateway device could be a router or a firewall, as shown in the figure. For example, the Cisco Adaptive Security Appliance (ASA) shown on the right side of the figure is a standalone firewall device that combines firewall, VPN concentrator, and intrusion prevention functionality into one software image.

The VPN gateway encapsulates and encrypts outbound traffic for all traffic from a particular site. It then sends the traffic through a VPN tunnel over the internet to a VPN gateway at the target site. Upon receipt, the receiving VPN gateway strips the headers, decrypts the content, and relays the packet toward the target host inside its private network.

Site-to-site VPNs are typically created and secured using IP security (IPsec).

Refer to **Online Course** for Illustration

GRE over IPsec - 8.2.4

Generic Routing Encapsulation (GRE) is a non-secure site-to-site VPN tunneling protocol. It can encapsulate various network layer protocols. It also supports multicast and broadcast traffic which may be necessary if the organization requires routing protocols to operate over a VPN. However, GRE does not by default support encryption; and therefore, it does not provide a secure VPN tunnel.

A standard IPsec VPN (non-GRE) can only create secure tunnels for unicast traffic. Therefore, routing protocols will not exchange routing information over an IPsec VPN.

To solve this problem, we can encapsulate routing protocol traffic using a GRE packet, and then encapsulate the GRE packet into an IPsec packet to forward it securely to the destination VPN gateway.

The terms used to describe the encapsulation of GRE over IPsec tunnel are passenger protocol, carrier protocol, and transport protocol, as shown in the figure.

For example, in the figure displaying a topology, Branch and HQ would like to exchange OSPF routing information over an IPsec VPN. However, IPsec does not support multicast traffic. Therefore, GRE over IPsec is used to support the routing protocol traffic over the IPsec VPN. Specifically, the OSPF packets (i.e., passenger protocol) would be encapsulated by GRE (i.e., carrier protocol) and subsequently encapsulated in an IPsec VPN tunnel.

The Wireshark screen capture in the figure displays an OSPF Hello packet that was sent using GRE over IPsec. In the example, the original OSPF Hello multicast packet (i.e., passenger protocol) was encapsulated with a GRE header (i.e., carrier protocol), which is subsequently encapsulated by another IP header (i.e., transport protocol). This IP header would then be forwarded over an IPsec tunnel.

Refer to **Online Course** for Illustration

Dynamic Multipoint VPNs - 8.2.5

Site-to-site IPsec VPNs and GRE over IPsec are adequate to use when there are only a few sites to securely interconnect. However, they are not sufficient when the enterprise adds many more sites. This is because each site would require static configurations to all other sites, or to a central site.

Dynamic Multipoint VPN (DMVPN) is a Cisco software solution for building multiple VPNs in an easy, dynamic, and scalable manner. Like other VPN types, DMVPN relies on IPsec to provide secure transport over public networks, such as the internet.

DMVPN simplifies the VPN tunnel configuration and provides a flexible option to connect a central site with branch sites. It uses a hub-and-spoke configuration to establish a full mesh topology. Spoke sites establish secure VPN tunnels with the hub site, as shown in the figure.

Each site is configured using Multipoint Generic Routing Encapsulation (mGRE). The mGRE tunnel interface allows a single GRE interface to dynamically support multiple IPsec tunnels. Therefore, when a new site requires a secure connection, the same configuration on the hub site would support the tunnel. No additional configuration would be required.

Spoke sites could also obtain information about remote sites from the central site. They can use this information to establish direct VPN tunnels, as shown in the figure.

Refer to **Online Course** for Illustration

IPsec Virtual Tunnel Interface - 8.2.6

Like DMVPNs, IPsec Virtual Tunnel Interface (VTI) simplifies the configuration process required to support multiple sites and remote access. IPsec VTI configurations are applied to a virtual interface instead of static mapping the IPsec sessions to a physical interface.

IPsec VTI is capable of sending and receiving both IP unicast and multicast encrypted traffic. Therefore, routing protocols are automatically supported without having to configure GRE tunnels.

IPsec VTI can be configured between sites or in a hub-and-spoke topology.

Refer to
Online Course
for Illustration

Service Provider MPLS VPNs - 8.2.7

Traditional service provider WAN solutions such as leased lines, Frame Relay, and ATM connections were inherently secure in their design. Today, service providers use MPLS in their core network. Traffic is forwarded through the MPLS backbone using labels that are previously distributed among the core routers. Like legacy WAN connections, traffic is secure because service provider customers cannot see each other's traffic.

MPLS can provide clients with managed VPN solutions; therefore, securing traffic between client sites is the responsibility of the service provider. There are two types of MPLS VPN solutions supported by service providers:

- **Layer 3 MPLS VPN** - The service provider participates in customer routing by establishing a peering between the customer's routers and the provider's routers. Then customer routes that are received by the provider's router are then redistributed through the MPLS network to the customer's remote locations.

- **Layer 2 MPLS VPN** - The service provider is not involved in the customer routing. Instead, the provider deploys a Virtual Private LAN Service (VPLS) to emulate an Ethernet multiaccess LAN segment over the MPLS network. No routing is involved. The customer's routers effectively belong to the same multiaccess network.

The figure shows a service provider that offers both Layer 2 and Layer 3 MPLS VPNs.

Go to the online course to take the quiz and exam.

Check Your Understanding - Types of VPNs - 8.2.8

IPsec - 8.3

Refer to **Video**
in online course

Video - IPsec Concepts - 8.3.1

In the previous topic you learned about types of VPNs. It is important to understand how IPsec works with a VPN.

Click Play in the figure for a video about IPsec.

Refer to
Online Course
for Illustration

IPsec Technologies - 8.3.2

IPsec is an IETF standard (RFC 2401-2412) that defines how a VPN can be secured across IP networks. IPsec protects and authenticates IP packets between source and destination. IPsec can protect traffic from Layer 4 through Layer 7.

Using the IPsec framework, IPsec provides these essential security functions:

- **Confidentiality** - IPsec uses encryption algorithms to prevent cybercriminals from reading the packet contents.

- **Integrity** - IPsec uses hashing algorithms to ensure that packets have not been altered between source and destination.

- **Origin authentication** - IPsec uses the Internet Key Exchange (IKE) protocol authenticate source and destination. Methods of authentication including using pre-shared keys (passwords), digital certificates, or RSA certificates.

- **Diffie-Hellman** - Secure key exchange typically various groups of the DH algorithm.

IPsec is not bound to any specific rules for secure communications. This flexibility of the framework allows IPsec to easily integrate new security technologies without updating the existing IPsec standards. The currently available technologies are aligned to their specific security function. The open slots shown in the IPsec framework in the figure can be filled with any of the choices that are available for that IPsec function to create a unique security association (SA).

The security functions are list in the table.

IPsec Function	Description
IPsec Protocol	The choices for IPsec Protocol include Authentication Header (AH) or Encapsulation Security Protocol (ESP). AH authenticates the Layer 3 packet. ESP encrypts the Layer 3 packet. **Note:** ESP+AH is rarely used as this combination will not successfully traverse a NAT device.
Confidentiality	Encryption ensures confidentiality of the Layer 3 packet. Choices include Data Encryption Standard (DES), Triple DES (3DES), Advanced Encryption Standard (AES), or Software-Optimized Encryption Algorithm (SEAL). No encryption is also an option.
Integrity	Ensures that data arrives unchanged at the destination using a hash algorithm, such as message-digest 5 (MD5) or Secure Hash Algorithm (SHA).
Authentication	IPsec uses Internet Key Exchange (IKE) to authenticate users and devices that can carry out communication independently. IKE uses several types of authentication, including username and password, one-time password, biometrics, pre-shared keys (PSKs), and digital certificates using the Rivest, Shamir, and Adleman (RSA) algorithm.
Diffie-Hellman	IPsec uses the DH algorithm to provide a public key exchange method for two peers to establish a shared secret key. There are several different groups to choose from including DH14, 15, 16 and DH 19, 20, 21 and 24. DH1, 2 and 5 are no longer recommended.

The figure shows examples of SAs for two different implementations. An SA is the basic building block of IPsec. When establishing a VPN link, the peers must share the same SA to negotiate key exchange parameters, establish a shared key, authenticate each other, and negotiate the encryption parameters. Notice that SA Example 1 is using no encryption.

Refer to
Online Course
for Illustration

IPsec Protocol Encapsulation - 8.3.3

Choosing the IPsec protocol encapsulation is the first building block of the framework. IPsec encapsulates packets using Authentication Header (AH) or Encapsulation Security Protocol (ESP).

The choice of AH or ESP establishes which other building blocks are available. Click each IPsec protocol in the figure for more information.

Refer to
Online Course
for Illustration

Confidentiality - 8.3.4

Confidentiality is achieved by encrypting the data, as shown in the figure. The degree of confidentiality depends on the encryption algorithm and the length of the key used in the encryption algorithm. If someone tries to hack the key through a brute-force attack, the number of possibilities to try is a function of the length of the key. The time to process all the possibilities is a function of the computer power of the attacking device. The shorter the key, the easier it is to break. A 64-bit key can take approximately one year to break with a relatively sophisticated computer. A 128-bit key with the same machine can take roughly 1019 or 10 quintillion years to decrypt.

The encryption algorithms highlighted in the figure are all symmetric key cryptosystems.

Refer to
Interactive Graphic
in online course

Refer to
Online Course
for Illustration

Integrity - 8.3.5

Data integrity means that the data that is received is exactly the same data that was sent. Potentially, data could be intercepted and modified. For example, in the figure, assume that a check for $100 is written to Alex. The check is then mailed to Alex, but it is intercepted by a threat actor. The threat actor changes the name on the check to Jeremy and the amount on the check to $1,000 and attempts to cash it. Depending on the quality of the forgery in the altered check, the attacker could be successful.

Because VPN data is transported over the public internet, a method of proving data integrity is required to guarantee that the content has not been altered. The Hashed Message Authentication Code (HMAC) is a data integrity algorithm that guarantees the integrity of the message using a hash value. The figure highlights the two most common HMAC algorithms. Click each algorithm for more information.

Note: Cisco now rates SHA-1 as legacy and recommends at least SHA-256 for integrity.

Refer to
Online Course
for Illustration

Authentication - 8.3.6

When conducting business long distance, you must know who is at the other end of the phone, email, or fax. The same is true of VPN networks. The device on the other end of the VPN tunnel must be authenticated before the communication path is considered secure. The figure highlights the two peer authentication methods.

The figure shows an example of PSK authentication. At the local device, the authentication key and the identity information are sent through a hash algorithm to form the hash for the local peer (Hash_L). One-way authentication is established by sending Hash_L to the remote device. If the remote device can independently create the same hash, the local device is authenticated. After the remote device authenticates the local device, the authentication process begins in the opposite direction, and all steps are repeated from the remote device to the local device.

The figure shows an example of RSA authentication. At the local device, the authentication key and identity information are sent through the hash algorithm to form the hash for the local peer (Hash_L). Then the Hash_L is encrypted using the local device's private encryption key. This creates a digital signature. The digital signature and a digital certificate are forwarded to the remote device. The public encryption key for decrypting the signature is included in the digital certificate. The remote device verifies the digital signature by decrypting it using the public encryption key. The result is Hash_L. Next, the remote device independently creates Hash_L from stored information. If the calculated Hash_L equals the decrypted Hash_L, the local device is authenticated. After the remote device authenticates the local device, the authentication process begins in the opposite direction and all steps are repeated from the remote device to the local device.

Refer to **Online Course** for Illustration

Secure Key Exchange with Diffie-Hellman - 8.3.7

Encryption algorithms require a symmetric, shared secret key to perform encryption and decryption. How do the encrypting and decrypting devices get the shared secret key? The easiest key exchange method is to use a public key exchange method, such as Diffie-Hellman (DH), as shown in the figure.

DH provides a way for two peers to establish a shared secret key that only they know, even though they are communicating over an insecure channel. Variations of the DH key exchange are specified as DH groups:

- DH groups 1, 2, and 5 should no longer be used. These groups support a key size of 768 bits, 1024 bits, and 1536 bits, respectively.

- DH groups 14, 15, and 16 use larger key sizes with 2048 bits, 3072 bits, and 4096 bits, respectively, and are recommended for use until 2030.

- DH groups 19, 20, 21 and 24 with respective key sizes of 256 bits, 384 bits, 521 bits, and 2048 bits support Elliptical Curve Cryptography (ECC), which reduces the time needed to generate keys. DH group 24 is the preferred next generation encryption.

The DH group you choose must be strong enough, or have enough bits, to protect the IPsec keys during negotiation. For example, DH group 1 is strong enough to support DES and 3DES encryption, but not AES. For example, if the encryption or authentication algorithms use a 128-bit key, use group 14, 19, 20 or 24. However, if the encryption or authentication algorithms use a 256-bit key or higher, use group 21 or 24.

Refer to **Video** in online course

Video - IPsec Transport and Tunnel Mode - 8.3.8

Click Play in the figure for a video about IPsec transport and tunnel modes.

Go to the online course to take the quiz and exam.

Check Your Understanding - IPsec - 8.3.9

Module Practice and Quiz - 8.4

What did I learn in this module? - 8.4.1

A VPN is virtual in that it carries information within a private network, but that information is actually transported over a public network. A VPN is private in that the traffic is encrypted to keep the data confidential while it is transported across the public network. Benefits of VPNs are cost savings, security, scalability, and compatibility. VPNs are commonly deployed in one of the following configurations: site-to-site or remote-access. VPNs can be managed and deployed as enterprise VPNs and service provider VPNs.

Remote-access VPNs let remote and mobile users securely connect to the enterprise by creating an encrypted tunnel. Remote access VPNs can be created using either IPsec or SSL. When a client negotiates an SSL VPN connection with the VPN gateway, it actually connects using TLS. SSL uses the public key infrastructure and digital certificates to authenticate peers. Site-to-site VPNs are used to connect networks across an untrusted network such as the internet. In a site-to-site VPN, end hosts send and receive normal unencrypted TCP/IP traffic through a VPN terminating device. The VPN terminating device is typically called a VPN gateway. A VPN gateway could be a router or a firewall. GRE is a non-secure site-to-site VPN tunneling protocol. DMVPN is a Cisco software solution for easily building multiple, dynamic, scalable VPNs. Like DMVPNs, IPsec VTI simplifies the configuration process required to support multiple sites and remote access. IPsec VTI configurations are applied to a virtual interface instead of static mapping the IPsec sessions to a physical interface. IPsec VTI can send and receive both IP unicast and multicast encrypted traffic. MPLS can provide clients with managed VPN solutions; therefore, securing traffic between client sites is the responsibility of the service provider. There are two types of MPLS VPN solutions supported by service providers, Layer 3 MPLS VPN and Layer 2 MPLS VPN.

IPsec protects and authenticates IP packets between source and destination. IPsec can protect traffic from Layer 4 through Layer 7. Using the IPsec framework, IPsec provides confidentiality, integrity, origin authentication, and Diffie-Hellman. Choosing the IPsec protocol encapsulation is the first building block of the framework. IPsec encapsulates packets using AH or ESP. The degree of confidentiality depends on the encryption algorithm and the length of the key used in the encryption algorithm. The HMAC is an algorithm that guarantees the integrity of the message using a hash value. The device on the other end of the VPN tunnel must be authenticated before the communication path is considered secure. A PSK value is entered into each peer manually. The PSK is combined with other information to form the authentication key. RSA authentication uses digital certificates to authenticate the peers. The local device derives a hash and encrypts it with its private key. The encrypted hash is attached to the message and is forwarded to the remote end and acts like a signature. DH provides a way for two peers to establish a shared secret key that only they know, even though they are communicating over an insecure channel.

Go to the online course to take the quiz and exam.

Chapter Quiz - VPN and IPsec Concepts

Your Chapter Notes

CHAPTER 9

QoS Concepts

Introduction - 9.0

Why should I take this module? - 9.0.1

Welcome to QoS Concepts!

Imagine driving on a heavily congested road and you are in a rush to meet a friend for dinner. You hear the siren and see the lights of an ambulance behind you. You need to move off the road to let the ambulance through. The ambulance getting to the hospital takes priority over you getting to the restaurant on time.

Much like the ambulance taking priority in the traffic on the highway, some forms of network traffic need priority over others. Why? Get started with this module to find out!

What will I learn to do in this module? - 9.0.2

Module Title: QoS Concepts

Module Objective: Explain how networking devices implement QoS.

Topic Title	Topic Objective
Network Transmission Quality	Explain how network transmission characteristics impact quality.
Traffic Characteristics	Describe minimum network requirements for voice, video, and data traffic.
Queuing Algorithms	Describe the queuing algorithms used by networking devices.
QoS Models	Describe the different QoS models.
QoS Implementation Techniques	Explain how QoS uses mechanisms to ensure transmission quality.

Network Transmission Quality - 9.1

Refer to **Video** in online course

Video Tutorial - The Purpose of QoS - 9.1.1

Click Play for a brief explanation of the purpose of QoS.

Refer to **Online Course** for Illustration

Prioritizing Traffic - 9.1.2

In the previous video, you learned about the purpose of Quality of Service (QoS). QoS is an ever-increasing requirement of networks today. New applications, such as voice and live video transmissions, create higher expectations for quality delivery among users.

Congestion occurs when multiple communication lines aggregate onto a single device such as a router, and then much of that data is placed on just a few outbound interfaces, or onto a slower interface. Congestion can also occur when large data packets prevent smaller packets from being transmitted in a timely manner.

When the volume of traffic is greater than what can be transported across the network, devices queue (hold) the packets in memory until resources become available to transmit them. Queuing packets causes delay because new packets cannot be transmitted until previous packets have been processed. If the number of packets to be queued continues to increase, the memory within the device fills up and packets are dropped. One QoS technique that can help with this problem is to classify data into multiple queues, as shown in the figure.

Note: A device implements QoS only when it is experiencing some type of congestion.

Refer to
Online Course
for Illustration

Bandwidth, Congestion, Delay, and Jitter - 9.1.3

Network bandwidth is measured in the number of bits that can be transmitted in a single second, or bits per second (bps). For example, a network device may be described as having the capability to perform at 10 gigabits per second (Gbps).

Network congestion causes delay. An interface experiences congestion when it is presented with more traffic than it can handle. Network congestion points are ideal candidates for QoS mechanisms. The figure shows three examples of typical congestion points.

Delay or latency refers to the time it takes for a packet to travel from the source to the destination. Two types of delays are fixed and variable. A fixed delay is a specific amount of time a specific process takes, such as how long it takes to place a bit on the transmission media. A variable delay takes an unspecified amount of time and is affected by factors such as how much traffic is being processed.

The sources of delay are summarized in the table.

Sources of Delay

Delay	Description
Code delay	The fixed amount of time it takes to compress data at the source before transmitting to the first internetworking device, usually a switch.
Packetization delay	The fixed time it takes to encapsulate a packet with all the necessary header information.
Queuing delay	The variable amount of time a frame or packet waits to be transmitted on the link.
Serialization delay	The fixed amount of time it takes to transmit a frame onto the wire.
Propagation delay	The variable amount of time it takes for the frame to travel between the source and destination.
De-jitter delay	The fixed amount of time it takes to buffer a flow of packets and then send them out in evenly spaced intervals.

Jitter is the variation in the delay of received packets. At the sending side, packets are sent in a continuous stream with the packets spaced evenly apart. Due to network congestion, improper queuing, or configuration errors, the delay between each packet can vary instead of remaining constant. Both delay and jitter need to be controlled and minimized to support real-time and interactive traffic.

Refer to **Online Course** for Illustration

Packet Loss - 9.1.4

Without any QoS mechanisms in place, packets are processed in the order in which they are received. When congestion occurs, network devices such as routers and switches can drop packets. This means that time-sensitive packets, such as real-time video and voice, will be dropped with the same frequency as data that is not time-sensitive, such as email and web browsing.

When a router receives a Real-Time Protocol (RTP) digital audio stream for Voice over IP (VoIP), it must compensate for the jitter that is encountered. The mechanism that handles this function is the playout delay buffer. The playout delay buffer must buffer these packets and then play them out in a steady stream, as shown in the figure. The digital packets are later converted back to an analog audio stream.

If the jitter is so large that it causes packets to be received out of the range of this buffer, the out-of-range packets are discarded and dropouts are heard in the audio, as shown in the figure.

For losses as small as one packet, the digital signal processor (DSP) interpolates what it thinks the audio should be and no problem is audible to the user. However, when jitter exceeds what the DSP can do to make up for the missing packets, audio problems are heard.

Packet loss is a very common cause of voice quality problems on an IP network. In a properly designed network, packet loss should be near zero. The voice codecs used by the DSP can tolerate some degree of packet loss without a dramatic effect on voice quality. Network engineers use QoS mechanisms to classify voice packets for zero packet loss. Bandwidth is guaranteed for the voice calls by giving priority to voice traffic over traffic that is not sensitive to delays.

Go to the online course to take the quiz and exam.

Check Your Understanding - Network Transmission Quality - 9.1.5

Traffic Characteristics - 9.2

Refer to **Video** in online course

Video Tutorial - Traffic Characteristics - 9.2.1

Click Play for an overview of how QoS can be used to treat packets differently based on the characteristics of the traffic.

Network Traffic Trends - 9.2.2

In a previous topic, you learned about network transmission quality. In this topic you will learn about traffic characteristics (voice, video, and data). In the early 2000s, the predominant types of IP traffic were voice and data. Voice traffic has a predictable bandwidth need and known packet arrival times. Data traffic is not real-time and has unpredictable bandwidth need. Data traffic can temporarily burst, as when a large file is being downloaded. This bursting can consume the entire bandwidth of a link.

More recently, video traffic has become increasingly important to business communications and operations. According to the Cisco Visual Networking Index (VNI), video traffic represented 70% of all traffic in 2017. By 2022, video will represent 82% of all traffic. In addition, mobile video traffic will reach 60.9 exabytes per month by 2022, up from 6.8 exabytes per month in 2017. The type of demands that voice, video, and data traffic place on the network are very different.

Voice - 9.2.3

Voice traffic is predictable and smooth, as shown in the figure. However, voice is very sensitive to delays and dropped packets. It makes no sense to re-transmit voice if packets are lost; therefore, voice packets must receive a higher priority than other types of traffic. For example, Cisco products use the RTP port range 16384 to 32767 to prioritize voice traffic. Voice can tolerate a certain amount of latency, jitter, and loss without any noticeable effects. Latency should be no more than 150 milliseconds (ms). Jitter should be no more than 30 ms, and voice packet loss should be no more than 1%. Voice traffic requires at least 30 Kbps of bandwidth. The table gives a summary of voice traffic characteristics and requirements.

Voice Traffic Characteristics	One-Way Requirements
• Smooth	• Latency ≤ 150ms
• Benign	• Jitter ≤ 30ms
• Drop sensitive	• Loss ≤ 1% Bandwidth (30 - 128 Kbps)
• Delay sensitive	
• UDP priority	

Refer to **Online Course** for Illustration

Video - 9.2.4

Without QoS and a significant amount of extra bandwidth capacity, video quality typically degrades. The picture appears blurry, jagged, or in slow motion. The audio portion of the feed may become unsynchronized with the video.

Video traffic tends to be unpredictable, inconsistent, and bursty compared to voice traffic. Compared to voice, video is less resilient to loss and has a higher volume of data per packet. Notice in the figure how voice packets arrive every 20 ms and are a predictable 200 bytes each.

In contrast, the number and size of video packets varies every 33 ms based on the content of the video, as shown in the figure. For example, if the video stream consists of content that is not changing much from frame to frame, then the video packets will be small, and fewer are required to maintain acceptable user experience. However, if the video steam consists of content that is rapidly changing, such as an action sequence in a movie, then the video packets will be larger. More are required per the 33 ms time slot to maintain an acceptable user experience.

UDP ports such as 554, are used for the Real-Time Streaming Protocol (RSTP) and should be given priority over other, less delay-sensitive, network traffic. Similar to voice, video can tolerate a certain amount of latency, jitter, and loss without any noticeable effects. Latency should be no more than 400 milliseconds (ms). Jitter should be no more than 50 ms, and video packet loss should be no more than 1%. Video traffic requires at least 384 Kbps of bandwidth. The table gives a summary of video traffic characteristics and requirements.

Video Traffic Characteristics	One-Way Requirements
• Bursty	• Latency ≤ 200-400 ms
• Greedy	• Jitter ≤ 30-50 ms
• Drop sensitive	• Loss ≤ 0.1-1%
• Delay sensitive	• Bandwidth (384 Kbps - > 20 Mbps)
• UDP priority	

Data - 9.2.5

Most applications use either TCP or UDP. Unlike UDP, TCP performs error recovery. Data applications that have no tolerance for data loss, such as email and web pages, use TCP to ensure that, if packets are lost in transit, they will be resent. Data traffic can be smooth or bursty. Network control traffic is usually smooth and predictable. When there is a topology change, the network control traffic may burst for a few seconds. But the capacity of today's networks can easily handle the increase in network control traffic as the network converges.

However, some TCP applications can consume a large portion of network capacity. FTP will consume as much bandwidth as it can get when you download a large file, such as a movie or game. The table summarizes data traffic characteristics.

Data Traffic Characteristics
• Smooth/bursty
• Benign/greedy
• Drop insensitive
• Delay insensitive
• TCP retransmits

Although data traffic is relatively insensitive to drops and delays compared to voice and video, a network administrator still needs to consider the quality of the user experience, sometimes referred to as Quality of Experience or QoE. There are two main factors that a network administrator needs to ask about the flow of data traffic:

- Does the data come from an interactive application?

- Is the data mission critical?

The table compares these two factors.

Factors to Consider for Data Delay

Factor	Mission Critical	Not Mission Critical
Interactive	Prioritize for the lowest delay of all data traffic and strive for a 1 to 2 second response time.	Applications could benefit from lower delay.
Not inter-active	Delay can vary greatly as long as the necessary minimum bandwidth is supplied.	Gets any leftover bandwidth after all voice, video, and other data application needs are met.

Go to the online course to take the quiz and exam.

Check Your Understanding - Traffic Characteristics - 9.2.6

Queuing Algorithms - 9.3

Refer to **Video** in online course

Video Tutorial - QoS Algorithms - 9.3.1

Click Play for an overview of the different types of QoS queuing algorithms.

Queuing Overview - 9.3.2

The previous topic covered traffic characteristics. This topic will explain the queuing algorithms used to implement QoS. The QoS policy implemented by the network administrator becomes active when congestion occurs on the link. Queuing is a congestion management tool that can buffer, prioritize, and, if required, reorder packets before being transmitted to the destination.

A number of queuing algorithms are available. For the purposes of this course, we will focus on the following:

- First-In, First-Out (FIFO)
- Weighted Fair Queuing (WFQ)
- Class-Based Weighted Fair Queuing (CBWFQ)
- Low Latency Queuing (LLQ)

Refer to **Online Course** for Illustration

First In First Out - 9.3.3

In its simplest form, First In First Out (FIFO) queuing, also known as first-come, first-served queuing, buffers and forwards packets in the order of their arrival.

FIFO has no concept of priority or classes of traffic and consequently, makes no decision about packet priority. There is only one queue, and all packets are treated equally. Packets are sent out an interface in the order in which they arrive, as shown in the figure. Although some traffic may be more important or time-sensitive based on the priority classification, notice that the traffic is sent out in the order it is received.

When FIFO is used, important or time-sensitive traffic can be dropped when there is congestion on the router or switch interface. When no other queuing strategies are configured, all interfaces, except serial interfaces at E1 (2.048 Mbps) and below, use FIFO by default. (Serial interfaces at E1 and below use WFQ by default.)

FIFO, which is the fastest method of queuing, is effective for large links that have little delay and minimal congestion. If your link has very little congestion, FIFO queuing may be the only queuing you need to use.

Refer to
Online Course
for Illustration

Weighted Fair Queuing (WFQ) - 9.3.4

Weighted Fair Queuing (WFQ) is an automated scheduling method that provides fair bandwidth allocation to all network traffic. WFQ does not allow classification options to be configured. WFQ applies priority, or weights, to identified traffic and classifies it into conversations or flows, as shown in the figure.

WFQ then determines how much bandwidth each flow is allowed relative to other flows. The flow-based algorithm used by WFQ simultaneously schedules interactive traffic to the front of a queue to reduce response time. It then fairly shares the remaining bandwidth among high-bandwidth flows. WFQ allows you to give low-volume, interactive traffic, such as Telnet sessions and voice, priority over high-volume traffic, such as FTP sessions. When multiple file transfers flows are occurring simultaneously, the transfers are given comparable bandwidth.

WFQ classifies traffic into different flows based on packet header addressing, including such characteristics as source and destination IP addresses, MAC addresses, port numbers, protocol, and Type of Service (ToS) value. The ToS value in the IP header can be used to classify traffic.

Low-bandwidth traffic flows, which comprise the majority of traffic, receive preferential service which allows their entire offered loads to be sent in a timely fashion. High-volume traffic flows share the remaining capacity proportionally among themselves.

Limitations

WFQ is not supported with tunneling and encryption because these features modify the packet content information required by WFQ for classification.

Although WFQ automatically adapts to changing network traffic conditions, it does not offer the degree of precise control over bandwidth allocation that CBWFQ offers.

Refer to
Online Course
for Illustration

Class-Based Weighted Fair Queuing (CBWFQ) - 9.3.5

Class-Based Weighted Fair Queuing (CBWFQ) extends the standard WFQ functionality to provide support for user-defined traffic classes. With CBWFQ, you define traffic classes based on match criteria including protocols, access control lists (ACLs), and input interfaces. Packets satisfying the match criteria for a class constitute the traffic for that class. A FIFO queue is reserved for each class, and traffic belonging to a class is directed to the queue for that class, as shown in the figure.

When a class has been defined according to its match criteria, you can assign it characteristics. To characterize a class, you assign it bandwidth, weight, and maximum packet limit. The bandwidth assigned to a class is the guaranteed bandwidth delivered to the class during congestion.

To characterize a class, you also specify the queue limit for that class, which is the maximum number of packets allowed to accumulate in the queue for the class. Packets belonging to a class are subject to the bandwidth and queue limits that characterize the class.

After a queue has reached its configured queue limit, adding more packets to the class causes tail drop or packet drop to take effect, depending on how class policy is configured. Tail drop means a router simply discards any packet that arrives at the tail end of a queue that has completely used up its packet-holding resources. This is the default queuing response to congestion. Tail drop treats all traffic equally and does not differentiate between classes of service.

Refer to
Online Course
for Illustration

Low Latency Queuing (LLQ) - 9.3.6

The Low Latency Queuing (LLQ) feature brings strict priority queuing (PQ) to CBWFQ. Strict PQ allows delay-sensitive packets such as voice to be sent before packets in other queues. LLQ provides strict priority queuing for CBWFQ, reducing jitter in voice conversations, as shown in the figure.

Without LLQ, CBWFQ provides WFQ based on defined classes with no strict priority queue available for real-time traffic. The weight for a packet belonging to a specific class is derived from the bandwidth you assigned to the class when you configured it. Therefore, the bandwidth assigned to the packets of a class determines the order in which packets are sent. All packets are serviced fairly based on weight; no class of packets may be granted strict priority. This scheme poses problems for voice traffic that is largely intolerant of delay, especially variation in delay. For voice traffic, variations in delay introduce irregularities of transmission manifesting as jitter in the heard conversation.

LLQ allows delay-sensitive packets such as voice to be sent first (before packets in other queues), giving delay-sensitive packets preferential treatment over other traffic. Although it is possible to classify various types of real-time traffic to the strict priority queue, Cisco recommends that only voice traffic be directed to the priority queue.

Go to the online
course to take the
quiz and exam.

Check Your Understanding - Queuing Algorithms - 9.3.7

QoS Models - 9.4

Refer to **Video**
in online course

Video Tutorial - QoS Models - 9.4.1

Click Play for a brief explanation of the purpose of QoS.

Selecting an Appropriate QoS Policy Model - 9.4.2

How can QoS be implemented in a network? There are three models for implementing QoS:

- Best-effort model
- Integrated services (IntServ)
- Differentiated services (DiffServ)

The table summarizes these three models. QoS is implemented in a network using either IntServ or DiffServ. While IntServ provides the highest guarantee of QoS, it is very resource-intensive, and therefore, not easily scalable. In contrast, DiffServ is less resource-intensive and more scalable. The two are sometimes co-deployed in network QoS implementations.

Models for Implementing QoS

Model	Description
Best-effort model	• This is not really an implementation as QoS is not explicitly configured. • Use this when QoS is not required.
Integrated services (IntServ)	• IntServ provides very high QoS to IP packets with guaranteed delivery. • It defines a signaling process for applications to signal to the network that they require special QoS for a period and that bandwidth should be reserved. • IntServ can severely limit the scalability of a network.
Differentiated services (DiffServ)	• DiffServ provides high scalability and flexibility in implementing QoS. • Network devices recognize traffic classes and provide different levels of QoS to different traffic classes.

Best Effort - 9.4.3

The basic design of the internet is best-effort packet delivery and provides no guarantees. This approach is still predominant on the internet today and remains appropriate for most purposes. The best-effort model treats all network packets in the same way, so an emergency voice message is treated the same way that a digital photograph attached to an email is treated. Without QoS, the network cannot tell the difference between packets and, as a result, cannot treat packets preferentially.

The best-effort model is similar in concept to sending a letter using standard postal mail. Your letter is treated exactly the same as every other letter. With the best-effort model, the letter may never arrive, and, unless you have a separate notification arrangement with the letter recipient, you may never know that the letter did not arrive.

The table lists the benefits and drawbacks of the best effort model.

Benefits and Drawbacks of Best-Effort Model

Benefits	Drawbacks
The model is the most scalable.	There are no guarantees of delivery.
Scalability is only limited by available bandwidth, in which case all traffic is equally affected.	Packets will arrive whenever they can and in any order possible, if they arrive at all.
No special QoS mechanisms are required.	No packets have preferential treatment.
It is the easiest and quickest model to deploy.	Critical data is treated the same as casual email is treated.

Refer to
Online Course
for Illustration

Integrated Services - 9.4.4

The IntServ architecture model (RFC 1633, 2211, and 2212) was developed in 1994 to meet the needs of real-time applications, such as remote video, multimedia conferencing, data visualization applications, and virtual reality. IntServ is a multiple-service model that can accommodate many QoS requirements.

IntServ delivers the end-to-end QoS that real-time applications require. IntServ explicitly manages network resources to provide QoS to individual flows or streams, sometimes called microflows. It uses resource reservation and admission-control mechanisms as building blocks to establish and maintain QoS. This is similar to a concept known as "hard QoS." Hard QoS guarantees traffic characteristics, such as bandwidth, delay, and packet-loss rates, from end to end. Hard QoS ensures both predictable and guaranteed service levels for mission-critical applications.

The figure shows a simple illustration of the IntServ model.

IntServ uses a connection-oriented approach inherited from telephony network design. Each individual communication must explicitly specify its traffic descriptor and requested resources to the network. The edge router performs admission control to ensure that available resources are sufficient in the network. The IntServ standard assumes that routers along a path set and maintain the state for each individual communication.

In the IntServ model, the application requests a specific kind of service from the network before sending data. The application informs the network of its traffic profile and requests a particular kind of service that can encompass its bandwidth and delay requirements. IntServ uses the Resource Reservation Protocol (RSVP) to signal the QoS needs of an application's traffic along devices in the end-to-end path through the network. If network devices along the path can reserve the necessary bandwidth, the originating application can begin transmitting. If the requested reservation fails along the path, the originating application does not send any data.

The edge router performs admission control based on information from the application and available network resources. The network commits to meeting the QoS requirements of the application as long as the traffic remains within the profile specifications. The network fulfills its commitment by maintaining the per-flow state and then performing packet classification, policing, and intelligent queuing based on that state.

The table lists the benefits and drawbacks of the IntServ model.

Benefits and Drawbacks of IntServ Model

Benefits	Drawbacks
• Explicit end-to-end resource admission control	• Resource intensive due to the stateful architecture requirement for continuous signaling.
• Per-request policy admission control	• Flow-based approach not scalable to large implementations such as the internet.
• Signaling of dynamic port numbers	

Refer to
Online Course
for Illustration

Differentiated Services - 9.4.5

The differentiated services (DiffServ) QoS model specifies a simple and scalable mechanism for classifying and managing network traffic. For example, DiffServ can provide low-latency guaranteed service to critical network traffic such as voice or video, while providing simple best-effort traffic guarantees to non-critical services such as web traffic or file transfers.

The DiffServ design overcomes the limitations of both the best-effort and IntServ models. The DiffServ model is described in RFCs 2474, 2597, 2598, 3246, 4594. DiffServ can provide an "almost guaranteed" QoS while still being cost-effective and scalable.

The DiffServ model is similar in concept to sending a package using a delivery service. You request (and pay for) a level of service when you send a package. Throughout the package network, the level of service you paid for is recognized and your package is given either preferential or normal service, depending on what you requested.

DiffServ is not an end-to-end QoS strategy because it cannot enforce end-to-end guarantees. However, DiffServ QoS is a more scalable approach to implementing QoS. Unlike IntServ and hard QoS, in which the end-hosts signal their QoS needs to the network, DiffServ does not use signaling. Instead, DiffServ uses a "soft QoS" approach. It works on the provisioned-QoS model, where network elements are set up to service multiple classes of traffic each with varying QoS requirements.

The figure shows a simple illustration of the DiffServ model.

As a host forwards traffic to a router, the router classifies the flows into aggregates (classes) and provides the appropriate QoS policy for the classes. DiffServ enforces and applies QoS mechanisms on a hop-by-hop basis, uniformly applying global meaning to each traffic class to provide both flexibility and scalability. For example, DiffServ could be configured to group all TCP flows as a single class, and allocate bandwidth for that class, rather than for the individual flows as IntServ would do. In addition to classifying traffic, DiffServ minimizes signaling and state maintenance requirements on each network node.

Specifically, DiffServ divides network traffic into classes based on business requirements. Each of the classes can then be assigned a different level of service. As the packets traverse a network, each of the network devices identifies the packet class and services the packet according to that class. It is possible to choose many levels of service with DiffServ. For example, voice traffic from IP phones is usually given preferential treatment over all other application traffic, email is generally given best-effort service, and nonbusiness traffic can either be given very poor service or blocked entirely.

The figure lists the benefits and drawbacks of the DiffServ model.

Note: Modern networks primarily use the DiffServ model. However, due to the increasing volumes of delay- and jitter-sensitive traffic, IntServ and RSVP are sometimes co-deployed.

Benefits and Drawbacks of DiffServ Model

Benefits	Drawbacks
• Highly scalable	• No absolute guarantee of service quality
• Provides many different levels of quality	• Requires a set of complex mechanisms to work in concert throughout the network

Go to the online course to take the quiz and exam.

Check Your Understanding - QoS Models - 9.4.6

QoS Implementation Techniques - 9.5

Refer to **Video** in online course

Video Tutorial - QoS Implementation Techniques - 9.5.1

Click Play for an overview of classification, marking, trust boundaries, congestion avoidance, shaping and policing.

Avoiding Packet Loss - 9.5.2

Now that you have learned about traffic characteristics, queuing algorithms, and QoS models, it is time to learn about QoS implementation techniques.

Let's start with packet loss. Packet loss is usually the result of congestion on an interface. Most applications that use TCP experience slowdown because TCP automatically adjusts to network congestion. Dropped TCP segments cause TCP sessions to reduce their window sizes. Some applications do not use TCP and cannot handle drops (fragile flows).

The following approaches can prevent drops in sensitive applications:

- Increase link capacity to ease or prevent congestion.

- Guarantee enough bandwidth and increase buffer space to accommodate bursts of traffic from fragile flows. WFQ, CBWFQ, and LLQ can guarantee bandwidth and provide prioritized forwarding to drop-sensitive applications.

- Drop lower-priority packets before congestion occurs. Cisco IOS QoS provides queuing mechanisms, such as weighted random early detection (WRED), that start dropping lower-priority packets before congestion occurs.

Refer to **Online Course** for Illustration

QoS Tools - 9.5.3

There are three categories of QoS tools, as described in the table:

- Classification and marking tools

- Congestion avoidance tools

- Congestion management tools

Tools for Implementing QoS

QoS Tools	Description
Classification and marking tools	• Sessions, or flows, are analyzed to determine what traffic class they belong to. • When the traffic class is determined, the packets are marked.
Congestion avoidance tools	• Traffic classes are allotted portions of network resources, as defined by the QoS policy. • The QoS policy also identifies how some traffic may be selectively dropped, delayed, or re-marked to avoid congestion.

QoS Tools	Description
	• The primary congestion avoidance tool is WRED and is used to regulate TCP data traffic in a bandwidth-efficient manner before tail drops caused by queue overflows occur.
Congestion management tools	• When traffic exceeds available network resources, traffic is queued to await availability of resources. • Common Cisco IOS-based congestion management tools include CBWFQ and LLQ algorithms.

Refer to the figure to help understand the sequence of how these tools are used when QoS is applied to packet flows.

As shown in the figure, ingress packets (gray squares) are classified and their respective IP header is marked (colored squares). To avoid congestion, packets are then allocated resources based on defined policies. Packets are then queued and forwarded out the egress interface based on their defined QoS shaping and policing policy.

Note: Classification and marking can be done on ingress or egress, whereas other QoS actions such queuing and shaping are usually done on egress.

Classification and Marking - 9.5.4

Before a packet can have a QoS policy applied to it, the packet has to be classified. Classification and marking allows us to identify or "mark" types of packets. Classification determines the class of traffic to which packets or frames belong. Only after traffic is marked can policies be applied to it.

How a packet is classified depends on the QoS implementation. Methods of classifying traffic flows at Layer 2 and 3 include using interfaces, ACLs, and class maps. Traffic can also be classified at Layers 4 to 7 using Network Based Application Recognition (NBAR).

Note: NBAR is a classification and protocol discovery feature of Cisco IOS software that works with QoS features. NBAR is out of scope for this course.

Marking means that we are adding a value to the packet header. Devices receiving the packet look at this field to see if it matches a defined policy. Marking should be done as close to the source device as possible. This establishes the trust boundary.

How traffic is marked usually depends on the technology. The table in the figure describes some the marking fields used in various technologies. The decision of whether to mark traffic at Layers 2 or 3 (or both) is not trivial and should be made after consideration of the following points:

- Layer 2 marking of frames can be performed for non-IP traffic.

- Layer 2 marking of frames is the only QoS option available for switches that are not "IP aware".

- Layer 3 marking will carry the QoS information end-to-end.

Traffic Marking for QoS

QoS Tools	Layer	Marking Field	Width in Bits
Ethernet (802.1Q, 802.1p)	2	Class of Service (CoS)	3
802.11 (Wi-Fi)	2	Wi-Fi Traffic Identifier (TID)	3
MPLS	2	Experimental (EXP)	3
IPv4 and IPv6	3	IP Precedence (IPP)	3
IPv4 and IPv6	3	Differentiated Services Code Point (DSCP)	6

Refer to **Online Course** for Illustration

Marking at Layer 2 - 9.5.5

802.1Q is the IEEE standard that supports VLAN tagging at Layer 2 on Ethernet networks. When 802.1Q is implemented, two fields are added to the Ethernet Frame. As shown in the figure, these two fields are inserted into the Ethernet frame following the source MAC address field.

The 802.1Q standard also includes the QoS prioritization scheme known as IEEE 802.1p. The 802.1p standard uses the first three bits in the Tag Control Information (TCI) field. Known as the Priority (PRI) field, this 3-bit field identifies the Class of Service (CoS) markings. Three bits means that a Layer 2 Ethernet frame can be marked with one of eight levels of priority (values 0-7) as displayed in the figure.

Ethernet Class of Service (CoS) Values

CoS Value	CoS Binary Value	Description
0	000	Best-Effort Data
1	001	Medium-Priority Data
2	010	High-Priority Data
3	011	Call Signaling
4	100	Videoconferencing
5	101	Voice bearer (voice traffic)
6	110	Reserved
7	111	Reserved

Refer to **Online Course** for Illustration

Marking at Layer 3 - 9.5.6

IPv4 and IPv6 specify an 8-bit field in their packet headers to mark packets. As shown in the figure, both IPv4 and IPv6 support an 8-bit field for marking: the Type of Service (ToS) field for IPv4 and the Traffic Class field for IPv6.

Refer to
Online Course
for Illustration

Type of Service and Traffic Class Field - 9.5.7

The Type of Service (IPv4) and Traffic Class (IPv6) carry the packet marking as assigned by the QoS classification tools. The field is then referred to by receiving devices which forward the packets based on the appropriate assigned QoS policy.

The figure displays the contents of the 8-bit field. In RFC 791, the original IP standard specified the IP Precedence (IPP) field to be used for QoS markings. However, in practice, these three bits did not provide enough granularity to implement QoS.

RFC 2474 supersedes RFC 791 and redefines the ToS field by renaming and extending the IPP field. The new field, as shown in the figure, has 6-bits allocated for QoS. Called the Differentiated Services Code Point (DSCP) field, these six bits offer a maximum of 64 possible classes of service. The remaining two IP Extended Congestion Notification (ECN) bits can be used by ECN-aware routers to mark packets instead of dropping them. The ECN marking informs downstream routers that there is congestion in the packet flow.

Refer to
Online Course
for Illustration

DSCP Values - 9.5.8

The 64 DSCP values are organized into three categories:

- **Best-Effort (BE)** - This is the default for all IP packets. The DSCP value is 0. The per-hop behavior is normal routing. When a router experiences congestion, these packets will be dropped. No QoS plan is implemented.

- **Expedited Forwarding (EF)** - RFC 3246 defines EF as the DSCP decimal value 46 (binary **101**110). The first 3 bits (101) map directly to the Layer 2 CoS value 5 used for voice traffic. At Layer 3, Cisco recommends that EF only be used to mark voice packets.

- **Assured Forwarding (AF)** - RFC 2597 defines AF to use the 5 most significant DSCP bits to indicate queues and drop preference. The definition of AF is illustrated in the figure.

Refer to
Online Course
for Illustration

Class Selector Bits - 9.5.9

Because the first 3 most significant bits of the DSCP field indicate the class, these bits are also called the Class Selector (CS) bits. These 3 bits map directly to the 3 bits of the CoS field and the IPP field to maintain compatibility with 802.1p and RFC 791, as shown in the figure.

The table in the figure shows how the CoS values map to the Class Selectors and the corresponding DSCP 6-bit value. This same table can be used to map IPP values to the Class Selectors.

Refer to
Online Course
for Illustration

Trust Boundaries - 9.5.10

Where should markings occur? Traffic should be classified and marked as close to its source as technically and administratively feasible. This defines the trust boundary, as shown in the figure.

1. Trusted endpoints have the capabilities and intelligence to mark application traffic to the appropriate Layer 2 CoS and/or Layer 3 DSCP values. Examples of trusted endpoints include IP phones, wireless access points, videoconferencing gateways and systems, IP conferencing stations, and more.

2. Secure endpoints can have traffic marked at the Layer 2 switch.

3. Traffic can also be marked at Layer 3 switches / routers.

Re-marking traffic, for example, re-marking CoS values to IP Precedent or DSCP values, is typically necessary.

Congestion Avoidance - 9.5.11

Refer to **Online Course** for Illustration

Congestion management includes queuing and scheduling methods where excess traffic is buffered or queued (and sometimes dropped) while it waits to be sent out an egress interface. Congestion avoidance tools are simpler. They monitor network traffic loads in an effort to anticipate and avoid congestion at common network and internetwork bottlenecks before congestion becomes a problem. These tools can monitor the average depth of the queue, as represented in the figure. When the queue is below the minimum threshold, there are no drops. As the queue fills up to the maximum threshold, a small percentage of packets are dropped. When the maximum threshold is passed, all packets are dropped.

Some congestion avoidance techniques provide preferential treatment for which packets will get dropped. For example, Cisco IOS QoS includes weighted random early detection (WRED) as a possible congestion avoidance solution. The WRED algorithm allows for congestion avoidance on network interfaces by providing buffer management and allowing TCP traffic to decrease, or throttle back, before buffers are exhausted. Using WRED helps avoid tail drops and maximizes network use and TCP-based application performance. There is no congestion avoidance for User Datagram Protocol (UDP)-based traffic, such as voice traffic. In case of UDP-based traffic, methods such as queuing and compression techniques help to reduce and even prevent UDP packet loss.

Shaping and Policing - 9.5.12

Refer to **Online Course** for Illustration

Traffic shaping and traffic policing are two mechanisms provided by Cisco IOS QoS software to prevent congestion.

Traffic shaping retains excess packets in a queue and then schedules the excess for later transmission over increments of time. The result of traffic shaping is a smoothed packet output rate, as shown in the figure.

Shaping implies the existence of a queue and of sufficient memory to buffer delayed packets, while policing does not.

Ensure that you have sufficient memory when enabling shaping. In addition, shaping requires a scheduling function for later transmission of any delayed packets. This scheduling function allows you to organize the shaping queue into different queues. Examples of scheduling functions are CBWFQ and LLQ.

Shaping is an outbound concept; packets going out an interface get queued and can be shaped. In contrast, policing is applied to inbound traffic on an interface. When the traffic rate reaches the configured maximum rate, excess traffic is dropped (or remarked).

Policing is commonly implemented by service providers to enforce a contracted customer information rate (CIR). However, the service provider may also allow bursting over the CIR if the service provider's network is not currently experiencing congestion.

QoS Policy Guidelines - 9.5.13

Your QoS policy must consider the full path from source to destination. If one device in the path is using a different policy than desired, then the entire QoS policy is impacted. For example, the stutter in video playback could be the result of one switch in the path that does not have the CoS value set appropriately.

A few guidelines that help ensure the best experience for end users includes the following:

- Enable queuing at every device in the path between source and destination.
- Classify and mark traffic as close the source as possible.
- Shape and police traffic flows as close to their sources as possible.

Go to the online course to take the quiz and exam.

Check Your Understanding - QoS Implementation Techniques - 9.5.14

Module Practice and Quiz - 9.6

What did I learn in this module? - 9.6.1

Network Transmission Quality

Voice and live video transmissions create higher expectations for quality delivery among users, and create a need for Quality of Service (QoS). Congestion occurs when multiple communication lines aggregate onto a single device such as a router, and then much of that data is placed on just a few outbound interfaces, or onto a slower interface. Congestion can also occur when large data packets prevent smaller packets from being transmitted in a timely manner. Without any QoS mechanisms in place, packets are processed in the order in which they are received. When congestion occurs, network devices such as routers and switches can drop packets. This means that time-sensitive packets, such as real-time video and voice, will be dropped with the same frequency as data that is not time-sensitive, such as email and web browsing. When the volume of traffic is greater than what can be transported across the network, devices queue (hold) the packets in memory until resources become available to transmit them. Queuing packets causes delay because new packets cannot be transmitted until previous packets have been processed. One QoS technique that can help with this problem is to classify data into multiple queues. Network congestion points are ideal candidates for QoS mechanisms to mitigate delay and latency. Two types of delays are fixed and variable. Sources of delay are code delay, packetization delay, queuing delay, serialization delay, propagation delay, and de-jitter delay. Jitter is the variation in the delay of received packets. Due to network congestion, improper queuing, or configuration errors, the delay between each packet can vary instead of remaining constant. Both delay and jitter need to be controlled and minimized to support real-time and interactive traffic.

Traffic Characteristics

Voice and video traffic are two of the main reasons for QoS. Voice traffic is smooth and benign, but it is sensitive to drops and delays. Voice can tolerate a certain amount of latency, jitter, and loss without any noticeable effects. Latency should be no more than 150 milliseconds (ms). Jitter should be no more than 30 ms, and voice packet loss should be no more than 1%. Voice traffic requires at least 30 Kbps of bandwidth. Video traffic is more demanding than voice traffic because of the size of the packets it sends across the network. Video traffic is bursty, greedy, drop sensitive, and delay sensitive. Without QoS and a significant amount of extra bandwidth, video quality typically degrades. UDP ports such as 554, are used for the Real-Time Streaming Protocol (RSTP) and should be given priority over other, less delay-sensitive, network traffic. Similar to voice, video can tolerate a certain amount of latency, jitter, and loss without any noticeable effects. Latency should be no more than 400 milliseconds (ms). Jitter should be no more than 50 ms, and video packet loss should be no more than 1%. Video traffic requires at least 384 Kbps of bandwidth. Data traffic is not as demanding as voice and video traffic. Data packets often use TCP applications which can retransmit data and, therefore, are not sensitive to drops and delays. Although data traffic is relatively insensitive to drops and delays compared to voice and video, a network administrator still needs to consider the quality of the user experience, sometimes referred to as Quality of Experience (QoE). The two main factors that a network administrator needs to ask about the flow of data traffic are if the data comes from an interactive application and if the data is mission critical.

Queuing Algorithms

The QoS policy implemented by the network administrator becomes active when congestion occurs on the link. Queuing is a congestion management tool that can buffer, prioritize, and, if required, reorder packets before being transmitted to the destination. This course focuses on the following queuing algorithms: First-In, First-Out (FIFO), Weighted Fair Queuing (WFQ), Class-Based Weighted Fair Queuing (CBWFQ), and Low Latency Queuing (LLQ). FIFO queuing buffers and forwards packets in the order of their arrival. FIFO has no concept of priority or classes of traffic and consequently, makes no decision about packet priority. When FIFO is used, important or time-sensitive traffic can be dropped when there is congestion on the router or switch interface. WFQ is an automated scheduling method that provides fair bandwidth allocation to all network traffic. WFQ applies priority, or weights, to identified traffic and classifies it into conversations or flows. WFQ classifies traffic into different flows based on packet header addressing, including such characteristics as source and destination IP addresses, MAC addresses, port numbers, protocol, and Type of Service (ToS) value. The ToS value in the IP header can be used to classify traffic. CBWFQ extends the standard WFQ functionality to provide support for user-defined traffic classes. With CBWFQ, you define traffic classes based on match criteria including protocols, access control lists (ACLs), and input interfaces. LLQ feature brings strict priority queuing (PQ) to CBWFQ. Strict PQ allows delay-sensitive packets, such as voice, to be sent before packets in other queues, reducing jitter in voice conversations.

QoS Models

There are three models for implementing QoS: Best-effort model, Integrated services (IntServ), and Differentiated services (DiffServ). The Best-effort model is the most scalable but does not guarantee delivery and does not give any packet preferential treatment. The IntServ architecture model was developed to meet the needs of real-time applications, such as remote video, multimedia conferencing, data visualization applications, and virtual reality. IntServ is a multiple-service model that can accommodate many QoS require-ments. IntServ explicitly manages network resources to provide QoS to individual flows or streams, sometimes called microflows. It uses resource reservation and admission-control mechanisms as building blocks to establish and maintain QoS. The DiffServ QoS model specifies a simple and scalable mechanism for classifying and managing network traffic. The DiffServ design overcomes the limitations of both the best-effort and IntServ models. The DiffServ model can provide an "almost guaranteed" QoS, while still being cost-effective and scalable. DiffServ divides network traffic into classes based on business requirements. Each of the classes can then be assigned a different level of service. As the packets traverse a network, each of the network devices identifies the packet class and services the packet according to that class. It is possible to choose many levels of service with DiffServ.

QoS Implementation Techniques

There are three categories of QoS tools: classification and marking tools, congestion avoidance tools, and congestion management tools. Before a packet can have a QoS policy applied to it, the packet has to be classified. Classification and marking allows us to iden-tify or "mark" types of packets. Classification determines the class of traffic to which packets or frames belong. Methods of classifying traffic flows at Layer 2 and 3 include using interfaces, ACLs, and class maps. Traffic can also be classified at Layers 4 to 7 using Network Based Application Recognition (NBAR). The Type of Service (IPv4) and Traffic Class (IPv6) carry the packet marking as assigned by the QoS classification tools. The field is then referred to by receiving devices which forward the packets based on the appropriate assigned QoS policy. These fields have 6-bits allocated for QoS. Called the Differentiated Services Code Point (DSCP) field, these six bits offer a maximum of 64 possible classes of service. The field is then referred to by receiving devices which forward the packets based on the appropriate assigned QoS policy. The 64 DSCP values are organized into three cat-egories: Best-Effort (BE), Expedited Forwarding (EF), Assured Forwarding (AF). Because the first 3 most significant bits of the DSCP field indicate the class, these bits are also called the Class Selector (CS) bits. Traffic should be classified and marked as close to its source as technically and administratively feasible. This defines the trust boundary. Con-gestion management includes queuing and scheduling methods where excess traffic is buff-ered or queued (and sometimes dropped) while it waits to be sent out an egress interface. Congestion avoidance tools help to monitor network traffic loads in an effort to anticipate and avoid congestion at common network and internetwork bottlenecks before congestion becomes a problem. Cisco IOS QoS includes weighted random early detection (WRED) as a possible congestion avoidance solution. The WRED algorithm allows for congestion avoidance on network interfaces by providing buffer management and allowing TCP traf-fic to decrease, or throttle back, before buffers are exhausted. Traffic shaping and traffic policing are two mechanisms provided by Cisco IOS QoS software to prevent congestion.

Go to the online course to take the quiz and exam.

Chapter Quiz - QoS Concepts

Your Chapter Notes

CHAPTER 10

Network Management

Introduction - 10.0

Why should I take this module? - 10.0.1

Welcome to Network Management!

Imagine that you are at the helm of a spaceship. There are many, many components that work together to move this ship. There are multiple systems to manage these components. To get where you are going you would need to have a full understanding of the components and the systems that manage them. You would probably appreciate any tools that would make managing your spaceship - *while you are also flying it* - simpler.

Like a complex spaceship, networks also need to be managed. Happily, there are many tools that are designed to make network management simpler. This module introduces you to several tools and protocols to help you manage your network - *while your users are using it*. It also includes many Packet Tracer activities and Hands On Labs to test your skills. These are the tools of great network administrators, so you will definitely want to get started!

What will I learn to do in this module? - 10.0.2

Module Title: Network Management

Module Objective: Implement protocols to manage the network.

Topic Title	Topic Objective
Device Discovery with CDP	Use CDP to map a network topology.
Device Discovery with LLDP	Use LLDP to map a network topology.
NTP	Implement NTP between an NTP client and NTP server.
SNMP	Explain how SNMP operates.
Syslog	Explain syslog operation.
Router and Switch File Maintenance	Use commands to back up and restore an IOS configuration file.
IOS Image Management	Implement protocols to manage the network.

Device Discovery with CDP - 10.1

CDP Overview - 10.1.1

Refer to
Online Course
for Illustration

The first thing you want to know about your network is what is in it? Where are these components? How are they connected? Basically, you need a map. This topic explains how you can use Cisco Discovery Protocol (CDP) to create a map of your network.

CDP is a Cisco proprietary Layer 2 protocol that is used to gather information about Cisco devices which share the same data link. CDP is media and protocol independent and runs on all Cisco devices, such as routers, switches, and access servers.

The device sends periodic CDP advertisements to connected devices, as shown in the figure.

These advertisements share information about the type of device that is discovered, the name of the devices, and the number and type of the interfaces.

Because most network devices are connected to other devices, CDP can assist in network design decisions, troubleshooting, and making changes to equipment. CDP can also be used as a network discovery tool to determine the information about the neighboring devices. This information gathered from CDP can help build a logical topology of a network when documentation is missing or lacking in detail.

Configure and Verify CDP - 10.1.2

For Cisco devices, CDP is enabled by default. For security reasons, it may be desirable to disable CDP on a network device globally, or per interface. With CDP, an attacker can gather valuable insight about the network layout, such as IP addresses, IOS versions, and types of devices.

To verify the status of CDP and display information about CDP, enter the **show cdp** command, as displayed in the example.

```
Router# show cdp

Global CDP information:

        Sending CDP packets every 60 seconds

        Sending a holdtime value of 180 seconds

        Sending CDPv2 advertisements is enabled
```

To enable CDP globally for all the supported interfaces on the device, enter **cdp run** in the global configuration mode. CDP can be disabled for all the interfaces on the device with the **no cdp run** command in the global configuration mode.

```
Router(config)# no cdp run

Router(config) # exit

Router# show cdp

CDP is not enabled

Router# configure terminal

Router(config) # cdp run
```

To disable CDP on a specific interface, such as the interface facing an ISP, enter **no cdp enable** in the interface configuration mode. CDP is still enabled on the device; however, no more CDP advertisements will be sent out that interface. To enable CDP on the specific interface again, enter **cdp enable**, as shown in the example.

```
Switch(config) # interface gigabitethernet 0/0/1

Switch(config-if)# cdp enable
```

To verify the status of CDP and display a list of neighbors, use the **show cdp neighbors** command in the privileged EXEC mode. The **show cdp neighbors** command displays important information about the CDP neighbors. Currently, this device does not have any neighbors because it is not physically connected to any devices, as indicated by the results of the **show cdp neighbors** command displayed in the example.

```
Router# show cdp neighbors
Capability Codes: R - Router, T - Trans Bridge, B - Source Route Bridge
                  S - Switch, H - Host, I - IGMP, r - Repeater, P
                  - Phone,
 D - Remote, C - CVTA, M - Two-port Mac Relay


Device ID Local Intrfce Holdtme Capability Platform Port ID


Total cdp entries displayed : 0
```

Use the **show cdp interface** command to display the interfaces that are CDP-enabled on a device. The status of each interface is also displayed. The figure shows that five interfaces are CDP-enabled on the router with only one active connection to another device.

```
Router# show cdp interface
GigabitEthernet0/0/0 is administratively down, line protocol is down
   Encapsulation ARPA
   Sending CDP packets every 60 seconds
   Holdtime is 180 seconds
GigabitEthernet0/0/1 is up, line protocol is up
   Encapsulation ARPA
   Sending CDP packets every 60 seconds
   Holdtime is 180 seconds
GigabitEthernet0/0/2 is down, line protocol is down
   Encapsulation ARPA
   Sending CDP packets every 60 seconds
   Holdtime is 180 seconds
Serial0/1/0 is administratively down, line protocol is down
   Encapsulation HDLC
   Sending CDP packets every 60 seconds
   Holdtime is 180 seconds
Serial0/1/1 is administratively down, line protocol is down
   Encapsulation HDLC
   Sending CDP packets every 60 seconds
   Holdtime is 180 seconds
```

```
GigabitEthernet0 is down, line protocol is down
  Encapsulation ARPA
  Sending CDP packets every 60 seconds
  Holdtime is 180 seconds
cdp enabled interfaces  : 6
interfaces up          : 1
interfaces down        : 5
```

Refer to
Online Course
for Illustration

Discover Devices by Using CDP - 10.1.3

Consider the lack of documentation in the topology shown in the figure. The network administrator only knows that R1 is connected to another device.

With CDP enabled on the network, the **show cdp neighbors** command can be used to determine the network layout, as shown in the output.

```
R1# show cdp neighbors
Capability Codes: R - Router, T - Trans Bridge, B - Source Route Bridge
                  S - Switch, H - Host, I - IGMP, r - Repeater, P - Phone,
                  D - Remote, C - CVTA, M - Two-port Mac Relay


Device ID          Local Intrfce   Holdtme   Capability   Platform   Port ID
S1                 Gig 0/0/1       179          S I       WS-C3560-  Fas 0/5
```

No information is available regarding the rest of the network. The **show cdp neighbors** command provides helpful information about each CDP neighbor device, including the following:

■ **Device identifiers** - This is the host name of the neighbor device (S1).

■ **Port identifier** - This is the name of the local and remote port (G0/0/1 and F0/5, respectively).

■ **Capabilities list** - This shows whether the device is a router or a switch (S for switch; I for IGMP is beyond scope for this course)

■ **Platform** - This is the hardware platform of the device (WS-C3560 for Cisco 3560 switch).

The output shows that there is another Cisco device, S1, connected to the G0/0/1 interface on R1. Furthermore, S1 is connected through its F0/5, as shown in the updated topology.

The network administrator uses **show cdp neighbors detail** to discover the IP address for S1. As displayed in the output, the address for S1 is 192.168.1.2.

```
R1# show cdp neighbors detail

-------------------------

Device ID: S1
```

```
Entry address(es):

  IP address: 192.168.1.2

Platform: cisco WS-C3560-24TS,   Capabilities: Switch IGMP

Interface: GigabitEthernet0/0/1,   Port ID (outgoing port):
FastEthernet0/5

Holdtime : 136 sec


Version :

Cisco IOS Software, C3560 Software (C3560-LANBASEK9-M), Version 15.0(2)
SE7, R

RELEASE SOFTWARE (fc1)

Technical Support: http://www.cisco.com/techsupport

Copyright (c) 1986-2014 by Cisco Systems, Inc.

Compiled Thu 23-Oct-14 14:49 by prod_rel_team


advertisement version: 2

Protocol Hello:  OUI=0x00000C, Protocol ID=0x0112; payload len=27,

value=00000000FFFFFFFF010221FF000000000000002291210380FF0000

VTP Management Domain: ''

Native VLAN: 1

Duplex: full

Management address(es):

  IP address: 192.168.1.2


Total cdp entries displayed : 1
```

By accessing S1 either remotely through SSH, or physically through the console port, the network administrator can determine what other devices are connected to S1, as displayed in the output of the **show cdp neighbors** in the figure.

```
S1# show cdp neighbors
Capability Codes: R - Router, T - Trans Bridge, B - Source Route Bridge
                  S - Switch, H - Host, I - IGMP, r - Repeater, P
                  - Phone,
                  D - Remote, C - CVTA, M - Two-port Mac Relay

Device ID      Local Intrfce  Holdtme  Capability  Platform   Port ID
S2             Fas 0/1        150      S I         WS-C2960-  Fas 0/1
R1             Fas 0/5        179      R S I       ISR4331/K  Gig 0/0/1
```

Another switch, S2, is revealed in the output. S2 is using F0/1 to connect to the F0/1 interface on S1, as shown in the figure.

Again, the network administrator can use **show cdp neighbors detail** to discover the IP address for S2, and then remotely access it. After a successful login, the network administrator uses the **show cdp neighbors** command to discover if there are more devices.

```
S2# show cdp neighbors
Capability Codes: R - Router, T - Trans Bridge, B - Source Route Bridge
                  S - Switch, H - Host, I - IGMP, r - Repeater, P
                  - Phone,
                  D - Remote, C - CVTA, M - Two-port Mac Relay
Device ID          Local Intrfce   Holdtme   Capability Platform  Port ID
S1                 Fas 0/1            141           S I  WS-C3560- Fas 0/1
```

The only device connected to S2 is S1. Therefore, there are no more devices to discover in the topology. The network administrator can now update the documentation to reflect the discovered devices.

Refer to
Interactive Graphic
in online course

Syntax Checker - Configure and Verify CDP - 10.1.4

Practice configuring and verifying CDP.

Refer to **Packet Tracer Activity** for this chapter

Packet Tracer - Use CDP to Map a Network - 10.1.5

A senior network administrator requires you to map the Remote Branch Office network and discover the name of a recently installed switch that still needs an IPv4 address to be configured. Your task is to create a map of the branch office network. To map the network, you will use SSH for remote access and the Cisco Discovery Protocol (CDP) to discover information about neighboring network devices, like routers and switches.

Device Discovery with LLDP - 10.2

Refer to
Online Course
for Illustration

LLDP Overview - 10.2.1

The Link Layer Discovery Protocol (LLDP) does the same thing as CDP, but it is not specific to Cisco devices. As a bonus, you can still use it if you have Cisco devices. One way or another, you will get your network map.

LLDP is a vendor-neutral neighbor discovery protocol similar to CDP. LLDP works with network devices, such as routers, switches, and wireless LAN access points. This protocol advertises its identity and capabilities to other devices and receives the information from a physically-connected Layer 2 device.

Configure and Verify LLDP - 10.2.2

Depending on the device, LLDP may be enabled by default. To enable LLDP globally on a Cisco network device, enter the **lldp run** command in the global configuration mode. To disable LLDP, enter the **no lldp run** command in the global configuration mode.

Similar to CDP, LLDP can be configured on specific interfaces. However, LLDP must be configured separately to transmit and receive LLDP packets, as shown in the figure.

To verify LLDP has been enabled on the device, enter the **show lldp** command in privileged EXEC mode.

```
Switch# conf t

Enter configuration commands, one per line.  End with CNTL/Z.

Switch(config)# lldp run

Switch(config)# interface gigabitethernet 0/1

Switch(config-if)# lldp transmit

Switch(config-if)# lldp receive

Switch(config-if)# end

Switch# show lldp

Global LLDP Information:

    Status: ACTIVE

    LLDP advertisements are sent every 30 seconds

    LLDP hold time advertised is 120 seconds

    LLDP interface reinitialisation delay is 2 seconds
```

Refer to
Online Course
for Illustration

Discover Devices by Using LLDP - 10.2.3

Consider the lack of documentation in the topology shown in the figure. The network administrator only knows that S1 is connected to two devices.

With LLDP enabled, device neighbors can be discovered by using the **show lldp neighbors** command, as displayed in the output.

```
S1# show lldp neighbors

Capability codes:

    (R) Router, (B) Bridge, (T) Telephone, (C) DOCSIS Cable Device

    (W) WLAN Access Point, (P) Repeater, (S) Station, (O) Other

Device ID          Local Intf      Hold-time   Capability      Port ID

R1                 Fa0/5           117         R               Gi0/0/1

S2                 Fa0/1           112         B               Fa0/1

Total entries displayed: 2
```

The network administrator discovers that S1 has a router and a switch as a neighbors. For this output, the letter B for bridge also means switch.

From the results of **show lldp neighbors**, a topology from S1 can be constructed, as displayed in the figure.

When more details about the neighbors are needed, the **show lldp neighbors detail** command can provide information, such as the neighbor IOS version, IP address, and device capability.

S1# **show lldp neighbors detail**

```
-------------------------------------------------
Chassis id: 848a.8d44.49b0

Port id: Gi0/0/1

Port Description: GigabitEthernet0/0/1

System Name: R1


System Description:

Cisco IOS Software [Fuji], ISR Software (X86_64_LINUX_IOSD-
UNIVERSALK9-M), Version 16.9.4, RELEASE SOFTWARE (fc2)

Technical Support: http://www.cisco.com/techsupport

Copyright (c) 1986-2019 by Cisco Systems, Inc.

Compiled Thu 22-Aug-19 18:09 by mcpre


Time remaining: 111 seconds

System Capabilities: B,R

Enabled Capabilities: R

Management Addresses - not advertised

Auto Negotiation - not supported

Physical media capabilities - not advertised

Media Attachment Unit type - not advertised

Vlan ID: - not advertised


-------------------------------------------------
Chassis id: 0025.83e6.4b00

Port id: Fa0/1

Port Description: FastEthernet0/1

System Name: S2


System Description:

Cisco IOS Software, C2960 Software (C2960-LANBASEK9-M), Version 15.0(2)
SE4, RELEASE SOFTWARE (fc1)

Technical Support: http://www.cisco.com/techsupport

Copyright (c) 1986-2013 by Cisco Systems, Inc.
```

```
Compiled Wed 26-Jun-13 02:49 by prod_rel_team


Time remaining: 107 seconds

System Capabilities: B

Enabled Capabilities: B

Management Addresses - not advertised

Auto Negotiation - supported, enabled

Physical media capabilities:

    100base-TX(FD)

    100base-TX(HD)

    10base-T(FD)

    10base-T(HD)

Media Attachment Unit type: 16

Vlan ID: 1


Total entries displayed: 2
```

Refer to **Interactive Graphic** in online course

Syntax Checker - Configure and Verify LLDP - 10.2.4

Practice configuring and verifying LLDP.

Go to the online course to take the quiz and exam.

Check Your Understanding - Compare CDP and LLDP - 10.2.5

Refer to **Packet Tracer Activity** for this chapter

Packet Tracer - Use LLDP to Map a Network - 10.2.6

In this Packet Tracer activity, you will complete the following objectives:

- Build the Network and Configure Basic Device Settings
- Network Discovery with CDP
- Network Discovery with LLDP

NTP - 10.3

Time and Calendar Services - 10.3.1

Before you get really deep into network management, the one thing that will help keep you on track is ensuring that all of your components are set to the same time and date.

The software clock on a router or switch starts when the system boots. It is the primary source of time for the system. It is important to synchronize the time across all devices on the network because all aspects of managing, securing, troubleshooting, and planning networks require accurate timestamping. When the time is not synchronized between devices, it will be impossible to determine the order of the events and the cause of an event.

Typically, the date and time settings on a router or switch can be set by using one of two methods You can manually configure the date and time, as shown in the example, or configure the Network Time Protocol (NTP).

```
R1# clock set 20:36:00 nov 15 2019

R1#

*Nov 15 20:36:00.000: %SYS-6-CLOCKUPDATE: System clock has been

updated from 21:32:31 UTC Fri Nov 15 2019 to 20:36:00 UTC Fri Nov 15

2019, configured from console by console.
```

As a network grows, it becomes difficult to ensure that all infrastructure devices are operating with synchronized time. Even in a smaller network environment, the manual method is not ideal. If a router reboots, how will it get an accurate date and timestamp?

A better solution is to configure the NTP on the network. This protocol allows routers on the network to synchronize their time settings with an NTP server. A group of NTP clients that obtain time and date information from a single source have more consistent time settings. When NTP is implemented in the network, it can be set up to synchronize to a private master clock, or it can synchronize to a publicly available NTP server on the internet.

NTP uses UDP port 123 and is documented in RFC 1305.

Refer to
Online Course
for Illustration

NTP Operation - 10.3.2

NTP networks use a hierarchical system of time sources. Each level in this hierarchical system is called a stratum. The stratum level is defined as the number of hop counts from the authoritative source. The synchronized time is distributed across the network by using NTP. The figure displays a sample NTP network.

NTP servers are arranged in three levels showing the three strata. Stratum 1 is connected to Stratum 0 clocks.

Stratum 0

An NTP network gets the time from authoritative time sources. These authoritative time sources, also referred to as stratum 0 devices, are high-precision timekeeping devices assumed to be accurate and with little or no delay associated with them. Stratum 0 devices are represented by the clock in the figure.

Stratum 1

The stratum 1 devices are directly connected to the authoritative time sources. They act as the primary network time standard.

Stratum 2 and Lower

The stratum 2 servers are connected to stratum 1 devices through network connections. Stratum 2 devices, such as NTP clients, synchronize their time by using the NTP packets from stratum 1 servers. They could also act as servers for stratum 3 devices.

Smaller stratum numbers indicate that the server is closer to the authorized time source than larger stratum numbers. The larger the stratum number, the lower the stratum level. The max hop count is 15. Stratum 16, the lowest stratum level, indicates that a device is unsynchronized. Time servers on the same stratum level can be configured to act as a peer with other time servers on the same stratum level for backup or verification of time.

Refer to **Online Course** for Illustration

Configure and Verify NTP - 10.3.3

The figure shows the topology used to demonstrate NTP configuration and verification.

Before NTP is configured on the network, the **show clock** command displays the current time on the software clock, as shown in the example. With the **detail** option, notice that the time source is user configuration. That means the time was manually configured with the **clock** command.

```
R1# show clock detail

20:55:10.207 UTC Fri Nov 15 2019

Time source is user configuration
```

The **ntp server** *ip-address* command is issued in global configuration mode to configure 209.165.200.225 as the NTP server for R1. To verify the time source is set to NTP, use the **show clock detail** command. Notice that now the time source is NTP.

```
R1(config)# ntp server 209.165.200.225

R1(config)# end

R1# show clock detail

21:01:34.563 UTC Fri Nov 15 2019

Time source is NTP
```

In the next example, the **show ntp associations** and **show ntp status** commands are used to verify that R1 is synchronized with the NTP server at 209.165.200.225. Notice that R1 is synchronized with a stratum 1 NTP server at 209.165.200.225, which is synchronized with a GPS clock. The **show ntp status** command displays that R1 is now a stratum 2 device that is synchronized with the NTP server at 209.165.220.225.

Note: The highlight **st** stands for stratum.

```
R1# show ntp associations


  address        ref clock   st   when  poll reach delay  offset disp
*~209.165.200.225 .GPS.       1    61    64   377  0.481  7.480  4.261
```

```
   * sys.peer, # selected, + candidate, - outlyer, x falseticker, ~
configured
```

```
R1# show ntp status
```

```
Clock is synchronized, stratum 2, reference is 209.165.200.225

nominal freq is 250.0000 Hz, actual freq is 249.9995 Hz, precision is
2**19

ntp uptime is 589900 (1/100 of seconds), resolution is 4016

reference time is DA088DD3.C4E659D3 (13:21:23.769 PST Fri Nov 15 2019)

clock offset is 7.0883 msec, root delay is 99.77 msec

root dispersion is 13.43 msec, peer dispersion is 2.48 msec

loopfilter state is 'CTRL' (Normal Controlled Loop), drift is
0.000001803 s/s

system poll interval is 64, last update was 169 sec ago.
```

Next, the clock on S1 is configured to synchronize to R1 with the **ntp server** command and then the configuration is verified with the **show ntp associations** command, as displayed.

```
S1(config)# ntp server 192.168.1.1
```

```
S1(config)# end
```

```
S1# show ntp associations
```

```
  address       ref clock       st when  poll reach  delay offset  disp
*~192.168.1.1  209.165.200.225  2   12    64   377   1.066 13.616 3.840
  * sys.peer, # selected, + candidate, - outlyer, x falseticker, ~
configured
```

Output from the **show ntp associations** command verifies that the clock on S1 is now synchronized with R1 at 192.168.1.1 via NTP. R1 is a stratum 2 device and NTP server to S1. Now S1 is a stratum 3 device that can provide NTP service to other devices in the network, such as end devices.

```
S1# show ntp status
```

```
Clock is synchronized, stratum 3, reference is 192.168.1.1

nominal freq is 119.2092 Hz, actual freq is 119.2088 Hz, precision is
2**17

reference time is DA08904B.3269C655 (13:31:55.196 PST Tue Nov 15 2019)

clock offset is 18.7764 msec, root delay is 102.42 msec

root dispersion is 38.03 msec, peer dispersion is 3.74 msec
```

```
loopfilter state is 'CTRL' (Normal Controlled Loop), drift is
0.000003925 s/s

system poll interval is 128, last update was 178 sec ago.
```

Refer to **Packet Tracer Activity** for this chapter

Packet Tracer - Configure and Verify NTP - 10.3.4

NTP synchronizes the time of day among a set of distributed time servers and clients. While there are a number of applications that require synchronized time, this lab will focus on the need to correlate events when listed in the system logs and other time-specific events from multiple network devices.

SNMP - 10.4

Refer to **Online Course** for Illustration

Introduction to SNMP - 10.4.1

Now that your network is mapped and all of your components are using the same clock, it is time to look at how you can manage your network by using Simple Network Management Protocol (SNMP).

SNMP was developed to allow administrators to manage nodes such as servers, workstations, routers, switches, and security appliances, on an IP network. It enables network administrators to monitor and manage network performance, find and solve network problems, and plan for network growth.

SNMP is an application layer protocol that provides a message format for communication between managers and agents. The SNMP system consists of three elements:

- SNMP manager
- SNMP agents (managed node)
- Management Information Base (MIB)

To configure SNMP on a networking device, it is first necessary to define the relationship between the manager and the agent.

The SNMP manager is part of a network management system (NMS). The SNMP manager runs SNMP management software. As shown in the figure, the SNMP manager can collect information from an SNMP agent by using the "get" action and can change configurations on an agent by using the "set" action. In addition, SNMP agents can forward information directly to a network manager by using "traps".

The SNMP agent and MIB reside on SNMP client devices. Network devices that must be managed, such as switches, routers, servers, firewalls, and workstations, are equipped with an SMNP agent software module. MIBs store data about the device and operational statistics and are meant to be available to authenticated remote users. The SNMP agent is responsible for providing access to the local MIB.

SNMP defines how management information is exchanged between network management applications and management agents. The SNMP manager polls the agents and queries the MIB for SNMP agents on UDP port 161. SNMP agents send any SNMP traps to the SNMP manager on UDP port 162.

Refer to
Online Course
for Illustration

SNMP Operation - 10.4.2

SNMP agents that reside on managed devices collect and store information about the device and its operation. This information is stored by the agent locally in the MIB. The SNMP manager then uses the SNMP agent to access information within the MIB.

There are two primary SNMP manager requests, get and set. A get request is used by the NMS to query the device for data. A set request is used by the NMS to change configuration variables in the agent device. A set request can also initiate actions within a device. For example, a set can cause a router to reboot, send a configuration file, or receive a configuration file. The SNMP manager uses the get and set actions to perform the operations described in the table.

Operation	Description
get-request	Retrieves a value from a specific variable.
get-next-request	Retrieves a value from a variable within a table; the SNMP manager does not need to know the exact variable name. A sequential search is performed to find the needed variable from within a table.
get-bulk-request	Retrieves large blocks of data, such as multiple rows in a table, that would otherwise require the transmission of many small blocks of data. (Only works with SNMPv2 or later.)
get-response	Replies to a **get-request**, **get-next-request**, and **set-request** sent by an NMS.
set-request	Stores a value in a specific variable.

The SNMP agent responds to SNMP manager requests as follows:

- **Get an MIB variable** - The SNMP agent performs this function in response to a GetRequest-PDU from the network manager. The agent retrieves the value of the requested MIB variable and responds to the network manager with that value.

- **Set an MIB variable** - The SNMP agent performs this function in response to a SetRequest-PDU from the network manager. The SNMP agent changes the value of the MIB variable to the value specified by the network manager. An SNMP agent reply to a set request includes the new settings in the device.

The figure illustrates the use of an SNMP GetRequest to determine if interface G0/0/0 is up/up.

Refer to
Online Course
for Illustration

SNMP Agent Traps - 10.4.3

An NMS periodically polls the SNMP agents that are residing on managed devices using the get request. The NMS queries the device for data. Using this process, a network management application can collect information to monitor traffic loads and to verify the device configurations of managed devices. The information can be displayed via a GUI on the NMS. Averages, minimums, or maximums can be calculated. The data can be graphed, or thresholds can be set to trigger a notification process when the thresholds are exceeded. For example, an NMS can monitor CPU utilization of a Cisco router. The SNMP manager samples the value periodically and presents this information in a graph for the network administrator to use in creating a baseline, creating a report, or viewing real time information.

Periodic SNMP polling does have disadvantages. First, there is a delay between the time that an event occurs and the time that it is noticed (via polling) by the NMS. Second, there is a trade-off between polling frequency and bandwidth usage.

To mitigate these disadvantages, it is possible for SNMP agents to generate and send traps to inform the NMS immediately of certain events. Traps are unsolicited messages alerting the SNMP manager to a condition or event on the network. Examples of trap conditions include, but are not limited to, improper user authentication, restarts, link status (up or down), MAC address tracking, closing of a TCP connection, loss of connection to a neighbor, or other significant events. Trap-directed notifications reduce network and agent resources by eliminating the need for some of SNMP polling requests.

The figure illustrates the use of an SNMP trap to alert the network administrator that interface G0/0/0 has failed. The NMS software can send the network administrator a text message, pop up a window on the NMS software, or turn the router icon red in the NMS GUI.

The exchange of all SNMP messages is illustrated in the figure.

Refer to
Interactive Graphic
in online course

SNMP Versions - 10.4.4

Refer to
Online Course
for Illustration

There are several versions of SNMP:

- **SNMPv1** - This is the Simple Network Management Protocol, a Full Internet Standard, that is defined in RFC 1157.

- **SNMPv2c** - This is defined in RFCs 1901 to 1908. It uses a community-string-based Administrative Framework.

- **SNMPv3** - This is an interoperable standards-based protocol originally defined in RFCs 2273 to 2275. It provides secure access to devices by authenticating and encrypting packets over the network. It includes these security features: message integrity to ensure that a packet was not tampered with in transit, authentication to determine that the message is from a valid source, and encryption to prevent the contents of a message from being read by an unauthorized source.

All versions use SNMP managers, agents, and MIBs. Cisco IOS software supports the above three versions. Version 1 is a legacy solution and is not often encountered in networks today; therefore, this course focuses on versions 2c and 3.

Both SNMPv1 and SNMPv2c use a community-based form of security. The community of managers that is able to access the MIB of the agent is defined by a community string.

Unlike SNMPv1, SNMPv2c includes a bulk retrieval mechanism and more detailed error message reporting to management stations. The bulk retrieval mechanism retrieves tables and large quantities of information, minimizing the number of round-trips required. The SNMPv2c improved error-handling includes expanded error codes that distinguish different kinds of error conditions. These conditions are reported through a single error code in SNMPv1. Error return codes in SNMPv2c include the error type.

Note: SNMPv1 and SNMPv2c offer minimal security features. Specifically, SNMPv1 and SNMPv2c can neither authenticate the source of a management message nor provide encryption. SNMPv3 is most currently described in RFCs 3410 to 3415. It adds methods to ensure the secure transmission of critical data between managed devices.

SNMPv3 provides for both security models and security levels. A security model is an authentication strategy set up for a user and the group within which the user resides. A security level is the permitted level of security within a security model. A combination of the security level and the security model determine which security mechanism is used when handling an SNMP packet. Available security models are SNMPv1, SNMPv2c, and SNMPv3.

The table identifies the characteristics of the different combinations of security models and levels.

Click each button for more information about the characteristics of the different combinations of security models and levels.

SNMPv1

SNMPv2c

SNMPv3 noAuthNoPriv

SNMPv3 authNoPriv

SNMPv3 authPriv

A network administrator must configure the SNMP agent to use the SNMP version supported by the management station. Because an agent can communicate with multiple SNMP managers, it is possible to configure the software to support communications by using SNMPv1, SNMPv2c, or SNMPv3.

Go to the online course to take the quiz and exam.

Check Your Understanding - SNMP Versions - 10.4.5

Refer to
Interactive Graphic
in online course

Community Strings - 10.4.6

For SNMP to operate, the NMS must have access to the MIB. To ensure that access requests are valid, some form of authentication must be in place.

SNMPv1 and SNMPv2c use community strings that control access to the MIB. Community strings are plaintext passwords. SNMP community strings authenticate access to MIB objects.

There are two types of community strings:

- **Read-only (ro)** - This type provides access to the MIB variables, but does not allow these variables to be changed, only read. Because security is minimal in version 2c, many organizations use SNMPv2c in read-only mode.

- **Read-write (rw)** - This type provides read and write access to all objects in the MIB.

To view or set MIB variables, the user must specify the appropriate community string for read or write access.

Click Play to see an animation about how SNMP operates with the community string.

Note: Plaintext passwords are not considered a security mechanism. This is because plaintext passwords are highly vulnerable to man-in-the-middle attacks, in which they are compromised through the capture of packets.

Refer to
Online Course
for Illustration

MIB Object ID - 10.4.7

The MIB organizes variables hierarchically. MIB variables enable the management software to monitor and control the network device. Formally, the MIB defines each variable as an object ID (OID). OIDs uniquely identify managed objects in the MIB hierarchy. The MIB organizes the OIDs based on RFC standards into a hierarchy of OIDs, usually shown as a tree.

The MIB tree for any given device includes some branches with variables common to many networking devices and some branches with variables specific to that device or vendor.

RFCs define some common public variables. Most devices implement these MIB variables. In addition, networking equipment vendors, like Cisco, can define their own private branches of the tree to accommodate new variables specific to their devices.

The figure shows portions of the MIB structure defined by Cisco. Note how the OID can be described in words or numbers to help locate a particular variable in the tree. OIDs belonging to Cisco, are numbered as follows: .iso (1).org (3).dod (6).internet (1).private (4). enterprises (1).cisco (9). Therefore, the OID is 1.3.6.1.4.1.9.

Refer to
Online Course
for Illustration

SNMP Polling Scenario - 10.4.8

SNMP can be used to observe CPU utilization over a period of time by polling devices. CPU statistics can then be compiled on the NMS and graphed. This creates a baseline for the network administrator. Threshold values can then be set relative to this baseline. When CPU utilization exceeds this threshold, notifications are sent. The figure illustrates 5-minute samples of router CPU utilization over the period of a few weeks.

The data is retrieved via the snmpget utility, issued on the NMS. Using the snmpget utility, you can manually retrieve real-time data, or have the NMS run a report. This report would give you a period of time that you could use the data to get the average. The snmpget utility requires that the SNMP version, the correct community, the IP address of the network device to query, and the OID number are set. The figure demonstrates the use of the freeware snmpget utility, which allows quick retrieval of information from the MIB.

Refer to
Online Course
for Illustration

SNMP Object Navigator - 10.4.9

The snmpget utility gives some insight into the basic mechanics of how SNMP works. However, working with long MIB variable names like 1.3.6.1.4.1.9.2.1.58.0 can be problematic for the average user. More commonly, the network operations staff uses a network management product with an easy-to-use GUI, which makes the entire MIB data variable naming transparent to the user.

The Cisco SNMP Navigator on the http://www.cisco.com website allows a network administrator to research details about a particular OID. The figure displays an example of using the navigator to research the OID information for the **whyReload** object.

Refer to
Lab Activity
for this chapter

Lab - Research Network Monitoring Software - 10.4.10

In this lab, you will complete the following objectives:

- Part 1: Survey Your Understanding of Network Monitoring
- Part 2: Research Network Monitoring Tools
- Part 3: Select a Network Monitoring Tool

Syslog - 10.5

Refer to
Online Course
for Illustration

Introduction to Syslog - 10.5.1

Like a Check Engine light on your car dashboard, the components in your network can tell you if there is something wrong. The syslog protocol was designed to ensure that you can receive and understand these messages. When certain events occur on a network, networking devices have trusted mechanisms to notify the administrator with detailed system messages. These messages can be either non-critical or significant. Network administrators have a variety of options for storing, interpreting, and displaying these messages. They can also be alerted to those messages that could have the greatest impact on the network infrastructure.

The most common method of accessing system messages is to use a protocol called syslog.

Syslog is a term used to describe a standard. It is also used to describe the protocol developed for that standard. The syslog protocol was developed for UNIX systems in the 1980s but was first documented as RFC 3164 by IETF in 2001. Syslog uses UDP port 514 to send event notification messages across IP networks to event message collectors, as shown in the figure.

Many networking devices support syslog, including: routers, switches, application servers, firewalls, and other network appliances. The syslog protocol allows networking devices to send their system messages across the network to syslog servers.

There are several different syslog server software packages for Windows and UNIX. Many of them are freeware.

The syslog logging service provides three primary functions, as follows:

- The ability to gather logging information for monitoring and troubleshooting
- The ability to select the type of logging information that is captured
- The ability to specify the destinations of captured syslog messages

Refer to
Online Course
for Illustration

Syslog Operation - 10.5.2

On Cisco network devices, the syslog protocol starts by sending system messages and **debug** output to a local logging process that is internal to the device. How the logging process manages these messages and outputs is based on device configurations. For example, syslog messages may be sent across the network to an external syslog server. These messages can be retrieved without needing to access the actual device. Log messages and outputs stored on the external server can be pulled into various reports for easier reading.

Alternatively, syslog messages may be sent to an internal buffer. Messages sent to the internal buffer are only viewable through the CLI of the device.

Finally, the network administrator may specify that only certain types of system messages be sent to various destinations. For example, the device may be configured to forward all system messages to an external syslog server. However, debug-level messages are forwarded to the internal buffer and are only accessible by the administrator from the CLI.

As shown in the figure, popular destinations for syslog messages include the following:

- Logging buffer (RAM inside a router or switch)
- Console line
- Terminal line
- Syslog server

It is possible to remotely monitor system messages by viewing the logs on a syslog server, or by accessing the device through Telnet, SSH, or through the console port.

Syslog Message Format - 10.5.3

Cisco devices produce syslog messages as a result of network events. Every syslog message contains a severity level and a facility.

The smaller numerical levels are the more critical syslog alarms. The severity level of the messages can be set to control where each type of message is displayed (i.e. on the console or the other destinations). The complete list of syslog levels is shown in the table.

Severity Name	Severity Level	Explanation
Emergency	Level 0	System Unusable
Alert	Level 1	Immediate Action Needed
Critical	Level 2	Critical Condition
Error	Level 3	Error Condition
Warning	Level 4	Warning Condition
Notification	Level 5	Normal, but Significant Condition
Informational	Level 6	Informational Message
Debugging	Level 7	Debugging Message

Each syslog level has its own meaning:

- **Warning Level 4 - Emergency Level 0**: These messages are error messages about software or hardware malfunctions; these types of messages mean that the functionality of the device is affected. The severity of the issue determines the actual syslog level applied.

- **Notification Level 5:** This notifications level is for normal, but significant events. For example, interface up or down transitions, and system restart messages are displayed at the notifications level.

- **Informational Level 6:** This is a normal information message that does not affect device functionality. For example, when a Cisco device is booting, you might see the following informational message: %LICENSE-6-EULA_ACCEPT_ALL: The Right to Use End User License Agreement is accepted.

- **Debugging Level 7:** This level indicates that the messages are output generated from issuing various **debug** commands.

Syslog Facilities - 10.5.4

In addition to specifying the severity, syslog messages also contain information on the facility. Syslog facilities are service identifiers that identify and categorize system state data for error and event message reporting. The logging facility options that are available are specific to the networking device. For example, Cisco 2960 Series switches running Cisco IOS Release 15.0(2) and Cisco 1941 routers running Cisco IOS Release 15.2(4) support 24 facility options that are categorized into 12 facility types.

Some common syslog message facilities reported on Cisco IOS routers include:

- IP

- OSPF protocol

- SYS operating system

- IP security (IPsec)

- Interface IP (IF)

By default, the format of syslog messages on the Cisco IOS Software is as follows:

```
%facility-severity-MNEMONIC: description
```

For example, sample output on a Cisco switch for an EtherChannel link changing state to up is:

```
%LINK-3-UPDOWN: Interface Port-channel1, changed state to up
```

Here the facility is LINK and the severity level is 3, with a MNEMONIC of UPDOWN.

The most common messages are link up and down messages, and messages that a device produces when it exits from configuration mode. If ACL logging is configured, the device generates syslog messages when packets match a parameter condition.

Configure Syslog Timestamp - 10.5.5

By default, log messages are not timestamped. In the example, the R1 GigabitEthernet 0/0/0 interface is shutdown. The message logged to the console does not identify when the interface state was changed. Log messages should be timestamped so that when they are sent to another destination, such as a Syslog server, there is record of when the message was generated.

```
R1# configure terminal

R1(config)# interface g0/0/0

R1(config-if)# shutdown

%LINK-5-CHANGED: Interface GigabitEthernet0/0/0, changed state to
administratively down

%LINEPROTO-5-UPDOWN: Line protocol on Interface GigabitEthernet0/0/0,
changed state to down

R1(config-if)# exit

R1(config)# service timestamps log datetime

R1(config)# interface g0/0/0

R1(config-if)# no shutdown

*Mar  1 11:52:42: %LINK-3-UPDOWN: Interface GigabitEthernet0/0/0,
changed state to down

*Mar  1 11:52:45: %LINK-3-UPDOWN: Interface GigabitEthernet0/0/0,
changed state to up

*Mar  1 11:52:46: %LINEPROTO-5-UPDOWN: Line protocol on Interface
GigabitEthernet0/0/0,

changed state to up

R1(config-if)#
```

Use the command **service timestamps log datetime** to force logged events to display the date and time. As shown in the figure, when the R1 GigabitEthernet 0/0/0 interface is reactivated, the log messages now contain the date and time.

Note: When using the **datetime** keyword, the clock on the networking device must be set, either manually or through NTP, as previously discussed.

Go to the online course to take the quiz and exam.

Check Your Understanding - Syslog Operation - 10.5.6

Router and Switch File Maintenance - 10.6

Router File Systems - 10.6.1

If you are thinking that you cannot possibly remember how you configured every device in your network, you are not alone. In a large network, it would not be possible to manually configure every device. Fortunately, there are many ways to copy or update your configurations, and then simply paste them in. To do this, you will need to know how to view and manage your file systems.

The Cisco IOS File System (IFS) allows the administrator to navigate to different directories and list the files in a directory. The administrator can also create subdirectories in flash memory or on a disk. The directories available depend on the device.

The example displays the output of the **show file systems** command, which lists all of the available file systems on a Cisco 4221 router.

```
Router# show file systems

File Systems:
```

	Size(b)	Free(b)	Type	Flags	Prefixes
	-	-	opaque	rw	system:
	-	-	opaque	rw	tmpsys:
*	7194652672	6294822912	disk	rw	bootflash: flash:
	256589824	256573440	disk	rw	usb0:
	1804468224	1723789312	disk	ro	webui:
	-	-	opaque	rw	null:
	-	-	opaque	ro	tar:
	-	-	network	rw	tftp:
	-	-	opaque	wo	syslog:
	33554432	33539983	nvram	rw	nvram:
	-	-	network	rw	rcp:
	-	-	network	rw	ftp:
	-	-	network	rw	http:
	-	-	network	rw	scp:
	-	-	network	rw	sftp:
	-	-	network	rw	https:
	-	-	opaque	ro	cns:

```
Router#
```

This command provides useful information such as the amount of total and free memory, the type of file system, and its permissions. Permissions include read only (ro), write only (wo), and read and write (rw). The permissions are shown in the Flags column of the command output.

Although there are several file systems listed, of interest to us will be the tftp, flash, and nvram file systems.

Notice that the flash file system also has an asterisk preceding it. This indicates that flash is the current default file system. The bootable IOS is located in flash; therefore, the pound symbol (#) is appended to the flash listing, indicating that it is a bootable disk.

The Flash File System

The example displays the output from the **dir** (directory) command.

```
Router# dir

Directory of bootflash:/
    11  drwx            16384   Aug 2 2019 04:15:13 +00:00  lost+found
```

```
370945    drwx              4096    Oct 3 2019 15:12:10 +00:00  .installer

338689    drwx              4096    Aug 2 2019 04:15:55 +00:00  .ssh

217729    drwx              4096    Aug 2 2019 04:17:59 +00:00  core

379009    drwx              4096    Sep 26 2019 15:54:10 +00:00 .prst_sync

80641     drwx              4096    Aug 2 2019 04:16:09 +00:00
                            .rollback_timer

161281    drwx              4096    Aug 2 2019 04:16:11 +00:00  gs_script

112897    drwx              102400  Oct 3 2019 15:23:07 +00:00  tracelogs

362881    drwx              4096    Aug 23 2019 17:19:54 +00:00 .dbpersist

298369    drwx              4096    Aug 2 2019 04:16:41 +00:00
                            virtual-instance

    12    -rw-                30    Oct 3 2019 15:14:11 +00:00
                            throughput_monitor_params

  8065    drwx              4096    Aug 2 2019 04:17:55 +00:00  onep

    13    -rw-                34    Oct 3 2019 15:19:30 +00:00
                            pnp-tech-time

249985    drwx              4096    Aug 20 2019 17:40:11 +00:00 Archives

    14    -rw-             65037    Oct 3 2019 15:19:42 +00:00
                            pnp-tech-discovery-summary

    17    -rw-           5032908    Sep 19 2019 14:16:23 +00:00
                            isr4200_4300_rommon_1612_1r_SPA.pkg

    18    -rw-         517153193    Sep 21 2019 04:24:04 +00:00  isr4200-
                            universalk9_ias.16.09.04.SPA.bin

7194652672 bytes total (6294822912 bytes free)

Router#
```

Because flash is the default file system, the **dir** command lists the contents of flash. Several files are located in flash, but of specific interest is the last listing. This is the name of the current Cisco IOS file image that is running in RAM.

The NVRAM File System

To view the contents of NVRAM, you must change the current default file system by using the **cd** (change directory) command, as shown in the example.

```
Router#

Router# cd nvram:

Router# pwd

nvram:/

Router# dir

Directory of nvram:/

32769    -rw-              1024                    startup-config
```

```
32770  ----            61              private-config
32771  -rw-          1024              underlying-config
    1  ----             4              private-KS1
    2  -rw-          2945              cwmp_inventory
    5  ----           447              persistent-data
    6  -rw-          1237              ISR4221-2x1GE_0_0_0
    8  -rw-            17              ecfm_ieee_mib
    9  -rw-             0              ifIndex-table
   10  -rw-          1431              NIM-2T_0_1_0
   12  -rw-           820              IOS-Self-Sig#1.cer
   13  -rw-           820              IOS-Self-Sig#2.cer

33554432 bytes total (33539983 bytes free)
Router#
```

The present working directory command is **pwd**. This command verifies that we are viewing the NVRAM directory. Finally, the **dir** command lists the contents of NVRAM. Although there are several configuration files listed, of specific interest is the startup-configuration file.

Switch File Systems - 10.6.2

With the Cisco 2960 switch flash file system, you can copy configuration files, and archive (upload and download) software images.

The command to view the file systems on a Catalyst switch is the same as on a Cisco router: **show file systems**, as displayed in the example.

```
Switch# show file systems

File Systems:

       Size(b)      Free(b)     Type    Flags    Prefixes
*     32514048    20887552     flash      rw       flash:
          -            -       opaque     rw         vb:
          -            -       opaque     ro         bs:
          -            -       opaque     rw     system:
          -            -       opaque     rw     tmpsys:
         65536        48897     nvram     rw      nvram:
          -            -       opaque     ro     xmodem:
          -            -       opaque     ro     ymodem:
          -            -       opaque     rw       null:
          -            -       opaque     ro        tar:
```

```
        -              -        network     rw      tftp:

        -              -        network     rw       rcp:

        -              -        network     rw      http:

        -              -        network     rw       ftp:

        -              -        network     rw       scp:

        -              -        network     rw     https:

        -              -         opaque     ro       cns:

Switch#
```

Refer to **Online Course** for Illustration

Use a Text File to Back Up a Configuration - 10.6.3

Configuration files can be saved to a text file by using Tera Term, as shown in the figure.

Step 1.   On the File menu, click **Log.**

Step 2.   Choose the location to save the file. Tera Term will begin capturing text.

Step 3.   After capture has been started, execute the **show running-config** or **show startup-config** command at the privileged EXEC prompt. Text displayed in the terminal window will be directed to the chosen file.

Step 4.   When the capture is complete, select **Close** in the Tera Term: Log window.

Step 5.   View the file to verify that it was not corrupted.

Refer to **Online Course** for Illustration

Use a Text File to Restore a Configuration - 10.6.4

A configuration can be copied from a file and then directly pasted to a device. The IOS executes each line of the configuration text as a command. This means that the file will require editing to ensure that encrypted passwords are in plaintext, and that non-command text such as **--More--** and IOS messages are removed. In addition, you may want to add **enable** and **configure terminal** to the beginning of the file or enter global configuration mode before pasting the configuration. This process is discussed in the lab later in this topic.

Instead of copying and pasting, a configuration can be restored from a text file by using Tera Term, as shown in the figure.

When using Tera Term, the steps are as follows:

Step 1.   On the File menu, click **Send** file.

Step 2.   Locate the file to be copied into the device and click **Open.**

Step 3.   Tera Term will paste the file into the device.

The text in the file will be applied as commands in the CLI and become the running configuration on the device.

Use TFTP to Back Up and Restore a Configuration - 10.6.5

Use TFTP to Back Up a Configuration

Copies of configuration files should be stored as backup files in the event of a problem. Configuration files can be stored on a Trivial File Transfer Protocol (TFTP) server, or a USB drive. A configuration file should also be included in the network documentation.

To save the running configuration or the startup configuration to a TFTP server, use either the **copy running-config tftp** or **copy startup-config tftp** command, as shown in the example.

```
R1# copy running-config tftp
Remote host []?192.168.10.254
Name of the configuration file to write[R1-config]? R1-Jan-2019
Write file R1-Jan-2019 to 192.168.10.254? [confirm]
Writing R1-Jan-2019 !!!!!! [OK]
```

Follow these steps to back up the running configuration to a TFTP server:

Step 1.   Enter the **copy running-config tftp** command.

Step 2.   Enter the IP address of the host where the configuration file will be stored.

Step 3.   Enter the name to assign to the configuration file.

Step 4.   Press Enter to confirm each choice.

Use TFTP to Restore a Configuration

To restore the running configuration or the startup configuration from a TFTP server, use either the **copy tftp running-config** or **copy tftp startup-config** command. Use the following steps to restore the running configuration from a TFTP server:

Step 1.   Enter the **copy tftp running-config** command.

Step 2.   Enter the IP address of the host where the configuration file is stored.

Step 3.   Enter the name to assign to the configuration file.

Step 4.   Press **Enter** to confirm each choice.

Refer to
Online Course
for Illustration

USB Ports on a Cisco Router - 10.6.6

The Universal Serial Bus (USB) storage feature enables certain models of Cisco routers to support USB flash drives. The USB flash feature provides an optional secondary storage capability and an additional boot device. Images, configurations, and other files can be copied to or from the Cisco USB flash memory with the same reliability as storing and retrieving files by using the Compact Flash card. In addition, modular integrated services routers can boot any Cisco IOS Software image saved on USB flash memory. Ideally, USB flash can hold multiple copies of the Cisco IOS and multiple router configurations. The USB ports of a Cisco 4321 Router are shown in the figure.

Use the **dir** command to view the contents of the USB flash drive, as shown in the example.

```
Router# dir usbflash0:

Directory of usbflash0:/

1 -rw- 30125020 Dec 22 2032 05:31:32 +00:00 c3825-entservicesk9-mz.123-14.T

63158272 bytes total (33033216 bytes free)
```

Use USB to Back Up and Restore a Configuration - 10.6.7

When backing up to a USB port, it is a good idea to issue the **show file systems** command to verify that the USB drive is there and confirm the name, as shown in the example.

```
R1# show file systems

File Systems:
```

	Size(b)	Free(b)	Type	Flags	Prefixes
	-	-	opaque	rw	archive:
	-	-	opaque	rw	system:
	-	-	opaque	rw	tmpsys:
	-	-	opaque	rw	null:
	-	-	network	rw	tftp:
*	256487424	184819712	disk	rw	flash0: flash:#
	-	-	disk	rw	flash1:
	262136	249270	nvram	rw	nvram:
	-	-	opaque	wo	syslog:
	-	-	opaque	rw	xmodem:
	-	-	opaque	rw	ymodem:
	-	-	network	rw	rcp:
	-	-	network	rw	http:
	-	-	network	rw	ftp:
	-	-	network	rw	scp:
	-	-	opaque	ro	tar:
	-	-	network	rw	https:
	-	-	opaque	ro	cns:
	4050042880	3774152704	usbflash	rw	usbflash0:

```
R1#
```

Notice the last line of output shows the USB port and name: "usbflash0:".

Next, use the **copy run usbflash0:/** command to copy the configuration file to the USB flash drive. Be sure to use the name of the flash drive, as indicated in the file system. The slash is optional but indicates the root directory of the USB flash drive.

The IOS will prompt for the filename. If the file already exists on the USB flash drive, the router will prompt to overwrite, as shown in the examples.

When copying to USB flash drive, with no pre-existing file will display the following output.

```
R1# copy running-config usbflash0:
Destination filename [running-config]? R1-Config
5024 bytes copied in 0.736 secs (6826 bytes/sec)
```

When copying to USB flash drive, with the same configuration file already on the drive will display the following output.

```
R1# copy running-config usbflash0:
Destination filename [running-config]? R1-Config
%Warning:There is a file already existing with this name
Do you want to over write? [confirm]
5024 bytes copied in 1.796 secs (2797 bytes/sec)
R1#
```

Use the **dir** command to see the file on the USB drive and use the **more** command to see the contents, as shown in the example.

```
R1# dir usbflash0:/
Directory of usbflash0:/
    1   drw-       0   Oct 15 2010 16:28:30 +00:00   Cisco
   16   -rw-   5024   Jan 7 2013 20:26:50 +00:00   R1-Config
4050042880 bytes total (3774144512 bytes free)
R1#
R1# more usbflash0:/R1-Config
!
! Last configuration change at 20:19:54 UTC Mon Jan 7 2013 by
admin version 15.2
service timestamps debug datetime msec
service timestamps log datetime msec
no service password-encryption
!
hostname R1
!
boot-start-marker
```

```
boot-end-marker
!
logging buffered 51200 warnings
!
no aaa new-model
!
no ipv6 cef
R1#
```

Restore Configurations with a USB Flash Drive

To copy the file back, it will be necessary to edit the USB R1-Config file with a text editor. Assuming the file name is **R1-Config**, use the command **copy usbflash0:/R1-Config** *running-config* to restore a running configuration.

Password Recovery Procedures - 10.6.8

Passwords on devices are used to prevent unauthorized access. For encrypted passwords, such as the enable secret passwords, the passwords must be replaced after recovery. Depending on the device, the detailed procedure for password recovery varies. However, all the password recovery procedures follow the same principle:

Step 1. Enter the ROMMON mode.

Step 2. Change the configuration register.

Step 3. Copy the startup-config to the running-config.

Step 4. Change the password.

Step 5. Save the running-config as the new startup-config.

Step 6. Reload the device.

Console access to the device through a terminal or terminal emulator software on a PC is required for password recovery. The terminal settings to access the device are:

- 9600 baud rate
- No parity
- 8 data bits
- 1 stop bit
- No flow control

Refer to **Interactive Graphic** in online course

Password Recovery Example - 10.6.9

Click each step for an example of completing a password recovery.

Step 1. Enter the ROMMON mode.

With console access, a user can access the ROMMON mode by using a break sequence during the boot up process or removing the external flash memory when the device is powered off. When successful, the **rommon 1 >** prompt displays, as shown in the example.

Note: The break sequence for PuTTY is Ctrl+Break. A list of standard break key sequences for other terminal emulators and operating systems can be found by searching the internet.

```
Readonly ROMMON initialized


monitor: command "boot" aborted due to user interrupt
rommon 1 >
```

Step 2. Change the configuration register.

The ROMMON software supports some basic commands, such as **confreg**. The **confreg 0x2142** command allows the user to set the configuration register to 0x2142. With the configuration register at 0x2142, the device will ignore the startup config file during startup. The startup config file is where the forgotten passwords are stored. After setting the configuration register to 0x2142, type **reset** at the prompt to restart the device. Enter the break sequence while the device is rebooting and decompressing the IOS. The example displays the terminal output of a 1941 router in the ROMMON mode after using a break sequence during the boot up process.

```
rommon 1 > confreg 0x2142
rommon 2 > reset


System Bootstrap, Version 15.0(1r)M9, RELEASE SOFTWARE (fc1)
Technical Support: http://www.cisco.com/techsupport
Copyright (c) 2010 by cisco Systems, Inc.
(output omitted)
```

Step 3. Copy the startup-config to the running-config.

After the device has finished reloading, copy the startup config to the running config by using the **copy startup-config running-config** command, as displayed in the example. Notice that the router prompt changed to **R1#** because the hostname is set to R1 in the startup-config.

Caution: Do not enter **copy running-config startup-config**. This command erases your original startup configuration.

```
Router# copy startup-config running-config
Destination filename [running-config]?
```

```
1450 bytes copied in 0.156 secs (9295 bytes/sec)

R1#
```

Step 4. Change the password.

Because you are in privileged EXEC mode, you can now configure all the necessary passwords, as shown in the example.

Note: The password **cisco** is not a strong password and is used here only as an example.

```
R1# configure terminal

Enter configuration commands, one per line. End with CNTL/Z.

R1(config)# enable secret cisco
```

Step 5. Save the running-config as the new startup-config.

After the new passwords are configured, change the configuration register back to 0x2102 by using the **config-register 0x2102** command in the global configuration mode. Save the running-config to startup-config, as shown in the example.

```
R1(config)# config-register 0x2102

R1(config)# end

R1# copy running-config startup-config

Destination filename [startup-config]?

Building configuration...

[OK]

R1#
```

Step 6. Reload the device.

Reload the device, as shown in the example. The device now uses the newly configured passwords for authentication. Be sure to use **show** commands to verify that all the configurations are still in place. For example, verify that the appropriate interfaces are not shut down after password recovery.

To find detailed instructions for password recovery procedures for a specific device, search the internet.

```
R1# reload
```

Refer to **Packet Tracer Activity** for this chapter

Packet Tracer - Back Up Configuration Files - 10.6.10

In this activity you will restore a configuration from a backup and then perform a new backup. Due to an equipment failure, a new router has been put in place. Fortunately, backup configuration files have been saved to a Trivial File Transfer Protocol (TFTP) Server. You are required to restore the files from the TFTP Server to get the router back online as quickly as possible.

Refer to
Lab Activity
for this chapter

Lab - Use Tera Term to Manage Router Configuration Files - 10.6.11

In this lab, you will complete the following objectives:

- Part 1: Configure Basic Device Settings
- Part 2: Use Terminal Emulation Software to Create a Backup Configuration File
- Part 3: Use a Backup Configuration File to Restore a Router

Refer to
Lab Activity
for this chapter

Lab - Use TFTP, Flash, and USB to Manage Configuration Files - 10.6.12

In this lab, you will complete the following objectives:

- Part 1: Build the Network and Configure Basic Device Settings
- Part 2: (Optional) Download TFTP Server Software
- Part 3: Use TFTP to Back Up and Restore the Switch Running Configuration
- Part 4: Use TFTP to Back Up and Restore the Router Running Configuration
- Part 5: Back Up and Restore Running Configurations Using Router Flash Memory
- Part 6: (Optional) Use a USB Drive to Back Up and Restore the Running Configuration

Refer to
Lab Activity
for this chapter

Lab - Research Password Recovery Procedures - 10.6.13

In this lab, you will complete the following objectives:

- Part 1: Research the Configuration Register
- Part 2: Document the Password Recovery Procedure for a Specific Cisco Router

IOS Image Management - 10.7

Refer to Video
in online course

Video - Managing Cisco IOS Images - 10.7.1

Click Play in the figure to view a demonstration of managing Cisco IOS images.

Refer to
Online Course
for Illustration

TFTP Servers as a Backup Location - 10.7.2

In the previous topic you learned the ways to copy and paste a configuration. This topic takes that idea one step further with IOS software images. As a network grows, Cisco IOS Software images and configuration files can be stored on a central TFTP server, as shown in the figure. This helps to control the number of IOS images and the revisions to those IOS images, as well as the configuration files that must be maintained.

Production internetworks usually span wide areas and contain multiple routers. For any network, it is good practice to keep a backup copy of the Cisco IOS Software image in case the system image on the router becomes corrupted or accidentally erased.

Widely distributed routers need a source or backup location for Cisco IOS Software images. Using a network TFTP server allows image and configuration uploads and downloads over the network. The network TFTP server can be another router, a workstation, or a host system.

Refer to **Interactive Graphic** in online course

Refer to **Online Course** for Illustration

Backup IOS Image to TFTP Server Example - 10.7.3

To maintain network operations with minimum down time, it is necessary to have procedures in place for backing up Cisco IOS images. This allows the network administrator to quickly copy an image back to a router in case of a corrupted or erased image.

In the figure, the network administrator wants to create a backup of the current image file on the router (isr4200-universalk9_ias.16.09.04.SPA.bin) to the TFTP server at 172.16.1.100.

Click each button for the steps to create a backup of the Cisco IOS image to a TFTP server.

Step 1. Ping the TFTP server.

Ensure that there is access to the network TFTP server. Ping the TFTP server to test connectivity, as shown in the example.

```
R1# ping 172.16.1.100

Type escape sequence to abort.

Sending 5, 100-byte ICMP Echos to 172.16.1.100, timeout is 2 seconds:

!!!!!

Success rate is 100 percent (5/5),

round-trip min/avg/max = 56/56/56 ms
```

Step 2. Verify image size in flash.

Verify that the TFTP server has sufficient disk space to accommodate the Cisco IOS Software image. Use the **show flash0:** command on the router to determine the size of the Cisco IOS image file. The file in the example is 517153193 bytes long.

```
R1# show flash0:

-# - --length-- -----date/time------ path

8 517153193     Apr 2 2019 21:29:58 +00:00

                        isr4200-universalk9_ias.16.09.04.SPA.bin

(output omitted)
```

Step 3. Copy the image to the TFTP server.

Copy the image to the TFTP server by using the **copy** *source-url destination-url* command. After issuing the command by using the specified source and destination URLs, the user is prompted for the source file name, IP address of the remote host, and destination file name. Typically, you will press **Enter** to accept the source filename as the destination file name. The transfer will then begin.

```
R1# copy flash: tftp:
Source filename []? isr4200-universalk9_ias.16.09.04.SPA.bin
Address or name of remote host []? 172.16.1.100
Destination filename [isr4200-universalk9_ias.16.09.04.SPA.bin]?
Writing isr4200-universalk9_ias.16.09.04.SPA.bin...
!!!!!!!!!!!!!!!!!!!!!!!!!!!!!!!!!!!!!!!!!!!!
(output omitted)
517153193 bytes copied in 863.468 secs (269058 bytes/sec)
```

Refer to
Interactive Graphic
in online course

Refer to
Online Course
for Illustration

Copy an IOS Image to a Device Example - 10.7.4

Cisco consistently releases new Cisco IOS software versions to resolve caveats and provide new features. This example uses IPv6 for the transfer to show that TFTP can also be used across IPv6 networks.

The figure illustrates copying a Cisco IOS software image from a TFTP server. A new image file (isr4200-universalk9_ias.16.09.04.SPA.bin) will be copied from the TFTP server at 2001:DB8:CAFE:100::99 to the router.

Select a Cisco IOS image file that meets the requirements in terms of platform, features, and software. Download the file from cisco.com and transfer it to the TFTP server. Click each button for the steps to upgrade the IOS image on the Cisco router.

Step 1. Ping the TFTP server.

Ensure that there is access to the network TFTP server. Ping the TFTP server to test connectivity, as shown in the example.

```
R1# ping 2001:db8:cafe:100::99
Type escape sequence to abort.
Sending 5, 100-byte ICMP Echos to 2001:DB8:CAFE:100::99,
timeout is 2 seconds:
!!!!!
Success rate is 100 percent (5/5),
round-trip min/avg/max = 56/56/56 ms
```

Step 2. Verify the amount of free flash.

Ensure that there is sufficient flash space on the router that is being upgraded. The amount of free flash can be verified by using the **show flash:** command. Compare the free flash space with the new image file size. The **show flash:** command in the example is used to verify free flash size. Free flash space in the example is 6294806528 bytes.

```
R1# show flash:
-# - --length-- -----date/time------ path
(output omitted)
6294806528 bytes available (537251840 bytes used)
R1#
```

Step 3. Copy the new IOS image to flash.

Copy the IOS image file from the TFTP server to the router by using the **copy** command, shown in the example. After issuing this command with specified source and destination URLs, the user will be prompted for the IP address of the remote host, source file name, and destination file name. Typically, you will press **Enter** to accept the source filename as the destination file name. The transfer of the file will begin.

```
R1# copy tftp: flash:
Address or name of remote host []?2001:DB8:CAFE:100::99
Source filename []? isr4200-universalk9_ias.16.09.04.SPA.bin
Destination filename [isr4200-universalk9_ias.16.09.04.SPA.bin]?
Accessing tftp://2001:DB8:CAFE:100::99/ isr4200-
universalk9_ias.16.09.04.SPA.bin...
Loading isr4200-universalk9_ias.16.09.04.SPA.bin
from 2001:DB8:CAFE:100::99 (via
GigabitEthernet0/0/0): !!!!!!!!!!!!!!!!!!!!!!

[OK - 517153193 bytes]
517153193 bytes copied in 868.128 secs (265652 bytes/sec)
```

The boot system Command - 10.7.5

To upgrade to the copied IOS image after that image is saved on the flash memory of the router, configure the router to load the new image during bootup by using the **boot system** command, as shown in the example. Save the configuration. Reload the router to boot the router with new image.

```
R1# configure terminal
R1(config)# boot system flash0:isr4200-universalk9_ias.16.09.04.SPA.bin
R1(config)# exit
R1# copy running-config startup-config
R1# reload
```

During startup, the bootstrap code parses the startup configuration file in NVRAM for the **boot system** commands that specify the name and location of the Cisco IOS Software image to load. Several **boot system** commands can be entered in sequence to provide a fault-tolerant boot plan.

If there are no **boot system** commands in the configuration, the router defaults to loading the first valid Cisco IOS image in flash memory and runs it.

After the router has booted, to verify that the new image has loaded, use the **show version** command, as displayed in the example.

```
R1# show version

Cisco IOS XE Software, Version 16.09.04

Cisco IOS Software [Fuji], ISR Software (X86_64_LINUX_IOSD-UNIVERSALK9_
IAS-M), Version 16.9.4, RELEASE SOFTWARE (fc2)

Technical Support: http://www.cisco.com/techsupport

Copyright (c) 1986-2019 by Cisco Systems, Inc.

Compiled Thu 22-Aug-19 18:09 by mcpre

Cisco IOS-XE software, Copyright (c) 2005-2019 by cisco Systems, Inc.

All rights reserved. Certain components of Cisco IOS-XE software are

licensed under the GNU General Public License ("GPL") Version 2.0. The

software code licensed under GPL Version 2.0 is free software that comes

with ABSOLUTELY NO WARRANTY. You can redistribute and/or modify such

GPL code under the terms of GPL Version 2.0. For more details, see the

documentation or "License Notice" file accompanying the IOS-XE software,

or the applicable URL provided on the flyer accompanying the IOS-XE

software.


ROM: IOS-XE ROMMON

Router uptime is 2 hours, 19 minutes

Uptime for this control processor is 2 hours, 22 minutes

System returned to ROM by PowerOn

System image file is "flash:isr4200-universalk9_ias.16.09.04.SPA.bin"

(output omitted)
```

Refer to **Packet Tracer Activity** for this chapter

Packet Tracer - Use a TFTP Server to Upgrade a Cisco IOS Image - 10.7.6

A TFTP server can help manage the storage of IOS images and revisions to IOS images. For any network, it is good practice to keep a backup copy of the Cisco IOS Software image in case the system image in the router becomes corrupted or accidentally erased. A TFTP server can also be used to store new upgrades to the IOS and then deployed throughout the network where it is needed. In this activity, you will upgrade the IOS images on Cisco devices by using a TFTP server. You will also backup an IOS image with the use of a TFTP server.

Module Practice and Quiz - 10.8

Refer to **Packet Tracer Activity** for this chapter

Packet Tracer - Configure CDP, LLDP, and NTP - 10.8.1

In this Packet Tracer activity, you will complete the following objectives:

- Build the Network and Configure Basic Device Settings
- Network Discovery with CDP
- Network Discovery with LLDP
- Configure and Verify NTP

Refer to **Lab Activity** for this chapter

Lab - Configure CDP, LLDP, and NTP - 10.8.2

In this lab, you will complete the following objectives:

- Build the Network and Configure Basic Device Settings
- Network Discovery with CDP
- Network Discovery with LLDP
- Configure and Verify NTP

What did I learn in this module? - 10.8.3

Device Discovery with CDP

Cisco Discovery Protocol (CDP) is a Cisco proprietary Layer 2 protocol that is used to gather information about Cisco devices which share the same data link. The device sends periodic CDP advertisements to connected devices. CDP can be used as a network discovery tool to determine the information about the neighboring devices. This information gathered from CDP can help build a logical topology of a network when documentation is missing or lacking in detail. CDP can assist in network design decisions, troubleshooting, and making changes to equipment. On Cisco devices, CDP is enabled by default. To verify the status of CDP and display information about CDP, enter the **show cdp** command. To enable CDP globally for all the supported interfaces on the device, enter **cdp run** in the global configuration mode. To enable CDP on the specific interface, enter the **cdp enable** command. To verify the status of CDP and display a list of neighbors, use the **show cdp neighbors** command in the privileged EXEC mode. The **show cdp neighbors** command provides helpful information about each CDP neighbor device, including device identifiers, port identifier, capabilities list, and platform. Use the **show cdp interface** command to display the interfaces that are CDP enabled on a device.

Device Discovery with LLDP

Cisco devices also support Link Layer Discovery Protocol (LLDP), which is a vendor-neutral neighbor discovery protocol similar to CDP. This protocol advertises its identity and capabilities to other devices and receives the information from a physically connected Layer 2 device. To enable LLDP globally on a Cisco network device, enter the **lldp run** command in the global configuration mode. To verify LLDP has been enabled on the device, enter the **show lldp** command in privileged EXEC mode. With LLDP enabled, device neighbors can be discovered by using the **show lldp neighbors** command. When more details about the neighbors are needed, the **show lldp neighbors detail** command can provide information, such as the neighbor IOS version, IP address, and device capability.

NTP

The software clock on a router or switch starts when the system boots and is the primary source of time for the system. When the time is not synchronized between devices, it will be impossible to determine the order of the events and the cause of an event. You can manually configure the date and time, or you can configure the NTP. This protocol allows routers on the network to synchronize their time settings with an NTP server. When NTP is implemented in the network, it can be set up to synchronize to a private master clock or it can synchronize to a publicly available NTP server on the Internet. NTP networks use a hierarchical system of time sources and each level in this system is called a stratum. The synchronized time is distributed across the network by using NTP. Authoritative time sources, also referred to as stratum 0 devices, are high-precision timekeeping devices. Stratum 1 devices are directly connected to the authoritative time sources. Stratum 2 devices, such as NTP clients, synchronize their time by using the NTP packets from stratum 1 servers. The **ntp server** *ip-address* command is issued in global configuration mode to configure a device as the NTP server. To verify the time source is set to NTP, use the **show clock detail** command. The **show ntp associations** and **show ntp status** commands are used to verify that a device is synchronized with the NTP server.

SNMP

SNMP allows administrators to manage servers, workstations, routers, switches, and security appliances, on an IP network. SNMP is an application layer protocol that provides a message format for communication between managers and agents. The SNMP system consists of three elements: SNMP manager, SNMP agents, and the MIB. To configure SNMP on a networking device, you must define the relationship between the manager and the agent. The SNMP manager is part of an NMS. The SNMP manager can collect information from an SNMP agent by using the "get" action and can change configurations on an agent by using the "set" action. SNMP agents can forward information directly to a network manager by using "traps". The SNMP agent responds to SNMP manager GetRequest-PDUs (to get an MIB variable) and SetRequest-PDUs (to set an MIB variable). An NMS periodically uses the get request to poll the SNMP agents by querying the device for data. A network management application can collect information to monitor traffic loads and to verify device configurations of managed devices.

SNMPv1, SNMPv2c, and SNMPv3 are all versions of SNMP. SNMPv1 is a legacy solution. Both SNMPv1 and SNMPv2c use a community-based form of security. The community of managers that is able to access the agent's MIB is defined by a community string. SNMPv2c includes a bulk retrieval mechanism and more detailed error message reporting. SNMPv3 provides for both security models and security levels. SNMP community strings are read-only (ro) and read-write (rw). They are used to authenticate access to MIB objects. The MIB organizes variables hierarchically. MIB variables enable the management software to monitor and control the network device. OIDs uniquely identify managed objects in the MIB hierarchy. The snmpget utility gives some insight into the basic mechanics of how SNMP works. The Cisco SNMP Navigator on the http://www.cisco.com website allows a network administrator to research details about a particular OID.

Syslog

The most common method of accessing system messages is to use a protocol called syslog. The syslog protocol uses UDP port 514 to allow networking devices to send their system messages across the network to syslog servers. The syslog logging service provides three primary functions: gather logging information for monitoring and troubleshooting, select the type of logging information that is captured, and specify the destinations of captured syslog messages. Destinations for syslog messages include the logging buffer (RAM inside a router or switch), console line, terminal line, and syslog server. This table shows syslog levels:

Severity Name	Severity Level	Explanation
Emergency	Level 0	System Unusable
Alert	Level 1	Immediate Action Needed
Critical	Level 2	Critical Condition
Error	Level 3	Error Condition
Warning	Level 4	Warning Condition
Notification	Level 5	Normal, but Significant Condition
Informational	Level 6	Informational Message
Debugging	Level 7	Debugging Message

Syslog facilities identify and categorize system state data for error and event message reporting. Common syslog message facilities reported on Cisco IOS routers include: IP, OSPF protocol, SYS operating system, IPsec, and IF. The default format of syslog messages on Cisco IOS software is: %facility-severity-MNEMONIC: description. Use the command **service timestamps log datetime** to force logged events to display the date and time.

Router and Switch File Maintenance

The Cisco IFS lets the administrator navigate to different directories and list the files in a directory, and to create subdirectories in flash memory or on a disk. Use the **show file systems command** to display lists all of the available file systems on a Cisco router. Use the directory command **dir** to display the directory of bootflash. Use the change directory command **cd** to view the contents of NVRAM. Use the present working directory command **pwd** to that you are viewing the current directory. Use the **show file systems** command to view the file systems on a Catalyst switch or a Cisco router. Configuration files can be saved to a text file by using Tera Term. A configuration can be copied from a file and then directly pasted to a device. Configuration files can be stored on a TFTP server, or a USB drive. To save the running configuration or the startup configuration to a TFTP server, use either the **copy running-config tftp** or **copy startup-config tftp** command. Use the **dir** command to view the contents of the USB flash drive. Use the **copy run usbflash0:/** command to copy the configuration file to the USB flash drive. Use the **dir** command to see the file on the USB drive. Use the **more** command to see the contents of the drive. For encrypted passwords, such as the enable secret passwords, the passwords must be replaced after recovery.

IOS Image Management

Cisco IOS Software images and configuration files can be stored on a central TFTP server to control the number of IOS images and the revisions to those IOS images, as well as the configuration files that must be maintained. Select a Cisco IOS image file that meets the requirements in terms of platform, features, and software. Download the file from cisco.com and transfer it to the TFTP server. Ping the TFTP server. Verify the amount of free flash. The amount of free flash can be verified by using the **show flash:** command. If there is enough free flash to hold the new IOS image, copy the new IOS image to flash. To upgrade to the copied IOS image after that image is saved on the router's flash memory, configure the router to load the new image during bootup by using the **boot system** command. Save the configuration. Reload the router to boot the router with new image. After the router has booted, to verify the new image has loaded, use the **show version** command.

Go to the online course to take the quiz and exam.

Chapter Quiz - Network Management

Your Chapter Notes

CHAPTER 11

Network Design

Introduction - 11.0

Why should I take this module? - 11.0.1

Welcome to Network Design!

You are a sought after spaceship designer! You have been asked to design a new spaceship. Your first questions are, "What will this ship be used for? How large is the crew? Will it be a war ship? A cargo ship? A science and exploration vessel?" What if the answer is, "The crew can be as few as 50 people, but it must be able to hold as many as 500. It will be used in a variety of ways."? How do you design a ship like this? You must design the size and configuration of the ship, and the power it requires, wisely.

Designing a network to meet current requirements and to adapt to future requirements is a complex task. But it can be done, thanks to hierarchical and scalable network designs that use the right components. You know you want to learn about this. Even if you have not designed your current network, knowing about network design will increase your value to the organization as a great network administrator! And who doesn't want that?

What will I learn to do in this module? - 11.0.2

Module Title: Network Design

Module Objective: Explain the characteristics of scalable network architectures.

Topic Title	Topic Objective
Hierarchical Networks	Explain how data, voice, and video are converged in a switched network.
Scalable Networks	Explain considerations for designing a scalable network.
Switch Hardware	Explain how switch hardware features support network requirements.
Router Hardware	Describe the types of routers available for small to-medium-sized business networks.

Hierarchical Networks - 11.1

Refer to **Video** in online course

Video - Three-Layer Network Design - 11.1.1

Refer to **Interactive Graphic** in online course

The Need to Scale the Network - 11.1.2

Our digital world is changing. The ability to access the internet and the corporate network is no longer confined to physical offices, geographical locations, or time zones. In today's globalized workplace, employees can access resources from anywhere in the world and information must be available at any time, and on any device. These requirements drive the need to build next-generation networks that are secure, reliable, and highly available.

These next-generation networks must not only support current expectations and equipment but must also be able to integrate legacy platforms. Businesses increasingly rely on their network infrastructure to provide mission-critical services. As businesses grow and evolve, they hire more employees, open branch offices, and expand into global markets. These changes directly affect the requirements of a network which must be able to scale to meet the needs of business.

Click Play in the figure to view an animation of a small network expanding into a larger network.

A network must support the exchange of various types of network traffic, including data files, email, IP telephony, and video applications for multiple business units. All enterprise networks must be able to do the following:

- Support critical applications

- Support converged network traffic

- Support diverse business needs

- Provide centralized administrative control

The LAN is the networking infrastructure that provides access to network communication services and resources for end users and devices. The end users and devices may be spread over a single floor or building. You create a campus network by interconnecting a group of LANs that are spread over a small geographic area. Campus network designs include small networks that use a single LAN switch, up to very large networks with thousands of connections.

Refer to
Online Course
for Illustration

Borderless Switched Networks - 11.1.3

With the increasing demands of the converged network, the network must be developed with an architectural approach that embeds intelligence, simplifies operations, and is scalable to meet future demands. One of the more recent developments in network design is the Cisco Borderless Network.

The Cisco Borderless Network is a network architecture that combines innovation and design. It allows organizations to support a borderless network that can connect anyone, anywhere, anytime, on any device; securely, reliably, and seamlessly. This architecture is designed to address IT and business challenges, such as supporting the converged network and changing work patterns.

The Cisco Borderless Network provides the framework to unify wired and wireless access, including policy, access control, and performance management across many different device types. Using this architecture, the borderless network, shown in the figure, is built on a hierarchical infrastructure of hardware that is scalable and resilient.

By combining this hardware infrastructure with policy-based software solutions, the Cisco Borderless Network provides two primary sets of services: network services, and user and endpoint services under the umbrella of an integrated management solution. It enables different network elements to work together, and allows users to access resources from any place, at any time, while providing optimization, scalability, and security.

Refer to
Interactive Graphic
in online course

Refer to
Online Course
for Illustration

Hierarchy in the Borderless Switched Network - 11.1.4

Creating a borderless switched network requires that sound network design principles are used to ensure maximum availability, flexibility, security, and manageability. The borderless switched network must deliver on current requirements and future required services and technologies. Borderless switched network design guidelines are built upon the following principles:

- **Hierarchical** - The design facilitates understanding the role of each device at every tier, simplifies deployment, operation, and management, and reduces fault domains at every tier.

- **Modularity** - The design allows seamless network expansion and integrated service enablement on an on-demand basis.

- **Resiliency** - The design satisfies user expectations for keeping the network always on.

- **Flexibility** - The design allows intelligent traffic load sharing by using all network resources.

These are not independent principles. Understanding how each principle fits in the context of the others is critical. Designing a borderless switched network in a hierarchical fashion creates a foundation that allows network designers to overlay security, mobility, and unified communication features. Two time-tested and proven hierarchical design frameworks for campus networks are the three-tier layer and the two-tier layer models.

The three critical layers within these tiered designs are the access, distribution, and core layers. Each layer can be seen as a well-defined, structured module with specific roles and functions in the campus network. Introducing modularity into the campus hierarchical design further ensures that the campus network remains resilient and flexible enough to provide critical network services. Modularity also helps to allow for growth and changes that occur over time.

Click each button for an example of each design.

Three-Tier Model

The figure shows an example of the three-tier model. At the top there are two clouds depicting the internet. There are redundant links connecting to two firewall routers. The routers have redundant links to two core layer multilayer switches. The switches have an EtherChannel between each other with four links. They also have redundant links to two distribution layer multilayer switches. The distribution layer switches have redundant links to three access layer switches. Two of the switches have links to access points. Both access points have connections to tablets. The access layer switches are also connected to IP phones and PCs.

Two-Tier Model

The figure shows an example of the two-tier model. At the top there are two clouds depicting the internet. There are redundant links connecting to two firewall routers. The routers have redundant links to two core/distribution layer multilayer switches. The core/distribution layer switches have redundant links to three access layer switches. Two of the switches have links to access points. Both access points have connections to tablets. The access layer switches are also connected to IP phones and PCs.

Refer to
Interactive Graphic
in online course

Access, Distribution, and Core Layer Functions - 11.1.5

The access, distribution, and core layers perform specific functions in a hierarchical network design.

Click each button for a description of the functions of each layer.

Access Layer

The access layer represents the network edge, where traffic enters or exits the campus network. Traditionally, the primary function of an access layer switch is to provide network access to the user. Access layer switches connect to distribution layer switches, which implement network foundation technologies such as routing, quality of service, and security.

To meet network application and end-user demand, the next-generation switching platforms now provide more converged, integrated, and intelligent services to various types of endpoints at the network edge. Building intelligence into access layer switches allows applications to operate on the network more efficiently and securely.

Distribution Layer

The distribution layer interfaces between the access layer and the core layer to provide many important functions, including the following:

- Aggregating large-scale wiring closet networks

- Aggregating Layer 2 broadcast domains and Layer 3 routing boundaries

- Providing intelligent switching, routing, and network access policy functions to access the rest of the network

- Providing high availability through redundant distribution layer switches to the end user, and equal cost paths to the core

- Providing differentiated services to various classes of service applications at the edge of the network

Core Layer

The core layer is the network backbone. It connects several layers of the campus network. The core layer serves as the aggregator for all of the distribution layer devices and ties the campus together with the rest of the network. The primary purpose of the core layer is to provide fault isolation and high-speed backbone connectivity.

Refer to
Interactive Graphic
in online course

Three-Tier and Two-Tier Examples - 11.1.6

Click each button for an example and explanation of a three-tier and two-tier design.

Refer to
Online Course
for Illustration

Three-Tier Example

The figure shows a three-tier campus network design for organizations where the access, distribution, and core are each separate layers. To build a simplified, scalable, cost-effective, and efficient physical cable layout design, the recommendation is to build an extended-star physical network topology from a centralized building location to all other buildings on the same campus.

Two-Tier Example

In some cases where extensive physical or network scalability does not exist, maintaining separate distribution and core layers is not required. In smaller campus locations where there are fewer users accessing the network, or in campus sites consisting of a single building, separate core and distribution layers may not be needed. In this scenario, the recommendation is the alternate two-tier campus network design, also known as the collapsed core network design, as shown in the figure.

Refer to
Online Course
for Illustration

Role of Switched Networks - 11.1.7

The role of switched networks has evolved dramatically in the last two decades. It was not long ago that flat Layer 2 switched networks were the norm. Flat Layer 2 switched networks relied on the Ethernet and the widespread use of hub repeaters to propagate LAN traffic throughout an organization.

As shown in the figure, networks have fundamentally changed to switched LANs in a hierarchical network.

A switched LAN allows additional flexibility, traffic management, quality of service, and security. It also affords support for wireless networking and connectivity, and support for other technologies such as IP telephone and mobility services.

Go to the online course to take the quiz and exam.

Check Your Understanding - Hierarchical Networks - 11.1.8

Scalable Networks - 11.2

Refer to
Interactive Graphic
in online course

Design for Scalability - 11.2.1

You understand that your network is going to change. Its number of users will likely increase, they may be found anywhere, and they will be using a wide variety of devices. Your network must be able to change along with its users. Scalability is the term for a network that can grow without losing availability and reliability.

Refer to
Online Course
for Illustration

To support a large, medium or small network, the network designer must develop a strategy to enable the network to be available and to scale effectively and easily. Included in a basic network design strategy are the following recommendations:

- Use expandable, modular equipment, or clustered devices that can be easily upgraded to increase capabilities. Device modules can be added to the existing equipment to support new features and devices without requiring major equipment upgrades. Some devices can be integrated in a cluster to act as one device to simplify management and configuration.

- Design a hierarchical network to include modules that can be added, upgraded, and modified, as necessary, without affecting the design of the other functional areas of the network. For example, creating a separate access layer that can be expanded without affecting the distribution and core layers of the campus network.

- Create an IPv4 and IPv6 address strategy that is hierarchical. Careful address planning eliminates the need to re-address the network to support additional users and services.

- Choose routers or multilayer switches to limit broadcasts and filter other undesirable traffic from the network. Use Layer 3 devices to filter and reduce traffic to the network core.

Click each button for more information about advanced network design requirements

Redundant Links

Implement redundant links in the network between critical devices and between access layer and core layer devices.

Multiple Links

Implement multiple links between equipment, with either link aggregation (EtherChannel) or equal cost load balancing, to increase bandwidth. Combining multiple Ethernet links into a single, load-balanced EtherChannel configuration increases available bandwidth. EtherChannel implementations can be used when budget restrictions prohibit purchasing high-speed interfaces and fiber runs.

Scalable Routing Protocol

Use a scalable routing protocol and implement features within that routing protocol to isolate routing updates and minimize the size of the routing table.

Wireless Connectivity

Implement wireless connectivity to allow for mobility and expansion.

Refer to
Online Course
for Illustration

Plan for Redundancy - 11.2.2

For many organizations, the availability of the network is essential to supporting business needs. Redundancy is an important part of network design. It can prevent disruption of network services by minimizing the possibility of a single point of failure. One method of implementing redundancy is by installing duplicate equipment and providing failover services for critical devices.

Another method of implementing redundancy is redundant paths, as shown in the figure above. Redundant paths offer alternate physical paths for data to traverse the network. Redundant paths in a switched network support high availability. However, due to the operation of switches, redundant paths in a switched Ethernet network may cause logical Layer 2 loops. For this reason, Spanning Tree Protocol (STP) is required.

STP eliminates Layer 2 loops when redundant links are used between switches. It does this by providing a mechanism for disabling redundant paths in a switched network until the path is necessary, such as when a failure occurs. STP is an open standard protocol, used in a switched environment to create a loop-free logical topology.

Using Layer 3 in the backbone is another way to implement redundancy without the need for STP at Layer 2. Layer 3 also provides best path selection and faster convergence during failover.

Refer to
Interactive Graphic
in online course

Refer to
Online Course
for Illustration

Reduce Failure Domain Size - 11.2.3

A well-designed network not only controls traffic, but also limits the size of failure domains. A failure domain is the area of a network that is impacted when a critical device or network service experiences problems.

The function of the device that initially fails determines the impact of a failure domain. For example, a malfunctioning switch on a network segment normally affects only the hosts on that segment. However, if the router that connects this segment to others fails, the impact is much greater.

The use of redundant links and reliable enterprise-class equipment minimize the chance of disruption in a network. Smaller failure domains reduce the impact of a failure on company productivity. They also simplify the troubleshooting process, thereby, shortening the downtime for all users.

Click each button to see the failure domain of each associated device.

Edge Router

AP1

S1

S2

S3

Limiting the Size of Failure Domains

Because a failure at the core layer of a network can have a potentially large impact, the network designer often concentrates on efforts to prevent failures. These efforts can greatly increase the cost of implementing the network. In the hierarchical design model, it is easiest and usually least expensive to control the size of a failure domain in the distribution layer. In the distribution layer, network errors can be contained to a smaller area; thus, affecting fewer users. When using Layer 3 devices at the distribution layer, every router functions as a gateway for a limited number of access layer users.

Switch Block Deployment

Routers, or multilayer switches, are usually deployed in pairs, with access layer switches evenly divided between them. This configuration is referred to as a building, or departmental, switch block. Each switch block acts independently of the others. As a result, the failure of a single device does not cause the network to go down. Even the failure of an entire switch block does not affect a significant number of end users.

Refer to
Online Course
for Illustration

Increase Bandwidth - 11.2.4

In hierarchical network design, some links between access and distribution switches may need to process a greater amount of traffic than other links. As traffic from multiple links converges onto a single, outgoing link, it is possible for that link to become a bottleneck.

Link aggregation, such as EtherChannel, allows an administrator to increase the amount of bandwidth between devices by creating one logical link made up of several physical links.

EtherChannel uses the existing switch ports. Therefore, additional costs to upgrade the link to a faster and more expensive connection are not necessary. The EtherChannel is seen as one logical link using an EtherChannel interface. Most configuration tasks are done on the EtherChannel interface, instead of on each individual port, ensuring configuration consistency throughout the links. Finally, the EtherChannel configuration takes advantage of load balancing between links that are part of the same EtherChannel, and depending on the hardware platform, one or more load-balancing methods can be implemented.

Refer to
Online Course
for Illustration

Expand the Access Layer - 11.2.5

The network must be designed to be able to expand network access to individuals and devices, as needed. An increasingly important option for extending access layer connectivity is through wireless. Providing wireless connectivity offers many advantages, such as increased flexibility, reduced costs, and the ability to grow and adapt to changing network and business requirements.

To communicate wirelessly, end devices require a wireless NIC that incorporates a radio transmitter/receiver and the required software driver to make it operational. Additionally, a wireless router or a wireless access point (AP) is required for users to connect, as shown in the figure.

There are many considerations when implementing a wireless network, such as the types of wireless devices to use, wireless coverage requirements, interference considerations, and security considerations.

Refer to
Online Course
for Illustration

Tune Routing Protocols - 11.2.6

Advanced routing protocols, such as Open Shortest Path First (OSPF), are used in large networks.

OSPF is a link-state routing protocol. As shown in the figure, OSPF works well for larger hierarchical networks where fast convergence is important. OSPF routers establish and maintain neighbor adjacencies with other connected OSPF routers. OSPF routers synchronize their link-state database. When a network change occurs, link-state updates are sent, informing other OSPF routers of the change and establishing a new best path, if one is available.

Go to the online
course to take the
quiz and exam.

Check Your Understanding - Scalable Networks - 11.2.7

Switch Hardware - 11.3

Refer to
Interactive Graphic
in online course

Switch Platforms - 11.3.1

Refer to
Online Course
for Illustration

One simple way to create hierarchical and scalable networks is to use the right equipment for the job. There is a variety of switch platforms, form factors, and other features that you should consider before choosing a switch.

When designing a network, it is important to select the proper hardware to meet current network requirements, as well as to allow for network growth. Within an enterprise network, both switches and routers play a critical role in network communication.

Click each button for more information about the categories of switches for enterprise networks.

Campus LAN Switches

To scale network performance in an enterprise LAN, there are core, distribution, access, and compact switches. These switch platforms vary from fanless switches with eight fixed ports to 13-blade switches supporting hundreds of ports. Campus LAN switch platforms include the Cisco 2960, 3560, 3650, 3850, 4500, 6500, and 6800 Series.

Cloud-Managed Switches

The Cisco Meraki cloud-managed access switches enable virtual stacking of switches. They monitor and configure thousands of switch ports over the web, without the intervention of onsite IT staff.

Data Center Switches

A data center should be built based on switches that promote infrastructure scalability, operational continuity, and transport flexibility. The data center switch platforms include the Cisco Nexus Series switches.

Service Provider Switches

Service provider switches fall under two categories: aggregation switches and Ethernet access switches. Aggregation switches are carrier-grade Ethernet switches that aggregate traffic at the edge of a network. Service provider Ethernet access switches feature application intelligence, unified services, virtualization, integrated security, and simplified management.

Virtual Networking

Networks are becoming increasingly virtualized. Cisco Nexus virtual networking switch platforms provide secure multi-tenant services by adding virtualization intelligence technology to the data center network.

Refer to
Interactive Graphic
in online course

Refer to
Online Course
for Illustration

Switch Form Factors - 11.3.2

When selecting switches, network administrators must determine the switch form factors. This includes fixed configuration, modular configuration, stackable, or non-stackable

Click each button for more information about switch form factors.

Fixed configuration switches

Features and options on fixed configuration switches are limited to those that originally come with the switch.

Modular configuration switches

The chassis on modular switches accept field-replaceable line cards.

Stackable configuration switches

Special cables are used to connect stackable switches that allow them to effectively operate as one large switch.

Thickness

The thickness of the switch, which is expressed in the number of rack units, is also important for switches that are mounted in a rack. For example, the fixed configuration switches shown in the figure are all one rack units (1U) or 1.75 inches (44.45 mm) in height.

Refer to
Online Course
for Illustration

Port Density - 11.3.3

The port density of a switch refers to the number of ports available on a single switch. The figure shows the port density of three different switches.

Fixed configuration switches support a variety of port density configurations. The Cisco Catalyst 3850 come in 12, 24, 48 port configurations, as shown in the figure. The 48-port switch has an option for additional ports for small form-factor pluggable (SFP) devices.

Modular switches can support very high port densities through the addition of multiple switchport line cards. The modular Catalyst 9400 switch shown in the next figure supports 384 switchport interfaces.

Large networks that support many thousands of network devices require high density, modular switches to make the best use of space and power. Without using a high-density modular switch, the network would need many fixed configuration switches to accommodate the number of devices that need network access. This approach can consume many power outlets and a lot of closet space.

The network designer must also consider the issue of uplink bottlenecks. A series of fixed configuration switches may consume many additional ports for bandwidth aggregation between switches, for the purpose of achieving target performance. With a single modular switch, bandwidth aggregation is less of an issue because the backplane of the chassis can provide the necessary bandwidth to accommodate the devices connected to the switchport line cards.

Forwarding Rates - 11.3.4

Forwarding rates define the processing capabilities of a switch by rating how much data the switch can process per second. Switch product lines are classified by forwarding rates. Entry-level switches have lower forwarding rates than enterprise-level switches. Forwarding rates are important to consider when selecting a switch. If the switch forwarding rate is too low, it cannot accommodate full wire-speed communication across all of its switch ports. Wire speed is the data rate that each Ethernet port on the switch is capable of attaining. Data rates can be 100 Mbps, 1 Gbps, 10 Gbps, or 100 Gbps.

For example, a typical 48-port gigabit switch operating at full wire speed generates 48 Gbps of traffic. If the switch only supports a forwarding rate of 32 Gbps, it cannot run at full wire speed across all ports simultaneously. Fortunately, access layer switches typically do not need to operate at full wire speed, because they are physically limited by their uplinks to the distribution layer. This means that less expensive, lower performing switches can be used at the access layer, and more expensive, higher performing switches can be used at the distribution and core layers, where the forwarding rate has a greater impact on network performance.

Refer to
Interactive Graphic
in online course

Refer to
Online Course
for Illustration

Power over Ethernet - 11.3.5

Power over Ethernet (PoE) allows the switch to deliver power to a device over the existing Ethernet cabling. This feature can be used by IP phones and some wireless access points, allowing them to be installed anywhere that there is an Ethernet cable. A network administrator should ensure that the PoE features are actually required for a given installation, because switches that support PoE are expensive.

Click each button to view PoE ports on different devices.

Switch

PoE ports look the same as any other switch port. Check the model of the switch to determine if the port supports PoE.

IP Phone

WAP

PoE ports on wireless access points look the same as any other switch port. Check the model of the wireless access point to determine if the port supports PoE.

Cisco Catalyst 2960-C

The Cisco Catalyst 2960-C and 3560-C Series compact switches support PoE pass-through. PoE pass-through allows a network administrator to power PoE devices that are connected to the switch, as well as the switch itself, by drawing power from certain upstream switches.

Refer to
Online Course
for Illustration

Multilayer Switching - 11.3.6

Multilayer switches are typically deployed in the core and distribution layers of an organization's switched network. Multilayer switches are characterized by their ability to build a routing table, support a few routing protocols, and forward IP packets at a rate close to that of Layer 2 forwarding. Multilayer switches often support specialized hardware, such as application-specific integrated circuits (ASICs). ASICs along with dedicated software data structures can streamline the forwarding of IP packets independent of the CPU.

There is a trend in networking toward a pure Layer 3 switched environment. When switches were first used in networks, none of them supported routing. Now, almost all switches support routing. It is likely that soon all switches will incorporate a route processor because the cost of doing so is decreasing relative to other constraints.

The figure shows a Catalyst 2960. Catalyst 2960 switches illustrate the migration to a pure Layer 3 environment. With IOS versions prior to 15.x, these switches supported only one active switched virtual interface (SVI). With IOS 15.x, these switches now support multiple active SVIs. This means that the switch can be remotely accessed via multiple IP addresses on distinct networks.

Business Considerations for Switch Selection - 11.3.7

The following table highlights other common business considerations when selecting switch equipment.

Consideration	Description
Cost	The cost of a switch will depend on the number and speed of the interfaces, supported features, and expansion capability.
Port density	Network switches must support the appropriate number of devices on the network.
Power	It is now common to power access points, IP phones, and compact switches user Power over Ethernet (PoE). In addition to PoE considerations, some chassis-based switches support redundant power supplies.
Reliability	The switch should provide continuous access to the network.
Port speed	The speed of the network connection is of primary concern to end users.
Frame buffers	The ability of the switch to store frames is important in a network where there may be congested ports to servers or other areas of the network.
Scalability	The number of users on a network typically grows over time; therefore, the switch should provide the opportunity for growth.

Go to the online course to take the quiz and exam.

Check Your Understanding - Switch Hardware - 11.3.8

Router Hardware - 11.4

Router Requirements - 11.4.1

Switches are not the only component of a network that come with a variety of features. Your choice of router is another very important decision. Routers play a critical role in networking by connecting homes and businesses to the internet, interconnecting multiple sites within an enterprise network, providing redundant paths, and connecting ISPs on the internet. Routers can also act as a translator between different media types and protocols. For example, a router can accept packets from an Ethernet network and re-encapsulate them for transport over a serial network.

Routers use the network portion (prefix) of the destination IP address to route packets to the proper destination. They select an alternate path if a link goes down. All hosts on a local network specify the IP address of the local router interface in their IP configuration. This router interface is the default gateway. The ability to route efficiently and recover from network link failures is critical to delivering packets to their destination.

Routers also serve other beneficial functions as follows:

- They provide broadcast containment by limiting broadcasts to the local network.

- They interconnect geographically separated locations.

- The group users logically by application or department within a company, who have command needs or require access to the same resources.

- They provide enhanced security by filtering unwanted traffic through access control lists.

Refer to
Interactive Graphic
in online course

Refer to
Online Course
for Illustration

Cisco Routers - 11.4.2

As the network grows, it is important to select the proper routers to meet its requirements. There are different categories of Cisco routers.

Click each button for more information about the categories of routers.

Branch Routers

Branch routers, shown in the figure, optimize branch services on a single platform while delivering an optimal application experience across branch and WAN infrastructures. Maximizing service availability at the branch requires networks designed for 24x7x365 uptime. Highly available branch networks must ensure fast recovery from typical faults, while minimizing or eliminating the impact on service, and provide simple network configuration and management. Shown are the Cisco Integrated Services Router (ISR) 4000 Series Routers.

Network Edge Routers

Network edge routers, shown in the figure, enable the network edge to deliver high-performance, highly secure, and reliable services that unite campus, data center, and branch networks. Customers expect a high-quality media experience and more types of content than ever before. Customers want interactivity, personalization, mobility, and control for all content. Customers also want to access content anytime and anyplace they choose, over any device, whether at home, at work, or on the go. Network edge routers must deliver enhanced quality of service and nonstop video and mobile capabilities. Shown are the Cisco Aggregation Services Routers (ASR) 9000 Series Routers.

Service Provider Routers

Service provider routers, shown in the figure, deliver end-to-end scalable solutions and subscriber-aware services. Operators must optimize operations, reduce expenses, and improve scalability and flexibility, to deliver next-generation internet experiences across all devices and locations. These systems are designed to simplify and enhance the operation and deployment of service-delivery networks. Shown are the Cisco Network Convergence System (NCS) 6000 Series Routers.

Industrial

Industrial routers, such as the ones shown in the figure, are designed to provide enterprise-class features in rugged and harsh environments. Their compact, modular, ruggedized design is excellent for mission-critical applications. Shown are the Cisco 1100 Series Industrial Integrated Services Routers.

Refer to
Interactive Graphic
in online course

Refer to
Online Course
for Illustration

Router Form Factors - 11.4.3

Like switches, routers also come in many form factors. Network administrators in an enterprise environment should be able to support a variety of routers, from a small desktop router to a rack-mounted or blade model.

Click each button for more information on various Cisco router platforms.

Cisco 900 Series

Cisco 900 Series

This is a small branch office router. It combines WAN, switching, security, and advanced connectivity options in a compact, fanless platform for small and medium-sized businesses.

ASR 9000 and 1000 Series

Cisco ASR 9000 and 1000 Series Aggregation Services Routers

These routers provide density and resiliency with programmability, for a scalable network edge.

5500 Series

Cisco Network Convergence System 5500 Series Routers

These routers are designed to efficiently scale between large data centers and large enterprise networks, web, and service provider WAN and aggregation networks.

Cisco 800

Cisco 800 Industrial Integrated Services Router

This router is compact and designed for harsh environments.

Routers can also be categorized as fixed configuration or modular. With the fixed configuration, the desired router interfaces are built-in. Modular routers come with multiple slots that allow a network administrator to change the interfaces on the router. Routers come with a variety of different interfaces, such as Fast Ethernet, Gigabit Ethernet, Serial, and Fiber-Optic.

A comprehensive list of Cisco routers can be found by searching Cisco's website www.cisco.com.

Go to the online course to take the quiz and exam.

Check Your Understanding - Router Hardware - 11.4.4

Module Practice and Quiz - 11.5

Refer to Packet Tracer Activity for this chapter

Packet Tracer - Compare Layer 2 and Layer 3 Devices - 11.5.1

In this Packet Tracer activity, you will use various commands to examine three different switching topologies and compare the similarities and differences between the 2960 and 3650 switches. You will also compare the routing table of a 4321 router with that of a 3650 switch.

What did I learn in this module? - 11.5.2

Hierarchical Networks

All enterprise networks must: support critical applications, support converged network traffic, support diverse business needs, and provide centralized administrative control. The Cisco Borderless Network provides the framework to unify wired and wireless access, including policy, access control, and performance management across many different device types. The borderless network is built on a hierarchical infrastructure of hardware that is scalable and resilient. Two proven hierarchical design frameworks for campus networks are the three-tier layer and the two-tier layer models. The three critical layers within these tiered designs are the access, distribution, and core layers. The access layer represents the network edge, where traffic enters or exits the campus network. Access layer switches connect to distribution layer switches, which implement network foundation technologies such as routing, quality of service, and security. The distribution layer interfaces between the access layer and the core layer. The primary purpose of the core layer is to provide fault isolation and high-speed backbone connectivity. Networks have fundamentally changed to switched LANs in a hierarchical network, providing QoS, security, support for wireless connectivity and IP telephony and mobility services.

Scalable Networks

A basic network design strategy includes the following recommendations: use expandable, modular equipment, or clustered devices; design a hierarchical network to include modules that can be added, upgraded, and modified; create a hierarchical IPv4 and IPv6 address strategy; and choose routers or multilayer switches to limit broadcasts and filter other undesirable traffic from the network. Implement redundant links in the network between critical devices and between access layer and core layer devices. Implement multiple links between equipment, with either link aggregation (EtherChannel) or equal cost load balancing, to increase bandwidth. Use a scalable routing protocol and implementing features within that routing protocol to isolate routing updates and minimize the size of the routing table. Implement wireless connectivity to allow for mobility and expansion. One method of implementing redundancy is by installing duplicate equipment and providing failover services for critical devices. Another method of implementing redundancy is to create redundant paths. A well-designed network not only controls traffic, but also limits the size of failure domains. Switch blocks act independently of the others, so the failure of a single device does not cause the network to go down. Link aggregation, such as Ether-Channel, allows an administrator to increase the amount of bandwidth between devices by creating one logical link made up of several physical links. Wireless connectivity expands the access layer. When implementing a wireless network, you must consider the types of wireless devices to use, wireless coverage requirements, interference considerations, and security. Link-state routing protocols such as OSPF, work well for larger hierarchical networks where fast convergence is important. OSPF routers establish and maintain neighbor adjacencies with other connected OSPF routers, they synchronize their link-state database. When a network change occurs, link state updates are sent, informing other OSPF routers of the change and establishing a new best path.

Switch Hardware

There are several categories of switches for enterprise networks including campus LAN, cloud-managed, data center, service provider, and virtual networking. Form factors for switches include fixed configuration, modular configuration, and stackable configuration. The thickness of a switch is expressed in number of rack units. The port density of a switch refers to the number of ports available on a single switch. Forwarding rates define the processing capabilities of a switch by rating how much data the switch can process per second. Power over Ethernet (PoE) allows the switch to deliver power to a device over the existing Ethernet cabling. Multilayer switches are typically deployed in the core and distribution layers of an organization's switched network. Multilayer switches are characterized by their ability to build a routing table, support a few routing protocols, and forward IP packets at a rate close to that of Layer 2 forwarding. Business considerations for switch selection include cost, port density, power, reliability, port speed, frame buffers, and scalability.

Router Hardware

Routers use the network portion (prefix) of the destination IP address to route packets to the proper destination. They select an alternate path if a link or path goes down. All hosts on a local network specify the IP address of the local router interface in their IP configuration. This router interface is the default gateway. Routers also serve other beneficial functions:

- They provide broadcast containment by limiting broadcasts to the local network.

- They interconnect geographically separated locations.

- They group users logically by application or department within a company, who have command needs or require access to the same resources.

- They provide enhanced security by filtering unwanted traffic through access control lists.

Cisco has several categories of routers including branch, network edge, service provider and industrial. Branch routers optimize branch services on a single platform while delivering an optimal application experience across branch and WAN infrastructures. Network edge routers deliver high-performance, highly secure, and reliable services that unite campus, data center, and branch networks. Service provider routers differentiate the service portfolio and increase revenues by delivering end-to-end scalable solutions and subscriber-aware services. Industrial routers are designed to provide enterprise-class features in rugged and harsh environments. Cisco router form factors include the Cisco 900 Series, the ASR 9000 and 1000 Series, the 5500 Series, and the Cisco 800. Routers can also be categorized as fixed configuration or modular. With the fixed configuration, the desired router interfaces are built-in. Modular routers come with multiple slots that allow a network administrator to change the interfaces on the router. Routers come with a variety of different interfaces, such as Fast Ethernet, Gigabit Ethernet, Serial, and Fiber-Optic.

Go to the online course to take the quiz and exam.

Chapter Quiz - Network Design

Your Chapter Notes

CHAPTER 12

Network Troubleshooting

Introduction - 12.0

Why should I take this module? - 12.0.1

Welcome to Network Troubleshooting!

Who is the best network administrator that you have ever seen? Why do you think this person is so good at it? Likely, it is because this person is really good at troubleshooting network problems. They are probably experienced administrators, but that is not the whole story. Good network troubleshooters generally go about this in a methodical fashion, and they use all of the tools available to them.

The truth is that the only way to become a good network troubleshooter is to always be troubleshooting. It takes time to get good at this. But luckily for you, there are many, many tips and tools that you can use. This module covers the different methods for network troubleshooting and all of the tips and tools you need to get started. This module also has two really good Packet Tracer activities to test your new skills and knowledge. Maybe your goal should be to become the best network administrator that someone else has ever seen!

What will I learn to do in this module? - 12.0.2

Module Title: Network Troubleshooting

Module Objective: Troubleshoot enterprise networks.

Topic Title	Topic Objective
Network Documentation	Explain how network documentation is developed and used to troubleshoot network issues.
Troubleshooting Process	Compare troubleshooting methods that use a systematic, layered approach.
Troubleshooting Tools	Describe different networking troubleshooting tools.
Symptoms and Causes of Network Problems	Determine the symptoms and causes of network problems using a layered model.
Troubleshooting IP Connectivity	Troubleshoot a network using the layered model.

Network Documentation - 12.1

Documentation Overview - 12.1.1

As with any complex activity like network troubleshooting, you will need to start with good documentation. Accurate and complete network documentation is required to effectively monitor and troubleshoot networks.

Common network documentation includes the following:

- Physical and logical network topology diagrams

- Network device documentation that records all pertinent device information

- Network performance baseline documentation

All network documentation should be kept in a single location, either as hard copy or on the network on a protected server. Backup documentation should be maintained and kept in a separate location.

Refer to
Interactive Graphic
in online course

Refer to
Online Course
for Illustration

Network Topology Diagrams - 12.1.2

Network topology diagrams keep track of the location, function, and status of devices on the network. There are two types of network topology diagrams: the physical topology and the logical topology.

Click each button for an example and explanation of physical and logical topologies.

Physical Topology

A physical network topology shows the physical layout of the devices connected to the network. You need to know how devices are physically connected to troubleshoot physical layer problems. Information recorded on the physical topology typically includes the following:

- Device name

- Device location (address, room number, rack location)

- Interface and ports used

- Cable type

The figure shows a sample physical network topology diagram.

Logical IPv4 Topology

A logical network topology illustrates how devices are logically connected to the network. This refers to how devices transfer data across the network when communicating with other devices. Symbols are used to represent network components, such as routers, switches, servers, and hosts. Additionally, connections between multiple sites may be shown, but do not represent actual physical locations.

Information recorded on a logical network topology may include the following:

- Device identifiers

- IP addresses and prefix lengths

- Interface identifiers

- Routing protocols / static routes

- Layer 2 information (i.e., VLANs, trunks, EtherChannels)

The figure displays a sample logical IPv4 network topology.

Logical IPv6 Topology

Although IPv6 addresses could also be displayed in the same IPv4 logical topology, for the sake of clarity, we have created a separate logical IPv6 network topology.

The figure displays a sample IPv6 logical topology.

Refer to
Interactive Graphic
in online course

Refer to
Online Course
for Illustration

Network Device Documentation - 12.1.3

Network device documentation should contain accurate, up-to-date records of the network hardware and software. Documentation should include all pertinent information about the network devices.

Many organizations create documents with tables or spreadsheets to capture relevant device information.

Click each button for examples of router, switch, and end device documentation.

Router Device Documentation

The table displays sample network device documentation for two interconnecting routers.

LAN Switch Device Documentation

This table displays sample device documentation for a LAN switch.

End-system Documentation Files

End-system documentation focuses on the hardware and software used in servers, network management consoles, and user workstations. An incorrectly configured end-system can have a negative impact on the overall performance of a network. For this reason, having access to end-system device documentation can be very useful when troubleshooting.

This table displays a sample of information that could be recorded in an end-system device document.

Establish a Network Baseline - 12.1.4

The purpose of network monitoring is to watch network performance in comparison to a predetermined baseline. A baseline is used to establish normal network or system performance to determine the "personality" of a network under normal conditions.

Establishing a network performance baseline requires collecting performance data from the ports and devices that are essential to network operation.

A network baseline should answer the following questions:

- How does the network perform during a normal or average day?
- Where are the most errors occurring?
- What part of the network is most heavily used?
- What part of the network is least used?
- Which devices should be monitored and what alert thresholds should be set?
- Can the network meet the identified policies?

Measuring the initial performance and availability of critical network devices and links allows a network administrator to determine the difference between abnormal behavior and proper network performance, as the network grows, or traffic patterns change. The baseline also provides insight into whether the current network design can meet business requirements. Without a baseline, no standard exists to measure the optimum nature of network traffic and congestion levels.

Analysis after an initial baseline also tends to reveal hidden problems. The collected data shows the true nature of congestion or potential congestion in a network. It may also reveal areas in the network that are underutilized, and quite often can lead to network redesign efforts, based on quality and capacity observations.

The initial network performance baseline sets the stage for measuring the effects of network changes and subsequent troubleshooting efforts. Therefore, it is important to plan for it carefully.

Step 1 - Determine What Types of Data to Collect - 12.1.5

When conducting the initial baseline, start by selecting a few variables that represent the defined policies. If too many data points are selected, the amount of data can be overwhelming, making analysis of the collected data difficult. Start out simply and fine-tune along the way. Some good starting variables are interface utilization and CPU utilization.

Refer to
Online Course
for Illustration

Step 2 - Identify Devices and Ports of Interest - 12.1.6

Use the network topology to identify those devices and ports for which performance data should be measured. Devices and ports of interest include the following:

- Network device ports that connect to other network devices

- Servers

- Key users

- Anything else considered critical to operations

A logical network topology can be useful in identifying key devices and ports to monitor. In the figure, the network administrator has highlighted the devices and ports of interest to monitor during the baseline test.

The devices of interest include PC1 (the Admin terminal), and the two servers (i.e., Srv1 and Svr2). The ports of interest typically include router interfaces and key ports on switches.

By shortening the list of ports that are polled, the results are concise, and the network management load is minimized. Remember that an interface on a router or switch can be a virtual interface, such as a switch virtual interface (SVI).

Refer to
Online Course
for Illustration

Step 3 - Determine the Baseline Duration - 12.1.7

The length of time and the baseline information being gathered must be long enough to determine a "normal" picture of the network. It is important that daily trends of network traffic are monitored. It is also important to monitor for trends that occur over a longer period, such as weekly or monthly. For this reason, when capturing data for analysis, the period specified should be, at a minimum, seven days long.

The figure displays examples of several screenshots of CPU utilization trends captured over a daily, weekly, monthly, and yearly period.

In this example, notice that the work week trends are too short to reveal the recurring utilization surge every weekend on Saturday evening, when a database backup operation consumes network bandwidth. This recurring pattern is revealed in the monthly trend. A yearly trend as shown in the example may be too long of a duration to provide meaningful baseline performance details. However, it may help identify long term patterns which should be analyzed further.

Typically, a baseline needs to last no more than six weeks, unless specific long-term trends need to be measured. Generally, a two-to-four-week baseline is adequate.

Baseline measurements should not be performed during times of unique traffic patterns, because the data would provide an inaccurate picture of normal network operations. Conduct an annual analysis of the entire network, or baseline different sections of the network on a rotating basis. Analysis must be conducted regularly to understand how the network is affected by growth and other changes.

Data Measurement - 12.1.8

When documenting the network, it is often necessary to gather information directly from routers and switches. Obvious useful network documentation commands include **ping**, **traceroute**, and **telnet**, as well as **show** commands.

The table lists some of the most common Cisco IOS commands used for data collection.

Command	Description
`show version`	Displays uptime, version information for device software and hardware.
`show ip interface [brief]` `show ipv6 interface [brief]`	• Displays all the configuration options that are set on an interface. • Use the **brief** keyword to only display up/down status of IP interfaces and the IP address of each interface.
`show interfaces`	• Displays detailed output for each interface. • To display detailed output for only a single interface, include the interface type and number in the command (e.g. Gigabit Ethernet 0/0/0).
`show ip route` `show ipv6 route`	• Displays the routing table content listing directly connected networks and learned remote networks. • Append **static**, **eigrp**, or **ospf** to display those routes only.
`show cdp neighbors detail`	Displays detailed information about directly connected Cisco neighbor devices.
`show arp` `show ipv6 neighbors`	Displays the contents of the ARP table (IPv4) and the neighbor table (IPv6).
`show running-config`	Displays current configuration.
`show vlan`	Displays the status of VLANs on a switch.
`show port`	Displays the status of ports on a switch.
`show tech-support`	• This command is useful for collecting a large amount of information about the device for troubleshooting purposes. • It executes multiple show commands which can be provided to technical support representatives when reporting a problem

Manual data collection using **show** commands on individual network devices is extremely time consuming and is not a scalable solution. Manual collection of data should be reserved for smaller networks or limited to mission-critical network devices. For simpler network designs, baseline tasks typically use a combination of manual data collection and simple network protocol inspectors.

Sophisticated network management software is typically used to baseline large and complex networks. These software packages enable administrators to automatically create and review reports, compare current performance levels with historical observations, automatically identify performance problems, and create alerts for applications that do not provide expected levels of service.

Establishing an initial baseline or conducting a performance-monitoring analysis may require many hours or days to accurately reflect network performance. Network management software or protocol inspectors and sniffers often run continuously over the course of the data collection process.

Go to the online course to take the quiz and exam.

Check Your Understanding - Network Documentation - 12.1.9

Troubleshooting Process - 12.2

Refer to **Online Course** for Illustration

General Troubleshooting Procedures - 12.2.1

Troubleshooting can be time consuming because networks differ, problems differ, and troubleshooting experience varies. However, experienced administrators know that using a structured troubleshooting method will shorten overall troubleshooting time.

Therefore, the troubleshooting process should be guided by structured methods. This requires well defined and documented troubleshooting procedures to minimize wasted time associated with erratic hit-and-miss troubleshooting. However, these methods are not static. The troubleshooting steps taken to solve a problem are not always the same or executed in the exact same order.

There are several troubleshooting processes that can be used to solve a problem. The figure displays the logic flowchart of a simplified three-stage troubleshooting process. However, a more detailed process may be more helpful to solve a network problem.

Refer to **Online Course** for Illustration

Seven-Step Troubleshooting Process - 12.2.2

Refer to **Interactive Graphic** in online course

The figure displays a more detailed seven-step troubleshooting process. Notice how some steps interconnect. This is because, some technicians may be able to jump between steps based on their level of experience.

Click each button for a detailed description of the steps to solve a network problem.

Define the Problem

The goal of this stage is to verify that there is a problem and then properly define what the problem is. Problems are usually identified by a symptom (e.g., the network is slow or has stopped working). Network symptoms may appear in many different forms, including alerts from the network management system, console messages, and user complaints.

While gathering symptoms, it is important to ask questions and investigate the issue in order to localize the problem to a smaller range of possibilities. For example, is the problem restricted to a single device, a group of devices, or an entire subnet or network of devices?

In an organization, problems are typically assigned to network technicians as trouble tickets. These tickets are created using trouble ticketing software that tracks the progress of each ticket. Trouble ticketing software may also include a self-service user portal to submit tickets, access to a searchable trouble tickets knowledge base, remote control capabilities to solve end-user issues, and more.

Gather Information

In this step, targets (i.e., hosts, devices) to be investigated must be identified, access to the target devices must be obtained, and information gathered. During this step, the technician may gather and document more symptoms, depending on the characteristics that are identified.

If the problem is outside the boundary of the organization's control (e.g., lost internet connectivity outside of the autonomous system), contact an administrator for the external system before gathering additional network symptoms.

Analyze information

Possible causes must be identified. The gathered information is interpreted and analyzed using network documentation, network baselines, searching organizational knowledge bases, searching the internet, and talking with other technicians.

Eliminate Possible Causes

If multiple causes are identified, then the list must be reduced by progressively eliminating possible causes to eventually identify the most probable cause. Troubleshooting experience is extremely valuable to quickly eliminate causes and identify the most probable cause.

Propose Hypothesis

When the most probable cause has been identified, a solution must be formulated. At this stage, troubleshooting experience is very valuable when proposing a plan.

Test Hypothesis

Before testing the solution, it is important to assess the impact and urgency of the problem. For instance, could the solution have an adverse effect on other systems or processes? The severity of the problem should be weighed against the impact of the solution. For example, if a critical server or router must be offline for a significant amount of time, it may be better to wait until the end of the workday to implement the fix. Sometimes, a workaround can be created until the actual problem is resolved.

Create a rollback plan identifying how to quickly reverse a solution. This may prove to be necessary if the solution fails.

Implement the solution and verify that it has solved the problem. Sometimes a solution introduces an unexpected problem. Therefore, it is important that a solution be thoroughly verified before proceeding to the next step.

If the solution fails, the attempted solution is documented and the changes are removed. The technician must now go back to the Gathering Information step and isolate the issue.

Solve the problem

When the problem is solved, inform the users and anyone involved in the troubleshooting process that the problem has been resolved. Other IT team members should be informed of the solution. Appropriate documentation of the cause and the fix will assist other support technicians in preventing and solving similar problems in the future.

Question End Users - 12.2.3

Many network problems are initially reported by an end user. However, the information provided is often vague or misleading. For example, users often report problems such as "the network is down", "I cannot access my email", or "my computer is slow".

In most cases, additional information is required to fully understand a problem. This usually involves interacting with the affected user to discover the "who", "what", and "when" of the problem.

The following recommendations should be employed when communicate with user:

- Speak at a technical level they can understand and avoid using complex terminology.

- Always listen or read carefully what the user is saying. Taking notes can be helpful when documenting a complex problem.

- Always be considerate and empathize with users while letting them know you will help them solve their problem. Users reporting a problem may be under stress and anxious to resolve the problem as quickly as possible.

When interviewing the user, guide the conversation and use effective questioning techniques to quickly ascertain the problem. For instance, use open questions (i.e., requires detailed response) and closed questions (i.e., yes, no, or single word answers) to discover important facts about the network problem.

The table provides some questioning guidelines and sample open ended end-user questions.

When done interviewing the user, repeat your understanding of the problem to the user to ensure that you both agree on the problem being reported.

Guidelines	Example Open Ended End-User Questions
Ask pertinent questions.	• What does not work? • What exactly is the problem? • What are you trying to accomplish?
Determine the scope of the problem.	• Who does this issue affect? Is it just you or others? • What device is this happening on?
Determine when the problem occurred / occurs.	• When exactly does the problem occur? • When was the problem first noticed? • Were there any error message(s) displayed?
Determine if the problem is constant or intermittent.	• Can you reproduce the problem? • Can you send me a screenshot or video of the problem?

Guidelines	Example Open Ended End-User Questions
Determine if anything has changed.	What has changed since the last time it did work?
Use questions to eliminate or discover possible problems.	• What works? • What does not work?

Gather Information - 12.2.4

To gather symptoms from suspected networking device, use Cisco IOS commands and other tools such as packet captures and device logs.

The table describes common Cisco IOS commands used to gather the symptoms of a network problem.

Command	Description
`ping {host	ip-address}`	• Sends an echo request packet to an address, then waits for a reply • The *host* or *ip-address* variable is the IP alias or IP address of the target system
`traceroute destination`	• Identifies the path a packet takes through the networks • The *destination* variable is the hostname or IP address of the target system
`telnet {host	ip-address}`	• Connects to an IP address using the Telnet application • Use SSH whenever possible instead of Telnet
`ssh -l user-id ip-address`	• Connects to an IP address using SSH • SSH is more secure than Telnet
`show ip interface brief` `show ipv6 interface brief`	• Displays a summary status of all interfaces on a device • Useful for quickly identifying IP addressing on all interfaces.
`show ip route` `show ipv6 route`	Displays the current IPv4 and IPv6 routing tables, which contains the routes to all known network destinations
`show protocols`	Displays the configured protocols and shows the global and interface-specific status of any configured Layer 3 protocol
`debug`	Displays a list of options for enabling or disabling debugging events

Note: Although the **debug** command is an important tool for gathering symptoms, it generates a large amount of console message traffic and the performance of a network device can be noticeably affected. If the **debug** must be performed during normal working hours, warn network users that a troubleshooting effort is underway, and that network performance may be affected. Remember to disable debugging when you are done.

Refer to
Online Course
for Illustration

Troubleshooting with Layered Models - 12.2.5

The OSI and TCP/IP models can be applied to isolate network problems when trouble-shooting. For example, if the symptoms suggest a physical connection problem, the network technician can focus on troubleshooting the circuit that operates at the physical layer.

The figure shows some common devices and the OSI layers that must be examined during the troubleshooting process for that device.

Notice that routers and multilayer switches are shown at Layer 4, the transport layer. Although routers and multilayer switches usually make forwarding decisions at Layer 3, ACLs on these devices can be used to make filtering decisions using Layer 4 information.

Refer to
Interactive Graphic
in online course

Structured Troubleshooting Methods - 12.2.6

Refer to
Online Course
for Illustration

There are several structured troubleshooting approaches that can be used. Which one to use will depend on the situation. Each approach has its advantages and disadvantages. This topic describes methods and provides guidelines for choosing the best method for a specific situation.

Click each button for a description of the different troubleshooting approaches that can be used.

Bottom-Up

In bottom-up troubleshooting, you start with the physical components of the network and move up through the layers of the OSI model until the cause of the problem is identified, as shown in the figure.

Bottom-up troubleshooting is a good approach to use when the problem is suspected to be a physical one. Most networking problems reside at the lower levels, so implementing the bottom-up approach is often effective.

The disadvantage with the bottom-up troubleshooting approach is it requires that you check every device and interface on the network until the possible cause of the problem is found. Remember that each conclusion and possibility must be documented so there can be a lot of paper work associated with this approach. A further challenge is to determine which devices to start examining first.

Top-Down

In the figure, top-down troubleshooting starts with the end-user applications and moves down through the layers of the OSI model until the cause of the problem has been identified.

End-user applications of an end system are tested before tackling the more specific networking pieces. Use this approach for simpler problems, or when you think the problem is with a piece of software.

The disadvantage with the top-down approach is it requires checking every network application until the possible cause of the problem is found. Each conclusion and possibility must be documented. The challenge is to determine which application to start examining first.

Divide-and-Conquer

The figure shows the divide-and-conquer approach to troubleshooting a networking problem.

The network administrator selects a layer and tests in both directions from that layer.

In divide-and-conquer troubleshooting, you start by collecting user experiences of the problem, document the symptoms and then, using that information, make an informed guess as to which OSI layer to start your investigation. When a layer is verified to be functioning properly, it can be assumed that the layers below it are functioning. The administrator can work up the OSI layers. If an OSI layer is not functioning properly, the administrator can work down the OSI layer model.

For example, if users cannot access the web server, but they can ping the server, then the problem is above Layer 3. If pinging the server is unsuccessful, then the problem is likely at a lower OSI layer.

Follow-the-Path

This is one of the most basic troubleshooting techniques. The approach first discovers the actual traffic path all the way from source to destination. The scope of troubleshooting is reduced to just the links and devices that are in the forwarding path. The objective is to eliminate the links and devices that are irrelevant to the troubleshooting task at hand. This approach usually complements one of the other approaches.

Substitution

This approach is also called swap-the-component because you physically swap the problematic device with a known, working one. If the problem is fixed, then the problem is with the removed device. If the problem remains, then the cause may be elsewhere.

In specific situations, this can be an ideal method for quick problem resolution, such as with a critical single point of failure. For example, a border router goes down. It may be more beneficial to simply replace the device and restore service, rather than to troubleshoot the issue.

If the problem lies within multiple devices, it may not be possible to correctly isolate the problem.

Comparison

This approach is also called the spot-the-differences approach and attempts to resolve the problem by changing the nonoperational elements to be consistent with the working ones. You compare configurations, software versions, hardware, or other device properties, links, or processes between working and nonworking situations and spot significant differences between them.

The weakness of this method is that it might lead to a working solution, without clearly revealing the root cause of the problem.

Educated Guess

This approach is also called the shoot-from-the-hip troubleshooting approach. This is a less-structured troubleshooting method that uses an educated guess based on the symptoms of the problem. Success of this method varies based on your troubleshooting experience and ability. Seasoned technicians are more successful because they can rely on their extensive knowledge and experience to decisively isolate and solve network issues. With a less-experienced network administrator, this troubleshooting method may be more like random troubleshooting.

Refer to **Online Course** for Illustration

Guidelines for Selecting a Troubleshooting Method - 12.2.7

To quickly resolve network problems, take the time to select the most effective network troubleshooting method.

The figure illustrates which method could be used when a certain type of problem is discovered.

For instance, software problems are often solved using a top-down approach while hardware-based problem are solved using the bottom-up approach. New problems may be solved by an experienced technician using the divide-and-conquer method. Otherwise, the bottom-up approach may be used.

Troubleshooting is a skill that is developed by doing it. Every network problem you identify and solve gets added to your skill set.

Go to the online course to take the quiz and exam.

Check Your Understanding - Troubleshooting Process - 12.2.8

Troubleshooting Tools - 12.3

Refer to **Interactive Graphic** in online course

Software Troubleshooting Tools - 12.3.1

As you know, networks are made up of software and hardware. Therefore, both software and hardware have their respective tools for troubleshooting. This topic discusses the troubleshooting tools available for both.

A wide variety of software and hardware tools are available to make troubleshooting easier. These tools may be used to gather and analyze symptoms of network problems. They often provide monitoring and reporting functions that can be used to establish the network baseline.

Click each button for a detailed description of common software troubleshooting tools.

Network Management System Tools

Network management system (NMS) tools include device-level monitoring, configuration, and fault-management tools. These tools can be used to investigate and correct network problems. Network monitoring software graphically displays a physical view of network devices, allowing network managers to monitor remote devices continuously and automatically. Device management software provides dynamic device status, statistics, and configuration information for key network devices. Search the internet for "NMS Tools" for more information.

Knowledge Bases

Online network device vendor knowledge bases have become indispensable sources of information. When vendor-based knowledge bases are combined with internet search engines, a network administrator has access to a vast pool of experience-based information.

For example, the **Cisco Tools & Resources** page can be found at http://www.cisco.com under the **Support** menu. This page provides tools that can be used for Cisco hardware and software.

Baselining Tools

Many tools for automating the network documentation and baselining process are available. Baselining tools help with common documentation tasks. For example, they can draw network diagrams, help keep network software and hardware documentation up-to-date, and help to cost-effectively measure baseline network bandwidth use. Search the internet for "Network Performance Monitoring Tools" for more information.

Refer to
Online Course
for Illustration

Protocol Analyzers - 12.3.2

Protocol analyzers can investigate packet content while flowing through the network. A protocol analyzer decodes the various protocol layers in a recorded frame and presents this information in a relatively easy to use format. The figure shows a screen capture of the Wireshark protocol analyzer.

The information displayed by a protocol analyzer includes the physical layer bit data, data link layer information, protocols, and descriptions for each frame. Most protocol analyzers can filter traffic that meets certain criteria so that all traffic to and from a device can be captured. Protocol analyzers such as Wireshark can help troubleshoot network performance problems. It is important to have both a good understanding of TCP/IP and how to use a protocol analyzer to inspect information at each TCP/IP layer.

Refer to
Interactive Graphic
in online course

Hardware Troubleshooting Tools - 12.3.3

There are multiple types of hardware troubleshooting tools.

Click each button for a detailed description of common hardware troubleshooting tools.

Refer to
Online Course
for Illustration

Digital Multimeters

Digital multimeters (DMMs), such as the Fluke 179 shown in the figure, are test instruments that are used to directly measure electrical values of voltage, current, and resistance.

In network troubleshooting, most tests that would need a multimeter involve checking power supply voltage levels and verifying that network devices are receiving power.

Cable Testers

Cable testers are specialized, handheld devices designed for testing the various types of data communication cabling. The figure displays the Fluke LinkRunner AT Network Auto-Tester.

Cable testers can be used to detect broken wires, crossed-over wiring, shorted connections, and improperly paired connections. These devices can be inexpensive continuity testers, moderately priced data cabling testers, or expensive time-domain reflectometers (TDRs). TDRs are used to pinpoint the distance to a break in a cable. These devices send signals along the cable and wait for them to be reflected. The time between sending the signal and receiving it back is converted into a distance measurement. The TDR function is normally packaged with data cabling testers. TDRs used to test fiber-optic cables are known as optical time-domain reflectometers (OTDRs).

Cable Analyzers

Cable analyzers, such as the Fluke DTX Cable Analyzer in the figure, are multifunctional handheld devices that are used to test and certify copper and fiber cables for different services and standards.

The more sophisticated tools include advanced troubleshooting diagnostics that measure the distance to a performance defect such as near-end crosstalk (NEXT) or return loss (RL), identify corrective actions, and graphically display crosstalk and impedance behavior. Cable analyzers also typically include PC-based software. After field data is collected, the data from the handheld device can be uploaded so that the network administrator can create up-to-date reports.

Portable Network Analyzers

Portable devices like the Fluke OptiView, shown in the figure, are used for troubleshooting switched networks and VLANs.

By plugging the network analyzer in anywhere on the network, a network engineer can see the switch port to which the device is connected, and the average and peak utilization. The analyzer can also be used to discover VLAN configuration, identify top network talkers (hosts generating the most traffic), analyze network traffic, and view interface details. The device can typically output to a PC that has network monitoring software installed for further analysis and troubleshooting.

Cisco Prime NAM

The Cisco Prime Network Analysis Module (NAM) portfolio, shown in the figure, includes hardware and software for performance analysis in switching and routing environments. It includes an embedded browser-based interface that generates reports on the traffic that consumes critical network resources. In addition, the NAM can capture and decode packets and track response times to pinpoint an application problem to a network or server.

Syslog Server as a Troubleshooting Tool - 12.3.4

Syslog is a simple protocol used by an IP device known as a syslog client, to send text-based log messages to another IP device, the syslog server. Syslog is currently defined in RFC 5424.

Implementing a logging facility is an important part of network security and for network troubleshooting. Cisco devices can log information regarding configuration changes, ACL violations, interface status, and many other types of events. Cisco devices can send log messages to several different facilities. Event messages can be sent to one or more of the following:

- **Console** - Console logging is on by default. Messages log to the console and can be viewed when modifying or testing the router or switch using terminal emulation software while connected to the console port of the network device.

- **Terminal lines** - Enabled EXEC sessions can be configured to receive log messages on any terminal lines. Like console logging, this type of logging is not stored by the network device and, therefore, is only valuable to the user on that line.

- **Buffered logging** - Buffered logging is a little more useful as a troubleshooting tool because log messages are stored in memory for a time. However, log messages are cleared when the device is rebooted.

- **SNMP traps** - Certain thresholds can be preconfigured on routers and other devices. Router events, such as exceeding a threshold, can be processed by the router and forwarded as SNMP traps to an external SNMP network management station. SNMP traps are a viable security logging facility but require the configuration and maintenance of an SNMP system.

- **Syslog** - Cisco routers and switches can be configured to forward log messages to an external syslog service. This service can reside on any number of servers or workstations, including Microsoft Windows and Linux-based systems. Syslog is the most popular message logging facility, because it provides long-term log storage capabilities and a central location for all router messages.

Cisco IOS log messages fall into one of eight levels, as shown in the table.

	Level	Keyword	Description	Definition
Highest Level	0	Emergencies	System is unusable	LOG_EMERG
	1	Alerts	Immediate action is needed	LOG_ALERT
	2	Critical	Critical conditions exist	LOG_CRIT
	3	Errors	Error conditions exist	LOG_ERR
	4	Warnings	Warning conditions exist	LOG_WARNING
Lowest Level	5	Notifications	Normal (but significant) condition	LOG_NOTICE
	6	Informational	Informational messages only	LOG_NFO
	7	Debugging	Debugging messages	LOG_DEBUG

The lower the level number, the higher the severity level. By default, all messages from level 0 to 7 are logged to the console. While the ability to view logs on a central syslog server is helpful in troubleshooting, sifting through a large amount of data can be an overwhelming task. The **logging trap** *level* command limits messages logged to the syslog server based on severity. The level is the name or number of the severity level. Only messages equal to or numerically lower than the specified level are logged.

In the command output, system messages from level 0 (emergencies) to 5 (notifications) are sent to the syslog server at 209.165.200.225.

```
R1(config)# logging host 209.165.200.225

R1(config)# logging trap notifications

R1(config)# logging on

R1(config)#
```

Go to the online
course to take the
quiz and exam.

Check Your Understanding - Troubleshooting Tools - 12.3.5

Symptoms and Causes of Network Problems - 12.4

Refer to
Online Course
for Illustration

Physical Layer Troubleshooting - 12.4.1

Now that you have your documentation, some knowledge of troubleshooting methods and the software and hardware tools to use to diagnose problems, you are ready to start troubleshooting! This topic covers the most common issues that you will find when troubleshooting a network.

Issues on a network often present as performance problems. Performance problems mean that there is a difference between the expected behavior and the observed behavior, and the system is not functioning as could be reasonably expected. Failures and suboptimal conditions at the physical layer not only inconvenience users but can impact the productivity of the entire company. Networks that experience these kinds of conditions usually shut down. Because the upper layers of the OSI model depend on the physical layer to function, a network administrator must have the ability to effectively isolate and correct problems at this layer.

The figure summarizes the symptoms and causes of physical layer network problems.

The table lists common symptoms of physical layer network problems.

Symptom	Description
Performance lower than baseline	• Requires previous baselines for comparison. • The most common reasons for slow or poor performance include overloaded or underpowered servers, unsuitable switch or router configurations, traffic congestion on a low-capacity link, and chronic frame loss.
Loss of connectivity	• Loss of connectivity could be due to a failed or disconnected cable. • Can be verified using a simple ping test. • Intermittent connectivity loss can indicate a loose or oxidized connection.
Network bottlenecks or congestion	• If a router, interface, or cable fails, routing protocols may redirect traffic to other routes that are not designed to carry the extra capacity. • This can result in congestion or bottlenecks in parts of the network.
High CPU utilization rates	• High CPU utilization rates are a symptom that a device, such as a router, switch, or server, is operating at or exceeding its design limits. • If not addressed quickly, CPU overloading can cause a device to shut down or fail.
Console error messages	• Error messages reported on the device console could indicate a physical layer problem. • Console messages should be logged to a central syslog server.

The next table lists issues that commonly cause network problems at the physical layer.

Problem Cause	Description
Power-related	• This is the most fundamental reason for network failure. • Check the operation of the fans and ensure that the chassis intake and exhaust vents are clear. • If other nearby units have also powered down, suspect a power failure at the main power supply.
Hardware faults	• Faulty network interface cards (NICs) can be the cause of network transmission errors due to late collisions, short frames, and jabber. • Jabber is often defined as the condition in which a network device continually transmits random, meaningless data onto the network. • Other likely causes of jabber are faulty or corrupt NIC driver files, bad cabling, or grounding problems.
Cabling faults	• Many problems can be corrected by simply reseating cables that have become partially disconnected. • When performing a physical inspection, look for damaged cables, improper cable types, and poorly crimped RJ-45 connectors. • Suspect cables should be tested or exchanged with a known functioning cable.
Attenuation	• Attenuation can be caused if a cable length exceeds the design limit for the media, or when there is a poor connection resulting from a loose cable, or dirty or oxidized contacts. • If attenuation is severe, the receiving device cannot always successfully distinguish one bit in the data stream from another bit.
Noise	• Local electromagnetic interference (EMI) is commonly known as noise. • Noise can be generated by many sources, such as FM radio stations, police radio, building security, and avionics for automated landing, crosstalk (noise induced by other cables in the same pathway or adjacent cables), nearby electric cables, devices with large electric motors, or anything that includes a transmitter more powerful than a cell phone.
Interface configuration errors	• Many things can be misconfigured on an interface to cause it to go down, such as incorrect clock rate, incorrect clock source, and interface not being turned on. • This causes a loss of connectivity with attached network segments.
Exceeding design limits	• A component may be operating sub-optimally at the physical layer because it is being utilized beyond specifications or configured capacity. • When troubleshooting this type of problem, it becomes evident that resources for the device are operating at or near the maximum capacity and there is an increase in the number of interface errors.
CPU overload	• Symptoms include processes with high CPU utilization percentages, input queue drops, slow performance, SNMP timeouts, no remote access, or services such as DHCP, Telnet, and ping are slow or fail to respond. • On a switch the following could occur: spanning tree reconvergence, EtherChannel links bounce, UDLD flapping, IP SLAs failures. • For routers, there could be no routing updates, route flapping, or HSRP flapping. • One of the causes of CPU overload in a router or switch is high traffic. • If one or more interfaces are regularly overloaded with traffic, consider redesigning the traffic flow in the network or upgrading the hardware.

Refer to
Online Course
for Illustration

Data Link Layer Troubleshooting - 12.4.2

Troubleshooting Layer 2 problems can be a challenging process. The configuration and operation of these protocols are critical to creating a functional, well-tuned network. Layer 2 problems cause specific symptoms that, when recognized, will help identify the problem quickly.

The figure summarizes the symptoms and causes of data link layer network problems.

The table lists common symptoms of data link layer network problems.

Symptom	Description
No functionality or connectivity at the network layer or above	• Some Layer 2 problems can stop the exchange of frames across a link, while others only cause network performance to degrade.
Network is operating below baseline performance levels	• There are two distinct types of suboptimal Layer 2 operation that can occur in a network. • First, the frames take a suboptimal path to their destination but do arrive causing the network to experience unexpected high-bandwidth usage on links. • Second, some frames are dropped as identified through error counter statistics and console error messages that appear on the switch or router. • An extended or continuous ping can help reveal if frames are being dropped.
Excessive broadcasts	• Operating systems use broadcasts and multicasts extensively to discover network services and other hosts. • Generally, excessive broadcasts are the result of a poorly programmed or configured applications, a large Layer 2 broadcast domain, or an underlying network problem (e.g., STP loops or route flapping).
Console messages	• A router recognizes that a Layer 2 problem has occurred and sends alert messages to the console. • Typically, a router does this when it detects a problem with interpreting incoming frames (encapsulation or framing problems) or when keepalives are expected but do not arrive. • The most common console message that indicates a Layer 2 problem is a line protocol down message

The table lists issues that commonly cause network problems at the data link layer.

Problem Cause	Description
Encapsulation errors	• An encapsulation error occurs because the bits placed in a field by the sender are not what the receiver expects to see. • This condition occurs when the encapsulation at one end of a WAN link is configured differently from the encapsulation used at the other end.

Problem Cause	Description
Address mapping errors	• In topologies, such as point-to-multipoint or broadcast Ethernet, it is essential that an appropriate Layer 2 destination address be given to the frame. This ensures its arrival at the correct destination.
	• To achieve this, the network device must match a destination Layer 3 address with the correct Layer 2 address using either static or dynamic maps.
	• In a dynamic environment, the mapping of Layer 2 and Layer 3 information can fail because devices may have been specifically configured not to respond to ARP requests, the Layer 2 or Layer 3 information that is cached may have physically changed, or invalid ARP replies are received because of a misconfiguration or a security attack.
Framing errors	• Frames usually work in groups of 8-bit bytes.
	• A framing error occurs when a frame does not end on an 8-bit byte boundary.
	• When this happens, the receiver may have problems determining where one frame ends, and another frame starts.
	• Too many invalid frames may prevent valid keepalives from being exchanged.
	• Framing errors can be caused by a noisy serial line, an improperly designed cable (too long or not properly shielded), faulty NIC, duplex mismatch, or an incorrectly configured channel service unit (CSU) line clock.
STP failures or loops	• The purpose of the Spanning Tree Protocol (STP) is to resolve a redundant physical topology into a tree-like topology by blocking redundant ports.
	• Most STP problems are related to forwarding loops that occur when no ports in a redundant topology are blocked and traffic is forwarded in circles indefinitely.
	• This causes excessive flooding because of a high rate of STP topology changes.
	• A topology change should be a rare event in a well-configured network.
	• When a link between two switches goes up or down, there is eventually a topology change when the STP state of the port is changing to or from forwarding.
	• However, when a port is flapping (oscillating between up and down states), this causes repetitive topology changes and flooding, or slow STP convergence or re-convergence.
	• This can be caused by a mismatch between the real and documented topology, a configuration error, such as an inconsistent configuration of STP timers, an overloaded switch CPU during convergence, or a software defect.

Network Layer Troubleshooting - 12.4.3

Refer to
Online Course
for Illustration

Network layer problems include any problem that involves a Layer 3 protocol, such as IPv4, IPv6, EIGRP, OSPF, etc. The figure summarizes the symptoms and causes of network layer network problems.

The table lists common symptoms of network layer network problems.

Symptom	Description
Network failure	• Network failure is when the network is nearly or completely non-functional, affecting all users and applications on the network. • These failures are usually noticed quickly by users and network administrators and are obviously critical to the productivity of a company.
Suboptimal performance	• Network optimization problems usually involve a subset of users, applications, destinations, or a type of traffic. • Optimization issues can be difficult to detect and even harder to isolate and diagnose. • This is because they usually involve multiple layers, or even a single host computer. • Determining that the problem is a network layer problem can take time.

In most networks, static routes are used in combination with dynamic routing protocols. Improper configuration of static routes can lead to less than optimal routing. In some cases, improperly configured static routes can create routing loops which make parts of the network unreachable.

Troubleshooting dynamic routing protocols requires a thorough understanding of how the specific routing protocol functions. Some problems are common to all routing protocols, while other problems are particular to the individual routing protocol.

There is no single template for solving Layer 3 problems. Routing problems are solved with a methodical process, using a series of commands to isolate and diagnose the problem.

The table lists areas to explore when diagnosing a possible problem involving routing protocols.

Problem Cause	Description
General network issues	• Often a change in the topology, such as a down link, may have effects on other areas of the network that might not be obvious at the time. • This may include the installation of new routes, static or dynamic, or removal of other routes. • Determine whether anything in the network has recently changed, and if there is anyone currently working on the network infrastructure.
Connectivity issues	• Check for any equipment and connectivity problems, including power problems such as outages and environmental problems (for example, overheating). • Also check for Layer 1 problems, such as cabling problems, bad ports, and ISP problems.
Routing table	• Check the routing table for anything unexpected, such as missing routes or unexpected routes. • Use **debug** commands to view routing updates and routing table maintenance.
Neighbor issues	If the routing protocol establishes an adjacency with a neighbor, check to see if there are any problems with the routers forming neighbor adjacencies.
Topology database	If the routing protocol uses a topology table or database, check the table for anything unexpected, such as missing entries or unexpected entries.

Refer to
Online Course
for Illustration

Transport Layer Troubleshooting - ACLs - 12.4.4

Network problems can arise from transport layer problems on the router, particularly at the edge of the network where traffic is examined and modified. For instance, both access control lists (ACLs) and Network Address Translation (NAT) operate at the network layer and may involve operations at the transport layer, as shown in the figure.

The most common issues with ACLs are caused by improper configuration, as shown in the figure.

Problems with ACLs may cause otherwise working systems to fail. The table lists areas where misconfigurations commonly occur.

Misconfigurations	Description
Selection of traffic flow	• Traffic is defined by both the router interface through which the traffic is traveling and the direction in which this traffic is traveling. • An ACL must be applied to the correct interface, and the correct traffic direction must be selected to function properly.
Order of access control entries	• The entries in an ACL should be from specific to general. • Although an ACL may have an entry to specifically permit a type of traffic flow, packets never match that entry if they are being denied by another entry earlier in the list. • If the router is running both ACLs and NAT, the order in which each of these technologies is applied to a traffic flow is important. • Inbound traffic is processed by the inbound ACL before being processed by outside-to-inside NAT. • Outbound traffic is processed by the outbound ACL after being processed by inside-to-outside NAT.
Implicit deny any	• When high security is not required on the ACL, this implicit access control element can be the cause of an ACL misconfiguration.
Addresses and IPv4 wildcard masks	• Complex IPv4 wildcard masks provide significant improvements in efficiency but are more subject to configuration errors. • An example of a complex wildcard mask is using the IPv4 address 10.0.32.0 and wildcard mask 0.0.32.15 to select the first 15 host addresses in either the 10.0.0.0 network or the 10.0.32.0 network.
Selection of transport layer protocol	• When configuring ACLs, it is important that only the correct transport layer protocols be specified. • Many network administrators, when unsure whether a type of traffic flow uses a TCP port or a UDP port, configure both. • Specifying both opens a hole through the firewall, possibly giving intruders an avenue into the network. • It also introduces an extra element into the ACL, so the ACL takes longer to process, introducing more latency into network communications.

Misconfigurations	Description
Source and destination ports	• Properly controlling the traffic between two hosts requires symmetric access control elements for inbound and outbound ACLs. • Address and port information for traffic generated by a replying host is the mirror image of address and port information for traffic generated by the initiating host.
Use of the established keyword	• The **established** keyword increases the security provided by an ACL. • However, if the keyword is applied incorrectly, unexpected results may occur.
Uncommon protocols	• Misconfigured ACLs often cause problems for protocols other than TCP and UDP. • Uncommon protocols that are gaining popularity are VPN and encryption protocols.

The **log** keyword is a useful command for viewing ACL operation on ACL entries. This keyword instructs the router to place an entry in the system log whenever that entry condition is matched. The logged event includes details of the packet that matched the ACL element. The **log** keyword is especially useful for troubleshooting and provides information on intrusion attempts being blocked by the ACL.

Refer to **Online Course** for Illustration

Transport Layer Troubleshooting - NAT for IPv4 - 12.4.5

There are several problems with NAT, such as not interacting with services like DHCP and tunneling. These can include misconfigured NAT inside, NAT outside, or ACLs. Other issues include interoperability with other network technologies, especially those that contain or derive information from host network addressing in the packet.

The figure summarizes common interoperability areas with NAT.

The table lists common interoperability areas with NAT.

Symptom	Description
BOOTP and DHCP	• Both protocols manage the automatic assignment of IPv4 addresses to clients. • Recall that the first packet that a new client sends is a DHCP-Request broadcast IPv4 packet. • The DHCP-Request packet has a source IPv4 address of 0.0.0.0. • Because NAT requires both a valid destination and source IPv4 address, BOOTP and DHCP can have difficulty operating over a router running either static or dynamic NAT. • Configuring the IPv4 helper feature can help solve this problem.
DNS	• Because a router running dynamic NAT is changing the relationship between inside and outside addresses regularly as table entries expire and are recreated, a DNS server outside the NAT router does not have an accurate representation of the network inside the router. • Configuring the IPv4 helper feature can help solve this problem.

Symptom	Description
SNMP	• Like DNS packets, NAT is unable to alter the addressing information stored in the data payload of the packet.
	• Because of this, an SNMP management station on one side of a NAT router may not be able to contact SNMP agents on the other side of the NAT router.
	• Configuring the IPv4 helper feature can help solve this problem.
Tunneling and encryption protocols	• Encryption and tunneling protocols often require that traffic be sourced from a specific UDP or TCP port, or use a protocol at the transport layer that cannot be processed by NAT.
	• For example, IPsec tunneling protocols and generic routing encapsulation protocols used by VPN implementations cannot be processed by NAT.

Refer to **Online Course** for Illustration

Application Layer Troubleshooting - 12.4.6

Most of the application layer protocols provide user services. Application layer protocols are typically used for network management, file transfer, distributed file services, terminal emulation, and email. New user services are often added, such as VPNs and VoIP.

The figure shows the most widely known and implemented TCP/IP application layer protocols.

The table provides a short description of these application layer protocols.

Applications	Description
SSH/Telnet	Enables users to establish terminal session connections with remote hosts.
HTTP	Supports the exchanging of text, graphic images, sound, video, and other multimedia files on the web.
FTP	Performs interactive file transfers between hosts.
TFTP	Performs basic interactive file transfers typically between hosts and networking devices.
SMTP	Supports basic message delivery services.
POP	Connects to mail servers and downloads email.
SNMP	Collects management information from network devices.
DNS	Maps IP addresses to the names assigned to network devices.
Network File System (NFS)	Enables computers to mount drives on remote hosts and operate them as if they were local drives. Originally developed by Sun Microsystems, it combines with two other application layer protocols, external data representation (XDR) and remote-procedure call (RPC), to allow transparent access to remote network resources.

The types of symptoms and causes depend upon the actual application itself.

Application layer problems prevent services from being provided to application programs. A problem at the application layer can result in unreachable or unusable resources when the physical, data link, network, and transport layers are functional. It is possible to have full network connectivity, but the application simply cannot provide data.

Another type of problem at the application layer occurs when the physical, data link, network, and transport layers are functional, but the data transfer and requests for network services from a single network service or application do not meet the normal expectations of a user.

A problem at the application layer may cause users to complain that the network or an application that they are working with is sluggish or slower than usual when transferring data or requesting network services.

Go to the online course to take the quiz and exam.

Check Your Understanding - Symptoms and Causes of Network Problems - 12.4.7

Troubleshooting IP Connectivity - 12.5

Refer to **Online Course** for Illustration

Components of Troubleshooting End-to-End Connectivity - 12.5.1

This topic presents a single topology and the tools to diagnose, and in some cases solve, an end-to-end connectivity problem. Diagnosing and solving problems is an essential skill for network administrators. There is no single recipe for troubleshooting, and a problem can be diagnosed in many ways. However, by employing a structured approach to the troubleshooting process, an administrator can reduce the time it takes to diagnose and solve a problem.

Throughout this topic, the following scenario is used. The client host PC1 is unable to access applications on Server SRV1 or Server SRV2. The figure shows the topology of this network. PC1 uses SLAAC with EUI-64 to create its IPv6 global unicast address. EUI-64 creates the Interface ID using the Ethernet MAC address, inserting FFFE in the middle, and flipping the seventh bit.

When there is no end-to-end connectivity, and the administrator chooses to troubleshoot with a bottom-up approach, the following are common steps the administrator can take:

Step 1.   Check physical connectivity at the point where network communication stops. This includes cables and hardware. The problem might be with a faulty cable or interface, or involve misconfigured or faulty hardware.

Step 2.   Check for duplex mismatches.

Step 3.   Check data link and network layer addressing on the local network. This includes IPv4 ARP tables, IPv6 neighbor tables, MAC address tables, and VLAN assignments.

Step 4.   Verify that the default gateway is correct.

Step 5.   Ensure that devices are determining the correct path from the source to the destination. Manipulate the routing information if necessary.

Step 6.   Verify the transport layer is functioning properly. Telnet can also be used to test transport layer connections from the command line.

Step 7.   Verify that there are no ACLs blocking traffic.

Step 8.   Ensure that DNS settings are correct. There should be a DNS server that is accessible.

The outcome of this process is operational, end-to-end connectivity. If all the steps have been performed without any resolution, the network administrator may either want to repeat the previous steps or escalate the problem to a senior administrator.

Refer to
Online Course
for Illustration

Refer to
Interactive Graphic
in online course

End-to-End Connectivity Problem Initiates Troubleshooting - 12.5.2

Usually what initiates a troubleshooting effort is the discovery that there is a problem with end-to-end connectivity. Two of the most common utilities used to verify a problem with end-to-end connectivity are **ping** and **traceroute**, as shown in the figure.

Click each button to review the ping, traceroute, and tracert utilities.

IPv4 ping

Ping is probably the most widely-known connectivity-testing utility in networking and has always been part of Cisco IOS Software. It sends out requests for responses from a specified host address. The **ping** command uses a Layer 3 protocol that is a part of the TCP/IP suite called ICMP. Ping uses the ICMP echo request and ICMP echo reply packets. If the host at the specified address receives the ICMP echo request, it responds with an ICMP echo reply packet. Ping can be used to verify end-to-end connectivity for both IPv4 and IPv6. The command output shows a successful ping from PC1 to SRV1, at address 172.16.1.100.

```
C:\> ping 172.16.1.100

Pinging 172.16.1.100 with 32 bytes of data:

Reply from 172.16.1.100: bytes=32 time=199ms TTL=128

Reply from 172.16.1.100: bytes=32 time=193ms TTL=128

Reply from 172.16.1.100: bytes=32 time=194ms TTL=128

Reply from 172.16.1.100: bytes=32 time=196ms TTL=128

Ping statistics for 172.16.1.100:

     Packets: Sent = 4, Received = 4, Lost = 0 (0% loss),

Approximate round trip times in milli-seconds:

     Minimum = 193ms, Maximum = 199ms, Average = 195ms

C:\>
```

IPv4 traceroute

Like the **ping** command, the Cisco IOS **traceroute** command can be used for both IPv4 and IPv6. The **tracert** command is used with Windows operating systems. The trace generates a list of hops, router IP addresses and the destination IP address that are successfully reached along the path. This list provides important verification and troubleshooting information. If the data reaches the destination, the trace lists the interface on every router in the path. If the data fails at some hop along the way, the address of the last router that responded to the trace is known. This address is an indication of where the problem or security restrictions reside.

The **tracert** output illustrates the path the IPv4 packets take to reach their destination.

```
C:\> tracert 172.16.1.100

Tracing route to 172.16.1.100 over a maximum of 30 hops:

    1     1 ms     <1 ms     <1 ms    10.1.10.1

    2     2 ms      2 ms      1 ms    192.168.1.2

    3     2 ms      2 ms      1 ms    192.168.1.6

    4     2 ms      2 ms      1 ms    172.16.1.100

Trace complete.

C:\>
```

IPv6 ping and traceroute

When using these utilities, the Cisco IOS utility recognizes whether the address is an IPv4 or IPv6 address and uses the appropriate protocol to test connectivity. The command output shows the **ping** and **traceroute** commands on router R1 used to test IPv6 connectivity.

```
R1# ping 2001:db8:acad:4::100

Type escape sequence to abort.

Sending 5, 100-byte ICMP Echos to 2001:DB8:ACAD:4::100, timeout is 2 seconds:

!!!!!

Success rate is 100 percent (5/5), round-trip min/avg/max = 56/56/56 ms

R1#

R1# traceroute 2001:db8:acad:4::100

Type escape sequence to abort.

Tracing the route to 2001:DB8:ACAD:4::100

1.    2001:DB8:ACAD:2::2 20 msec 20 msec 20 msec

2.    2001:DB8:ACAD:3::2 44 msec 40 msec 40 msec

R1#
```

Note: The **traceroute** command is commonly performed when the **ping** command fails. If the **ping** succeeds, the **traceroute** command is commonly not needed because the technician knows that connectivity exists.

Refer to
Interactive Graphic
in online course

Step 1 - Verify the Physical Layer - 12.5.3

All network devices are specialized computer systems. At a minimum, these devices consist of a CPU, RAM, and storage space, allowing the device to boot and run the operating system and interfaces. This allows for the reception and transmission of network traffic. When a network administrator determines that a problem exists on a given device, and that problem might be hardware-related, it is worthwhile to verify the operation of these generic components. The most commonly used Cisco IOS commands for this purpose are **show processes cpu**, **show memory**, and **show interfaces**. This topic discusses the **show interfaces** command.

When troubleshooting performance-related issues and hardware is suspected to be at fault, the **show interfaces** command can be used to verify the interfaces through which the traffic passes.

Refer to the command output of the **show interfaces** command.

```
R1# show interfaces GigabitEthernet 0/0/0

GigabitEthernet0/0/0 is up, line protocol is up

Hardware is CN Gigabit Ethernet, address is d48c.b5ce.a0c0(bia d48c.
  b5ce.a0c0)

Internet address is 10.1.10.1/24

(Output omitted)

Input queue: 0/75/0/0 (size/max/drops/flushes); Total output drops: 0

Queueing strategy: fifo

 Output queue: 0/40 (size/max)

5 minute input rate 0 bits/sec, 0 packets/sec

5 minute output rate 0 bits/sec, 0 packets/sec

85 packets input, 7711 bytes, 0 no buffer

Received 25 broadcasts (0 IP multicasts)

0 runts, 0 giants, 0 throttles

0 input errors, 0 CRC, 0 frame, 0 overrun, 0 ignored

0 watchdog, 5 multicast, 0 pause input

10112 packets output, 922864 bytes, 0 underruns

0 output errors, 0 collisions, 1 interface resets

11 unknown protocol drops

0 babbles, 0 late collision, 0 deferred

0 lost carrier, 0 no carrier, 0 pause output

0 output buffer failures, 0 output buffers swapped out

R1#
```

Click each button for an explanation of the highlighted output.

Input queue drops

Input queue drops (and the related ignored and throttle counters) signify that at some point, more traffic was delivered to the router than it could process. This does not necessarily indicate a problem. That could be normal traffic during peak periods. However, it could be an indication that the CPU cannot process packets in time, so if this number is consistently high, it is worth trying to spot at which moments these counters are increasing and how this relates to CPU usage.

Output queue drops

Output queue drops indicate that packets were dropped due to congestion on the interface. Seeing output drops is normal for any point where the aggregate input traffic is higher than the output traffic. During peak traffic periods, packets are dropped if traffic is delivered to the interface faster than it can be sent out. However, even if this is considered normal behavior, it leads to packet drops and queuing delays, so applications that are sensitive to those, such as VoIP, might suffer from performance issues. Consistently seeing output queue drops can be an indicator that you need to implement an advanced queuing mechanism to implement or modify QoS.

Input errors

Input errors indicate errors that are experienced during the reception of the frame, such as CRC errors. High numbers of CRC errors could indicate cabling problems, interface hardware problems, or, in an Ethernet-based network, duplex mismatches.

Output errors

Output errors indicate errors, such as collisions, during the transmission of a frame. In most Ethernet-based networks today, full-duplex transmission is the norm, and half-duplex transmission is the exception. In full-duplex transmission, operation collisions cannot occur; therefore, collisions (especially late collisions) often indicate duplex mismatches.

Refer to
Online Course
for Illustration

Step 2 - Check for Duplex Mismatches - 12.5.4

Another common cause for interface errors is a mismatched duplex mode between two ends of an Ethernet link. In many Ethernet-based networks, point-to-point connections are now the norm, and the use of hubs and the associated half-duplex operation is becoming less common. This means that most Ethernet links today operate in full-duplex mode, and while collisions were normal for an Ethernet link, collisions today often indicate that duplex negotiation has failed, or the link is not operating in the correct duplex mode.

The IEEE 802.3ab Gigabit Ethernet standard mandates the use of autonegotiation for speed and duplex. In addition, although it is not strictly mandatory, practically all Fast Ethernet NICs also use autonegotiation by default. The use of autonegotiation for speed and duplex is the current recommended practice.

However, if duplex negotiation fails for some reason, it might be necessary to set the speed and duplex manually on both ends. Typically, this would mean setting the duplex mode to full-duplex on both ends of the connection. If this does not work, running half-duplex on both ends is preferred over a duplex mismatch.

Duplex configuration guidelines include the following:

- Autonegotiation of speed and duplex is recommended.

- If autonegotiation fails, manually set the speed and duplex on interconnecting ends.

- Point-to-point Ethernet links should always run in full-duplex mode.

- Half-duplex is uncommon and typically encountered only when legacy hubs are used.

Troubleshooting Example

In the previous scenario, the network administrator needed to add additional users to the network. To incorporate these new users, the network administrator installed a second switch and connected it to the first. Soon after S2 was added to the network, users on both switches began experiencing significant performance problems connecting with devices on the other switch, as shown in the figure.

The network administrator notices a console message on switch S2:

```
*Mar 1 00:45:08.756: %CDP-4-DUPLEX_MISMATCH: duplex mismatch discovered
on FastEthernet0/20 (not half duplex), with Switch FastEthernet0/20
(half duplex).
```

Using the **show interfaces fa 0/20** command, the network administrator examines the interface on S1 that is used to connect to S2 and notices it is set to full-duplex, as shown the command output.

```
S1# show interface fa 0/20

FastEthernet0/20 is up, line protocol is up (connected)

Hardware is Fast Ethernet, address is 0cd9.96e8.8a01 (bia
    0cd9.96e8.8a01)

MTU 1500 bytes, BW 10000 Kbit/sec, DLY 1000 usec, reliability 255/255,
txload 1/255, rxload 1/255

Encapsulation ARPA, loopback not set Keepalive set (10 sec)

Full-duplex, Auto-speed, media type is 10/100BaseTX


(Output omitted)


S1#
```

The network administrator now examines the other side of the connection, the port on S2. The command out shows that this side of the connection has been configured for half-duplex.

```
S2# show interface fa 0/20

FastEthernet0/20 is up, line protocol is up (connected)
```

```
Hardware is Fast Ethernet, address is 0cd9.96d2.4001 (bia
   0cd9.96d2.4001)
MTU 1500 bytes, BW 100000 Kbit/sec, DLY 100 usec, reliability 255/255,
txload 1/255, rxload 1/255
Encapsulation ARPA, loopback not set Keepalive set (10 sec)
Half-duplex, Auto-speed, media type is 10/100BaseTX


(Output omitted)


S2(config)# interface fa 0/20
S2(config-if)# duplex auto
S2(config-if)#
```

The network administrator corrects the setting to **duplex auto** to automatically negotiate the duplex. Because the port on S1 is set to full-duplex, S2 also uses full-duplex.

The users report that there are no longer any performance problems.

Refer to
Interactive Graphic
in online course

Step 3 - Verify Addressing on the Local Network - 12.5.5

When troubleshooting end-to-end connectivity, it is useful to verify mappings between destination IP addresses and Layer 2 Ethernet addresses on individual segments. In IPv4, this functionality is provided by ARP. In IPv6, the ARP functionality is replaced by the neighbor discovery process and ICMPv6. The neighbor table caches IPv6 addresses and their resolved Ethernet physical (MAC) addresses.

Click each button for an example and explanation of the command to verify Layer 2 and Layer 3 addressing.

Windows IPv4 ARP Table

The **arp** Windows command displays and modifies entries in the ARP cache that are used to store IPv4 addresses and their resolved Ethernet physical (MAC) addresses. As shown in the command output, the **arp** Windows command lists all devices that are currently in the ARP cache.

The information that is displayed for each device includes the IPv4 address, physical (MAC) address, and the type of addressing (static or dynamic).

The cache can be cleared by using the **arp -d** Windows command if the network administrator wants to repopulate the cache with updated information.

Note: The **arp** commands in Linux and MAC OS X have a similar syntax.

```
C:\> arp -a
Interface: 10.1.10.100 --- 0xd
```

```
Internet Address          Physical Address        Type

10.1.10.1                 d4-8c-b5-ce-a0-c0       dynamic

224.0.0.22                01-00-5e-00-00-16       static

224.0.0.251               01-00-5e-00-00-fb       static

239.255.255.250           01-00-5e-7f-ff-fa       static

255.255.255.255           ff-ff-ff-ff-ff-ff       static

C:\>
```

Windows IPv6 Neighbor Table

The **netsh interface ipv6 show neighbor** Windows command output lists all devices that are currently in the neighbor table.

The information that is displayed for each device includes the IPv6 address, physical (MAC) address, and the type of addressing. By examining the neighbor table, the network administrator can verify that destination IPv6 addresses map to correct Ethernet addresses. The IPv6 link-local addresses on all interfaces of R1 have been manually configured to FE80::1. Similarly, R2 has been configured with the link-local address of FE80::2 on its interfaces and R3 has been configured with the link-local address of FE80::3 on its interfaces. Remember, link-local addresses must be unique on the link or network.

Note: The neighbor table for Linux and MAC OS X can be displayed using **ip neigh show** command.

```
C:\> netsh interface ipv6 show neighbor

Internet Address                   Physical Address     Type

----------------------------       -----------------    -----------

fe80::9657:a5ff:fe0c:5b02          94-57-a5-0c-5b-02    Stale

fe80::1                            d4-8c-b5-ce-a0-c0    Reachable

(Router)

ff02::1                            33-33-00-00-00-01    Permanent

ff02::2                            33-33-00-00-00-02    Permanent

ff02::16                           33-33-00-00-00-16    Permanent

ff02::1:2                          33-33-00-01-00-02    Permanent

ff02::1:3                          33-33-00-01-00-03    Permanent

ff02::1:ff0c:5b02                  33-33-ff-0c-5b-02    Permanent

ff02::1:ff2d:a75e                  33-33-ff-2d-a7-5e    Permanent
```

IOS IPv6 Neighbor Table

The **show ipv6 neighbors** command output displays an example of the neighbor table on the Cisco IOS router.

Note: The neighbor states for IPv6 are more complex than the ARP table states in IPv4. Additional information is contained in RFC 4861.

```
R1# show ipv6 neighbors

IPv6 Address                         Age   Link-layer Addr   State Interface

FE80::21E:7AFF:FE79:7A81               8   001e.7a79.7a81    STALE  Gi0/0

2001:DB8:ACAD:1:5075:D0FF:FE8E:9AD8    0   5475.d08e.9ad8    REACH  Gi0/0
```

Switch MAC Address Table

When a destination MAC address is found in the switch MAC address table, the switch forwards the frame only to the port of the device that has that MAC address. To do this, the switch consults its MAC address table. The MAC address table lists the MAC address connected to each port. Use the **show mac address-table** command to display the MAC address table on the switch. An example of a switch MAC address table is shown in the command output.

Notice how the MAC address for PC1, a device in VLAN 10, has been discovered along with the S1 switch port to which PC1 attaches. Remember, the MAC address table of switch only contains Layer 2 information, including the Ethernet MAC address and the port number. IP address information is not included.

```
S1# show mac address-table

             Mac Address Table

-------------------------------------------

Vlan      Mac Address       Type     Ports

All       0100.0ccc.cccc    STATIC   CPU

All       0100.0ccc.cccd    STATIC   CPU

10        d48c.b5ce.a0c0    DYNAMIC  Fa0/4

10        000f.34f9.9201    DYNAMIC  Fa0/5

10        5475.d08e.9ad8    DYNAMIC  Fa0/13

Total Mac Addresses for this criterion: 5
```

Refer to Interactive Graphic in online course

Troubleshoot VLAN Assignment Example - 12.5.6

Another issue to consider when troubleshooting end-to-end connectivity is VLAN assignment. In the switched network, each port in a switch belongs to a VLAN. Each VLAN is considered a separate logical network, and packets destined for stations that do not belong to the VLAN must be forwarded through a device that supports routing. If a host in one VLAN sends a broadcast Ethernet frame, such as an ARP request, all hosts in the same VLAN receive the frame; hosts in other VLANs do not. Even if two hosts are in the same IP network, they will not be able to communicate if they are connected to ports assigned to two separate VLANs. Additionally, if the VLAN to which the port belongs is deleted, the port becomes inactive. All hosts attached to ports belonging to the VLAN that was deleted are unable to communicate with the rest of the network. Commands such as **show vlan** can be used to validate VLAN assignments on a switch.

Assume for example, that in an effort to improve the wire management in the wiring closet, your company has reorganized the cables connecting to switch S1. Almost immediately afterward, users started calling the support desk stating that they could no longer reach devices outside their own network.

Click each button for an explanation of the process used to troubleshoot this issue.

Check the ARP Table

An examination of PC1 ARP table using the **arp** Windows command shows that the ARP table no longer contains an entry for the default gateway 10.1.10.1, as shown in the command output.

```
C:\> arp -a

Interface: 10.1.10.100 --- 0xd
    Internet Address        Physical Address        Type
    224.0.0.22              01-00-5e-00-00-16        static
    224.0.0.251             01-00-5e-00-00-fb        static
    239.255.255.250         01-00-5e-7f-ff-fa        static
    255.255.255.255         ff-ff-ff-ff-ff-ff        static

C:\>
```

Check the Switch MAC Table

There were no configuration changes on the router, so S1 is the focus of the troubleshooting.

The MAC address table for S1, as shown in the command output, shows that the MAC address for R1 is on a different VLAN than the rest of the 10.1.10.0/24 devices, including PC1.

```
S1# show mac address-table
            Mac Address Table
-------------------------------------------

Vlan      Mac Address       Type      Ports
All       0100.0ccc.cccc    STATIC    CPU
All       0100.0ccc.cccd    STATIC    CPU
  1       d48c.b5ce.a0c0    DYNAMIC   Fa0/1
 10       000f.34f9.9201    DYNAMIC   Fa0/5
 10       5475.d08e.9ad8    DYNAMIC   Fa0/13
Total Mac Addresses for this criterion: 5
S1#
```

Correct the VLAN Assignment

During the re-cabling, the patch cable for R1 was moved from Fa 0/4 on VLAN 10 to Fa 0/1 on VLAN 1. After the network administrator configured the Fa 0/1 port of S1 to be on

VLAN 10, as shown in the command output, the problem was resolved. The MAC address table now shows VLAN 10 for the MAC address of R1 on port Fa 0/1.

```
S1(config)# interface fa0/1

S1(config-if)# switchport mode access

S1(config-if)# switchport access vlan 10

S1(config-if)# exit

S1#

S1# show mac address-table

              Mac Address Table

----------------------------------------------

Vlan       Mac Address       Type      Ports

All        0100.0ccc.cccc    STATIC    CPU

All        0100.0ccc.cccd    STATIC    CPU

10         d48c.b5ce.a0c0    DYNAMIC   Fa0/1

10         000f.34f9.9201    DYNAMIC   Fa0/5

10         5475.d08e.9ad8    DYNAMIC   Fa0/13

Total Mac Addresses for this criterion: 5

S1#
```

Refer to
Online Course
for Illustration

Refer to
Interactive Graphic
in online course

Step 4 - Verify Default Gateway - 12.5.7

If there is no detailed route on the router, or if the host is configured with the wrong default gateway, then communication between two endpoints in different networks does not work.

The figure illustrates how PC1 uses R1 as its default gateway. Similarly, R1 uses R2 as its default gateway or gateway of last resort. If a host needs access to resources beyond the local network, the default gateway must be configured. The default gateway is the first router on the path to destinations beyond the local network.

Troubleshooting IPv4 Default Gateway Example

In this example, R1 has the correct default gateway, which is the IPv4 address of R2. However, PC1 has the wrong default gateway. PC1 should have the default gateway of R1 10.1.10.1. This must be configured manually if the IPv4 addressing information was manually configured on PC1. If the IPv4 addressing information was obtained automatically from a DHCPv4 server, then the configuration on the DHCP server must be examined. A configuration problem on a DHCP server usually affects multiple clients.

Click each button to view the command output for R1 and PC1.

R1 Routing Table

The command output of the **show ip route** Cisco IOS command is used to verify the default gateway of R1

```
R1# show ip route | include Gateway|0.0.0.0
```

```
Gateway of last resort is 192.168.1.2 to network 0.0.0.0
S* 0.0.0.0/0 [1/0] via 192.168.1.2


R1#
```

PC1 Routing Table

On a Windows host, the **route print** Windows command is used to verify the presence of the IPv4 default gateway as shown in the command output.

```
C:\> route print
(Output omitted)


IPv4 Route Table
===========================================================================
Active Routes:
Network Destination        Netmask          Gateway       Interface   Metric
          0.0.0.0          0.0.0.0        10.1.10.1      10.1.10.10   11
(Output omitted)
```

Troubleshoot IPv6 Default Gateway Example - 12.5.8

In IPv6, the default gateway can be configured manually, using stateless autoconfiguration (SLAAC), or by using DHCPv6. With SLAAC, the default gateway is advertised by the router to hosts using ICMPv6 Router Advertisement (RA) messages. The default gateway in the RA message is the link-local IPv6 address of a router interface. If the default gateway is configured manually on the host, which is very unlikely, the default gateway can be set to either the global IPv6 address, or to the link-local IPv6 address.

Click each button for an example and explanation of troubleshooting an IPv6 default gateway issue.

R1 Routing Table

As shown in the command output, the **show ipv6 route** Cisco IOS command is used to check for the IPv6 default route on R1. R1 has a default route via R2.

```
R1# show ipv6 route


(Output omitted)


S ::/0 [1/0]

via 2001:DB8:ACAD:2::2

R1#
```

PC1 Addressing

The **ipconfig** Windows command is used to verify that a PC1 has an IPv6 default gateway. In the command output, PC1 is missing an IPv6 global unicast address and an IPv6 default gateway. PC1 is enabled for IPv6 because it has an IPv6 link-local address. The link-local address is automatically created by the device. Checking the network documentation, the network administrator confirms that hosts on this LAN should be receiving their IPv6 address information from the router using SLAAC.

Note: In this example, other devices on the same LAN using SLAAC would also experience the same problem receiving IPv6 address information.

```
C:\> ipconfig

Windows IP Configuration

    Connection-specific DNS Suffix . :

    Link-local IPv6 Address . . . . :  fe80::5075:d0ff:fe8e:9ad8%13

    IPv4 Address . . . . . . . . . :  10.1.10.10

    Subnet Mask  . . . . . . . . . :  255.255.255.0

    Default Gateway. . . . . . . . :  10.1.10.1

C:\>
```

Check R1 Interface Settings

The command output of the **show ipv6 interface GigabitEthernet 0/0/0** on R1 reveals that although the interface has an IPv6 address, it is not a member of the All-IPv6-Routers multicast group FF02::2. This means the router is not enabled as an IPv6 router. Therefore, it is not sending out ICMPv6 RAs on this interface.

```
R1# show ipv6 interface GigabitEthernet 0/0/0

GigabitEthernet0/0/0 is up, line protocol is up

  IPv6 is enabled, link-local address is FE80::1

  No Virtual link-local address(es):

  Global unicast address(es):

    2001:DB8:ACAD:1::1, subnet is 2001:DB8:ACAD:1::/64

  Joined group address(es):

      FF02:: 1

      FF02::1:FF00:1


(Output omitted)

R1#
```

Correct R1 IPv6 Routing

R1 is enabled as an IPv6 router using the **ipv6 unicast-routing** command. The **show ipv6 interface GigabitEthernet 0/0/0** command verifies that R1 is a member of ff02::2, the All-IPv6-Routers multicast group.

```
R1(config)# ipv6 unicast-routing

R1(config)# exit

R1# show ipv6 interface GigabitEthernet 0/0/0

GigabitEthernet0/0/0 is up, line protocol is up

  IPv6 is enabled, link-local address is FE80::1

  No Virtual link-local address(es):

  Global unicast address(es):

    2001:DB8:ACAD:1::1, subnet is 2001:DB8:ACAD:1::/64

  Joined group address(es):

      FF02:: 1

      FF02:: 2

      FF02::1:FF00:1

(Output omitted)

R1#
```

Verify PC1 Has an IPv6 Default Gateway

To verify that PC1 has the default gateway set, use the **ipconfig** command on the Microsoft Windows PC or the **ifconfig** command on Linux and Mac OS X. In the, PC1 has an IPv6 global unicast address and an IPv6 default gateway. The default gateway is set to the link-local address of router R1, fe80::1.

```
C:\> ipconfig

Windows IP Configuration

    Connection-specific DNS Suffix . :

    IPv6 Address. . . . . . . . . . : 2001:db8:acad:1:5075:d0ff:fe8e:
      9ad8

    Link-local IPv6 Address . . . . : fe80::5075:d0ff:fe8e:9ad8%13

    IPv4 Address . . . . . . . . . . : 10.1.10.10

    Subnet Mask . . . . . . . . . . : 255.255.255.0

    Default Gateway. . . . . . . . . : fe80::1

                                       10.1.10.1

C:\>
```

Refer to
Online Course
for Illustration

Refer to
Interactive Graphic
in online course

Step 5 - Verify Correct Path - 12.5.9

When troubleshooting, it is often necessary to verify the path to the destination network. The figure shows the reference topology indicating the intended path for packets from PC1 to SRV1.

The routers in the path make the routing decision based on information in the routing tables. Click each button to view the IPv4 and IPv6 routing tables for R1.

R1 IPv4 Routing Table

```
R1# show ip route | begin Gateway

Gateway of last resort is 192.168.1.2 to network 0.0.0.0

O*E2  0.0.0.0/0 [110/1] via 192.168.1.2, 00:00:13, Serial0/1/0
          10.0.0.0/8 is variably subnetted, 2 subnets, 2 masks
C         10.1.10.0/24 is directly connected, GigabitEthernet0/0/0
L         10.1.10.1/32 is directly connected, GigabitEthernet0/0/0
          172.16.0.0/24 is subnetted, 1 subnets
O         172.16.1.0 [110/100] via 192.168.1.2, 00:01:59, Serial0/1/0
          192.168.1.0/24 is variably subnetted, 3 subnets, 2 masks
C         192.168.1.0/30 is directly connected, Serial0/1/0
L         192.168.1.1/32 is directly connected, Serial0/1/0
O         192.168.1.4/30 [110/99] via 192.168.1.2, 00:06:25, Serial0/1/0
R1#
```

R1 IPv6 Routing Table

```
R1# show ipv6 route

IPv6 Routing Table - default - 8 entries

Codes:  C - Connected, L - Local, S - Static, U - Per-user Static route
        B - BGP, R - RIP, H - NHRP, I1 - ISIS L1
        I2 - ISIS L2, IA - ISIS interarea, IS - ISIS summary, D - EIGRP
        EX - EIGRP external, ND - ND Default, NDp - ND Prefix, DCE -
Destination
        NDr - Redirect, O - OSPF Intra, OI - OSPF Inter, OE1 - OSPF ext 1
        OE2 - OSPF ext 2, ON1 - OSPF NSSA ext 1, ON2 - OSPF NSSA ext 2
        a - Application

OE2 ::/0 [110/1], tag 1
     via FE80::2, Serial0/1/0
C   2001:DB8:ACAD:1::/64 [0/0]
     via GigabitEthernet0/0/0, directly connected
```

```
L    2001:DB8:ACAD:1::1/128  [0/0]

        via GigabitEthernet0/0/0, receive

C    2001:DB8:ACAD:2::/64  [0/0]

        via Serial0/1/0, directly connected

L    2001:DB8:ACAD:2::1/128  [0/0]

        via Serial0/1/0, receive

O    2001:DB8:ACAD:3::/64  [110/99]

        via FE80::2, Serial0/1/0

O    2001:DB8:ACAD:4::/64  [110/100]

        via FE80::2, Serial0/1/0

L    FF00::/8  [0/0]

        via Null0, receive

R1#
```

The IPv4 and IPv6 routing tables can be populated by the following methods:

- Directly connected networks
- Local host or local routes
- Static routes
- Dynamic routes
- Default routes

The process of forwarding IPv4 and IPv6 packets is based on the longest bit match or longest prefix match. The routing table process will attempt to forward the packet using an entry in the routing table with the greatest number of leftmost matching bits. The number of matching bits is indicated by the prefix length of the route.

The figure describes the process for both the IPv4 and IPv6 routing tables.

Troubleshooting Example

Devices are unable to connect to the server SRV1 at 172.16.1.100. Using the **show ip route** command, the administrator should check to see if a routing entry exists to network 172.16.1.0/24. If the routing table does not have a specific route to the SRV1 network, the network administrator must then check for the existence of a default or summary route entry in the direction of the 172.16.1.0/24 network. If none exists, then the problem may be with routing and the administrator must verify that the network is included within the dynamic routing protocol configuration or add a static route.

Step 6 - Verify the Transport Layer - 12.5.10

If the network layer appears to be functioning as expected, but users are still unable to access resources, then the network administrator must begin troubleshooting the upper layers. Two of the most common issues that affect transport layer connectivity include ACL configurations and NAT configurations. A common tool for testing transport layer functionality is the Telnet utility.

Caution: While Telnet can be used to test the transport layer, for security reasons, SSH should be used to remotely manage and configure devices.

Troubleshooting Example

A network administrator is troubleshooting a problem where they cannot connect to a router using HTTP. The administrator pings R2 as shown in the command output.

```
R1# ping 2001:db8:acad:2::2

Type escape sequence to abort.

Sending 5, 100-byte ICMP Echos to 2001:DB8:ACAD:2::2, timeout is 2 seconds:

!!!!!

Success rate is 100 percent (5/5), round-trip min/avg/max = 2/2/3 ms

R1#
```

R2 responds and confirms that the network layer, and all layers below the network layer are operational. The administrator knows the issue is with Layer 4 or up and must start troubleshooting those layers.

Next, the administrator verifies that they can Telnet to R2 as shown in the command output.

```
R1# telnet 2001:db8:acad:2::2

Trying 2001:DB8:ACAD:2::2 ... Open

User Access Verification

Password:

R2> exit

[Connection to 2001:db8:acad:2::2 closed by foreign host]

R1#
```

The administrator has confirmed that Telnet services is running on R2. Although the Telnet server application runs on its own well-known port number 23 and Telnet clients connect to this port by default, a different port number can be specified on the client to connect to any TCP port that must be tested. Using a different port other than TCP port 23 indicates whether the connection is accepted (as indicated by the word "Open" in the output), refused, or times out. From any of those responses, further conclusions can be made concerning the connectivity. Certain applications, if they use an ASCII-based session protocol, might even display an application banner, it may be possible to trigger some responses from the server by typing in certain keywords, such as with SMTP, FTP, and HTTP.

For example, the administrator attempts to Telnet to R2 using port 80.

```
R1# telnet 2001:db8:acad:2::2 80

Trying 2001:DB8:ACAD:2::2, 80 ... Open

^C

HTTP/1.1 400 Bad Request

Date: Mon, 04 Nov 2019 12:34:23 GMT
```

```
Server: cisco-IOS

Accept-Ranges: none

400 Bad Request

[Connection to 2001:db8:acad:2::2 closed by foreign host]

R1#
```

The output verifies a successful transport layer connection, but R2 is refusing the connection using port 80.

Refer to
Interactive Graphic
in online course

Refer to
Online Course
for Illustration

Step 7 - Verify ACLs - 12.5.11

On routers, there may be ACLs that prohibit protocols from passing through the interface in the inbound or outbound direction.

Use the **show ip access-lists** command to display the contents of all IPv4 ACLs and the **show ipv6 access-list** command to display the contents of all IPv6 ACLs configured on a router. The specific ACL can be displayed by entering the ACL name or number as an option for this command. The **show ip interfaces** and **show ipv6 interfaces** commands display IPv4 and IPv6 interface information that indicates whether any IP ACLs are set on the interface.

Troubleshooting Example

To prevent spoofing attacks, the network administrator decided to implement an ACL that is preventing devices with a source network address of 172.16.1.0/24 from entering the inbound S0/0/1 interface on R3, as shown in the figure. All other IP traffic should be allowed.

However, shortly after implementing the ACL, users on the 10.1.10.0/24 network were unable to connect to devices on the 172.16.1.0/24 network, including SRV1.

Click each button for an example of how to troubleshoot this issue.

show ip access-lists

The **show ip access-lists** command displays that the ACL is configured correctly, as shown in the command output.

```
R3# show ip access-lists
Extended IP access list 100
    10 deny ip 172.16.1.0 0.0.0.255 any (108 matches)
    20 permit ip any any (28 matches)
R3#
```

show ip interfaces

We can verify which interface has the ACL applied using the **show ip interfaces serial 0/1/1** command and the **show ip interfaces serial 0/0/0** command. The output reveals that the ACL was never applied to the inbound interface on Serial 0/0/1 but it was accidentally applied to the G0/0/0 interface, blocking all outbound traffic from the 172.16.1.0/24 network.

```
R3# show ip interface serial 0/1/1 | include access list
  Outgoing Common access list is not set
  Outgoing access list is not set
  Inbound Common access list is not set
  Inbound  access list is not set
R3#

R3# show ip interface gig 0/0/0 | include access list
  Outgoing Common access list is not set
  Outgoing access list is not set
  Inbound Common access list is not set
  Inbound  access list is 100
R3#
```

Correct the Issue

After correctly placing the IPv4 ACL on the Serial 0/0/1 inbound interface, as shown in the command output, devices can successfully connect to the server.

```
R3(config)# interface GigabitEthernet 0/0/0
R3(config-if)# no ip access-group 100 in
R3(config-if)# exit
R3(config)#
R3(config)# interface serial 0/1/1
R3(config-if)# ip access-group 100 in
R3(config-if)# end
R3#
```

Step 8 - Verify DNS - 12.5.12

The DNS protocol controls the DNS, a distributed database with which you can map hostnames to IP addresses. When you configure DNS on the device, you can substitute the hostname for the IP address with all IP commands, such as **ping** or **telnet**.

To display the DNS configuration information on the switch or router, use the **show running-config** command. When there is no DNS server installed, it is possible to enter names to IP mappings directly into the switch or router configuration. Use the **ip host** command to enter a name to be used instead of the IPv4 address of the switch or router, as shown in the command output.

```
R1(config)# ip host ipv4-server 172.16.1.100
R1(config)# exit
R1#
```

Now the assigned name can be used instead of using the IP address, as shown in the command output.

```
R1# ping ipv4-server
Type escape sequence to abort.
Sending 5, 100-byte ICMP Echos to 172.16.1.100, timeout is 2 seconds:
!!!!!
Success rate is 100 percent (5/5), round-trip min/avg/max = 4/5/7 ms
R1#
```

To display the name-to-IP-address mapping information on a Windows-based PC, use the **nslookup** command.

> Refer to **Packet Tracer Activity** for this chapter

Packet Tracer - Troubleshoot Enterprise Networks - 12.5.13

This activity uses a variety of technologies you have encountered during your CCNA studies, including routing, port security, EtherChannel, DHCP, and NAT. Your task is to review the requirements, isolate and resolve any issues, and then document the steps you took to verify the requirements.

Module Practice and Quiz - 12.6

> Refer to **Packet Tracer Activity** for this chapter

Packet Tracer - Troubleshooting Challenge - Document the Network - 12.6.1

In this Packet Tracer activity, you will document a network that is unknown to you.

- Test network connectivity.
- Compile host addressing information.
- Remotely access default gateway devices.
- Document default gateway device configurations.
- Discover devices on the network.
- Draw the network topology.

> Refer to **Packet Tracer Activity** for this chapter

Packet Tracer - Troubleshooting Challenge - Use Documentation to Solve Issues - 12.6.2

In this Packet Tracer activity, you use network documentation to identify and fix network communications problems.

- Use various techniques and tools to identify connectivity issues.
- Use documentation to guide troubleshooting efforts.

- Identify specific network problems.

- Implement solutions to network communication problems.

- Verify network operation.

What did I learn in this module? - 12.6.3

Network Documentation

Common network documentation includes: physical and logical network topologies, network device documentation recording all pertinent device information, and network performance baseline documentation. Information found on a physical topology typically includes the device name, device location (address, room number, rack location, etc.), interface and ports used, and cable type. Network device documentation for a router may include the interface, IPv4 address, IPv6 address, MAC address and routing protocol. Network device documentation for a switch may include the port, access, VLAN, trunk, EtherChannel, native, and enabled. Network device documentation for end-systems may include device name, OS, services, MAC address, IPv4 and IPv6 addresses, default gateway, and DNS. A network baseline should answer the following questions:

- How does the network perform during a normal or average day?

- Where are the most errors occurring?

- What part of the network is most heavily used?

- What part of the network is least used?

- Which devices should be monitored and what alert thresholds should be set?

- Can the network meet the identified policies?

When conducting the initial baseline, start by selecting a few variables that represent the defined policies, such as interface utilization and CPU utilization. A logical network topology diagram can be useful in identifying key devices and ports to monitor. The length of time and the baseline information being gathered must be long enough to determine a "normal" picture of the network. When documenting the network, gather information directly from routers and switches using the **show**, **ping**, **traceroute**, and **telnet** commands.

Troubleshooting Process

The troubleshooting process should be guided by structured methods. One method is the seven-step troubleshooting process: 1. Define the problem, 2. Gather information, 3. Analyze information, 4. Eliminate possible causes, 5. Propose hypothesis, 6. Test hypothesis, and 7. Solve the problem. When talking to end users about their network problems, ask both open and closed-ended questions. Use the **show**, **ping**, **traceroute**, and **telnet** commands to gather information from devices. Use the layered models to perform bottom-up, top-down, or divide-and-conquer troubleshooting. Other models include follow-the-path, substitution, comparison, and educated guess. Software problems are often solved using a top-down approach while hardware-based problems are solved using the bottom-up approach. New problems may be solved by an experienced technician using the divide-and-conquer method.

Troubleshooting Tools

Common software troubleshooting tools include NMS tools, knowledge bases, and base-lining tools. A protocol analyzer, such as Wireshark, decodes the various protocol layers in a recorded frame and presents this information in an easy to use format. Hardware trouble-shooting tools include digital multimeters, cable testers, cable analyzers, portable network analyzers, and Cisco Prime NAM. Syslog server can also be used as a troubleshooting tool. Implementing a logging facility for network troubleshooting. Cisco devices can log information regarding configuration changes, ACL violations, interface status, and many other types of events. Event messages can be sent to one or more of the following: console, terminal lines, buffered logging, SNMP traps, and syslog. The lower the level number, the higher the severity level. The **logging trap** *level* command limits messages logged to the syslog server based on severity. The level is the name or number of the severity level. Only messages equal to or numerically lower than the specified level are logged.

Symptoms and Causes of Network Problems

Failures and suboptimal conditions at the physical layer usually cause networks to shut down. Network administrators must have the ability to effectively isolate and correct problems at this layer. Symptoms include performance lower than baseline, loss of connectivity, congestion, high CPU utilization, and console error messages. The causes are usually power-related, hardware faults, cabling faults, attenuation, noise, interface configuration errors, exceeding component design limits, and CPU overload.

Data link layer problems cause specific symptoms that, when recognized, will help identify the problem quickly. Symptoms include no functionality/connectivity at Layer 2 or above, network operating below baseline levels, excessive broadcasts, and console messages. The causes are usually encapsulation errors, address mapping errors, framing errors, and STP failures or loops.

Network layer problems include any problem that involves a Layer 3 protocol, both routed protocols (such as IPv4 or IPv6) and routing protocols (such as EIGRP, OSPF, etc.). Symptoms include network failure and suboptimal performance. The causes are usually general network issues, connectivity issues, routing table problems, neighbor issues, and the topology database.

Transport layer problems can arise from transport layer problems on the router, particularly at the edge of the network where traffic is examined and modified. Symptoms include connectivity and access issues. Causes are likely to be misconfigured NAT or ACLs. ACL misconfigurations commonly occur at the selection of traffic flow, order of access control entries, implicit deny any, addresses and IPv4 wildcard masks, selection of transport layer protocol, source and destination ports, use of the established keyword, and uncommon protocols. There are several problems with NAT including misconfigured NAT inside, NAT outside, or ACL. Common interoperability areas with NAT include BOOTP and DHCP, DNS, SNMP, and tunneling and encryption protocols.

Application layer problems can result in unreachable or unusable resources when the physical, data link, network, and transport layers are functional. It is possible to have full network connectivity, but the application simply cannot provide data. Another type of problem at the application layer occurs when the physical, data link, network, and transport layers are functional, but the data transfer and requests for network services from a single network service or application do not meet the normal expectations of a user.

Troubleshooting IP Connectivity

Diagnosing and solving problems is an essential skill for network administrators. There is no single recipe for troubleshooting, and a problem can be diagnosed in many ways. However, by employing a structured approach to the troubleshooting process, an administrator can reduce the time it takes to diagnose and solve a problem.

End-to-end connectivity problems are usually what initiates a troubleshooting effort. Two of the most common utilities used to verify a problem with end-to-end connectivity are **ping** and **traceroute**. The **ping** command uses a Layer 3 protocol that is a part of the TCP/IP suite called ICMP. The **traceroute** command is commonly performed when the **ping** command fails.

Step 1. Verify the physical layer. The most commonly used Cisco IOS commands for this purpose are **show processes cpu**, **show memory**, and **show interfaces**.

Step 2. Check for duplex mismatches. Another common cause for interface errors is a mismatched duplex mode between two ends of an Ethernet link. In many Ethernet-based networks, point-to-point connections are now the norm, and the use of hubs and the associated half-duplex operation is becoming less common. Use the **show interfaces** *interface* command to diagnose this problem.

Step 3. Verify addressing on the local network. When troubleshooting end-to-end connectivity, it is useful to verify mappings between destination IP addresses and Layer 2 Ethernet addresses on individual segments. The **arp** Windows command displays and modifies entries in the ARP cache that are used to store IPv4 addresses and their resolved Ethernet physical (MAC) addresses. The **netsh interface ipv6 show neighbor** Windows command output lists all devices that are currently in the neighbor table. The **show ipv6 neighbors** command output displays an example of the neighbor table on the Cisco IOS router. Use the **show mac address-table** command to display the MAC address table on the switch.

VLAN assignment is another issue to consider when troubleshooting end-to-end connectivity. Use the **arp** Windows command to see the entry for a default gateway. Use the **show mac address-table** command to check the switch MAC table. This may show that not a VLAN assignments are correct.

Step 4. Verify the default gateway. The command output of the **show ip route** Cisco IOS command is used to verify the default gateway of a router. On a Windows host, the **route print** Windows command is used to verify the presence of the IPv4 default gateway.

In IPv6, the default gateway can be configured manually, using stateless autoconfiguration (SLAAC), or by using DHCPv6. The **show ipv6 route** Cisco IOS command is used to check for the IPv6 default route on a router. The **ipconfig** Windows command is used to verify if a PC1 has an IPv6 default gateway. The command output of the **show ipv6 interface** *interface* will tell you if a router is or is not enabled as an IPv6 router. Enable a router as an IPv6 router using the **ipv6 unicast-routing** command. To verify that a host has the default gateway set, use the **ipconfig** command on the Microsoft Windows PC or the **ifconfig** command on Linux and Mac OS X.

Step 5.   Verify correct path. The routers in the path make the routing decision based on information in the routing tables. Use the **show ip route | begin Gateway** command for an IPv4 routing table. Use the **show ipv6 route** command for an IPv6 routing table.

Step 6.   Verify the transport layer. Two of the most common issues that affect transport layer connectivity include ACL configurations and NAT configurations. A common tool for testing transport layer functionality is the Telnet utility.

Step 7.   Verify ACLs. Use the **show ip access-lists** command to display the contents of all IPv4 ACLs and the **show ipv6 access-list** command to show the contents of all IPv6 ACLs configured on a router. Verify which interface has the ACL applied using the **show ip interfaces** command.

Step 8.   Verify DNS. To display the DNS configuration information on the switch or router, use the **show running-config** command. Use the **ip host** command to enter name to IPv4 mapping to the switch or router as shown in the command output.

Go to the online
course to take the
quiz and exam.

Chapter Quiz - Network Troubleshooting

Your Chapter Notes

CHAPTER 13

Network Virtualization

Introduction - 13.0

Why should I take this module? - 13.0.1

Welcome to Network Virtualization!

Imagine you live in a two-bedroom house. You use the second bedroom for storage. The second bedroom is packed full of boxes, but you still have more to place in storage! You could consider building an addition on your house. It would be a costly endeavor and you may not need that much space forever. You decide to rent a storage unit for the overflow.

Similar to a storage unit, network virtualization and cloud services can provide a business with options other than adding servers into their own data center. In addition to storage, it offers other advantages. Get started with this module to learn more about what virtualization and cloud services can do!

What will I learn to do in this module? - 13.0.2

Module Title: Network Virtualization

Module Objective: Explain the purpose and characteristics of network virtualization.

Topic Title	Topic Objective
Cloud Computing	Explain the importance of cloud computing.
Virtualization	Explain the importance of virtualization.
Virtual Network Infrastructure	Describe the virtualization of network devices and services.
Software-Defined Networking	Describe software-defined networking.
Controllers	Describe controllers used in network programming.

Cloud Computing - 13.1

Refer to **Video** in online course

Video - Cloud and Virtualization - 13.1.1

Click Play for an overview of cloud computing and virtualization.

Refer to **Online Course** for Illustration

Cloud Overview - 13.1.2

In the previous video, an overview of cloud computing was explained. Cloud computing involves large numbers of computers connected through a network that can be physically located anywhere. Providers rely heavily on virtualization to deliver their cloud computing services. Cloud computing can reduce operational costs by using resources more efficiently. Cloud computing addresses a variety of data management issues:

- Enables access to organizational data anywhere and at any time

- Streamlines the organization's IT operations by subscribing only to needed services

- Eliminates or reduces the need for onsite IT equipment, maintenance, and management

- Reduces cost for equipment, energy, physical plant requirements, and personnel training needs

- Enables rapid responses to increasing data volume requirements

Cloud computing, with its "pay-as-you-go" model, allows organizations to treat computing and storage expenses more as a utility rather than investing in infrastructure. Capital expenditures are transformed into operating expenditures.

Refer to **Online Course** for Illustration

Cloud Services - 13.1.3

Cloud services are available in a variety of options, tailored to meet customer requirements. The three main cloud computing services defined by the National Institute of Standards and Technology (NIST) in their Special Publication 800-145 are as follows:

- **Software as a Service (SaaS)** - The cloud provider is responsible for access to applications and services, such as email, communication, and Office 365 that are delivered over the internet. The user does not manage any aspect of the cloud services except for limited user-specific application settings. The user only needs to provide their data.

- **Platform as a Service (PaaS)** - The cloud provider is responsible for providing users access to the development tools and services used to deliver the applications. These users are typically programmers and may have control over the configuration settings of the cloud provider's application hosting environment.

- **Infrastructure as a Service (IaaS)** - The cloud provider is responsible for giving IT managers access to the network equipment, virtualized network services, and supporting network infrastructure. Using this cloud service allows IT managers to deploy and run software code, which can include operating systems and applications.

Cloud service providers have extended this model to also provide IT support for each of the cloud computing services (ITaaS), as shown in the figure. For businesses, ITaaS can extend the capability of the network without requiring investment in new infrastructure, training new personnel, or licensing new software. These services are available on demand and delivered economically to any device anywhere in the world without compromising security or function.

Refer to **Online Course** for Illustration

Cloud Models - 13.1.4

There are four primary cloud models, as shown in the figure.

- **Public clouds** - Cloud-based applications and services offered in a public cloud are made available to the general population. Services may be free or are offered on a pay-per-use model, such as paying for online storage. The public cloud uses the internet to provide services.

- **Private clouds** - Cloud-based applications and services offered in a private cloud are intended for a specific organization or entity, such as the government. A private cloud can be set up using the organization's private network, though this can be expensive to build and maintain. A private cloud can also be managed by an outside organization with strict access security.

- **Hybrid clouds** - A hybrid cloud is made up of two or more clouds (example: part private, part public), where each part remains a separate object, but both are connected using a single architecture. Individuals on a hybrid cloud would be able to have degrees of access to various services based on user access rights.

- **Community clouds** - A community cloud is created for exclusive use by a specific community. The differences between public clouds and community clouds are the functional needs that have been customized for the community. For example, health-care organizations must remain compliant with policies and laws (e.g., HIPAA) that require special authentication and confidentiality.

Cloud Computing versus Data Center - 13.1.5

The terms data center and cloud computing are often used incorrectly. These are the correct definitions of data center and cloud computing:

- **Data center:** Typically, a data storage and processing facility run by an in-house IT department or leased offsite.

- **Cloud computing:** Typically, an off-premise service that offers on-demand access to a shared pool of configurable computing resources. These resources can be rapidly provisioned and released with minimal management effort.

Data centers are the physical facilities that provide the compute, network, and storage needs of cloud computing services. Cloud service providers use data centers to host their cloud services and cloud-based resources.

A data center can occupy one room of a building, one or more floors, or an entire building. Data centers are typically very expensive to build and maintain. For this reason, only large organizations use privately built data centers to house their data and provide services to users. Smaller organizations that cannot afford to maintain their own private data center can reduce the overall cost of ownership by leasing server and storage services from a larger data center organization in the cloud.

Go to the online course to take the quiz and exam.

Check Your Understanding - Cloud Computing - 13.1.6

Virtualization - 13.2

Refer to **Online Course** for Illustration

Cloud Computing and Virtualization - 13.2.1

In the previous topic, you learned about cloud services and cloud models. This topic will explain virtualization. The terms "cloud computing" and "virtualization" are often used interchangeably; however, they mean different things. Virtualization is the foundation of cloud computing. Without it, cloud computing, as it is most-widely implemented, would not be possible.

Virtualization separates the operating system (OS) from the hardware. Various providers offer virtual cloud services that can dynamically provision servers as required. For example, Amazon Web Services (AWS) provides a simple way for customers to dynamically provision the compute resources they need. These virtualized instances of servers are created on demand. As shown in the figure, the network administrator can deploy a variety of services from the AWS Management Console including virtual machines, web applications, virtual servers, and connections to IoT devices.

Refer to
Online Course
for Illustration

Dedicated Servers - 13.2.2

To fully appreciate virtualization, it is first necessary to understand some of the history of server technology. Historically, enterprise servers consisted of a server OS, such as Windows Server or Linux Server, installed on specific hardware, as shown in the figure. All of a server's RAM, processing power, and hard drive space were dedicated to the service provided (e.g., Web, email services, etc.).

The major problem with this configuration is that when a component fails, the service that is provided by this server becomes unavailable. This is known as a single point of failure. Another problem was that dedicated servers were underused. Dedicated servers often sat idle for long periods of time, waiting until there was a need to deliver the specific service they provide. These servers wasted energy and took up more space than was warranted by the amount of service provided. This is known as server sprawl.

Refer to
Online Course
for Illustration

Server Virtualization - 13.2.3

Server virtualization takes advantage of idle resources and consolidates the number of required servers. This also allows for multiple operating systems to exist on a single hardware platform.

For example, in the figure, the previous eight dedicated servers have been consolidated into two servers using hypervisors to support multiple virtual instances of the operating systems.

The use of virtualization normally includes redundancy to protect from a single point of failure. Redundancy can be implemented in different ways. If the hypervisor fails, the VM can be restarted on another hypervisor. Also, the same VM can run on two hypervisors concurrently, copying the RAM and CPU instructions between them. If one hypervisor fails, the VM continues running on the other hypervisor. The services running on the VMs are also virtual and can be dynamically installed or uninstalled, as needed.

The hypervisor is a program, firmware, or hardware that adds an abstraction layer on top of the physical hardware. The abstraction layer is used to create virtual machines which have access to all the hardware of the physical machine such as CPUs, memory, disk controllers, and NICs. Each of these virtual machines runs a complete and separate operating system. With virtualization, enterprises can now consolidate the number of servers they require. For example, it is not uncommon for 100 physical servers to be consolidated as virtual machines on top of 10 physical servers that are using hypervisors.

Advantages of Virtualization - 13.2.4

One major advantage of virtualization is overall reduced cost:

■ **Less equipment is required** - Virtualization enables server consolidation, which requires fewer physical servers, fewer networking devices, and less supporting infrastructure. It also means lower maintenance costs.

- **Less energy is consumed** - Consolidating servers lowers the monthly power and cooling costs. Reduced consumption helps enterprises to achieve a smaller carbon footprint.

- **Less space is required** - Server consolidation with virtualization reduces the overall footprint of the data center. Fewer servers, network devices, and racks reduce the amount of required floor space.

These are additional benefits of virtualization:

- **Easier prototyping** - Self-contained labs, operating on isolated networks, can be rapidly created for testing and prototyping network deployments. If a mistake is made, an administrator can simply revert to a previous version. The testing environments can be online, but isolated from end users. When testing is completed, the servers and systems can be deployed to end users.

- **Faster server provisioning** - Creating a virtual server is far faster than provisioning a physical server.

- **Increased server uptime** - Most server virtualization platforms now offer advanced redundant fault tolerance features, such as live migration, storage migration, high availability, and distributed resource scheduling.

- **Improved disaster recovery** - Virtualization offers advanced business continuity solutions. It provides hardware abstraction capability so that the recovery site no longer needs to have hardware that is identical to the hardware in the production environment. Most enterprise server virtualization platforms also have software that can help test and automate the failover before a disaster does happen.

- **Legacy support** - Virtualization can extend the life of OSs and applications providing more time for organizations to migrate to newer solutions.

Abstraction Layers - 13.2.5

Refer to
Online Course
for Illustration

To help explain how virtualization works, it helps to use layers of abstraction in computer architectures. A computer system consists of the following abstraction layers, as illustrated in the figure:

- Services

- OS

- Firmware

- Hardware

At each of these layers of abstraction, some type of programming code is used as an interface between the layer below and the layer above. For example, the C programming language is often used to program the firmware that accesses the hardware.

An example of virtualization is shown in the figure. A hypervisor is installed between the firmware and the OS. The hypervisor can support multiple instances of OSs.

Refer to
Online Course
for Illustration

Type 2 Hypervisors - 13.2.6

A Type 2 hypervisor is software that creates and runs VM instances. The computer, on which a hypervisor is supporting one or more VMs, is a host machine. Type 2 hypervisors are also called hosted hypervisors. This is because the hypervisor is installed on top of the existing OS, such as macOS, Windows, or Linux. Then, one or more additional OS instances are installed on top of the hypervisor, as shown in the figure.

A big advantage of Type 2 hypervisors is that management console software is not required.

Type 2 hypervisors are very popular with consumers and for organizations experimenting with virtualization. Common Type 2 hypervisors include:

- Virtual PC
- VMware Workstation
- Oracle VM VirtualBox
- VMware Fusion
- Mac OS X Parallels

Many of these Type 2 hypervisors are free. However, some hypervisors offer more advanced features for a fee.

Note: It is important to make sure that the host machine is robust enough to install and run the VMs, so that it does not run out of resources.

Go to the online
course to take the
quiz and exam.

Check Your Understanding - Virtualization - 13.2.7

Virtual Network Infrastructure - 13.3

Refer to
Online Course
for Illustration

Type 1 Hypervisors - 13.3.1

In the previous topic, you learned about virtualization. This topic will cover the virtual network infrastructure.

Type 1 hypervisors are also called the "bare metal" approach because the hypervisor is installed directly on the hardware. Type 1 hypervisors are usually used on enterprise servers and data center networking devices.

With Type 1 hypervisors, the hypervisor is installed directly on the server or networking hardware. Then, instances of an OS are installed on the hypervisor, as shown in the figure. Type 1 hypervisors have direct access to the hardware resources. Therefore, they are more efficient than hosted architectures. Type 1 hypervisors improve scalability, performance, and robustness.

Refer to
Online Course
for Illustration

Installing a VM on a Hypervisor - 13.3.2

When a Type 1 hypervisor is installed, and the server is rebooted, only basic information is displayed, such as the OS version, the amount of RAM, and the IP address. An OS instance cannot be created from this screen. Type 1 hypervisors require a "management console" to manage the hypervisor. Management software is used to manage multiple servers using the same hypervisor. The management console can automatically consolidate servers and power on or off servers as required.

For example, assume that Server1 in the figure becomes low on resources. To make more resources available, the network administrator uses the management console to move the Windows instance to the hypervisor on Server2. The management console can also be programmed with thresholds that will trigger the move automatically.

The management console provides recovery from hardware failure. If a server component fails, the management console automatically moves the VM to another server. The management console for the Cisco Unified Computing System (UCS) Manager is shown in the figure. Cisco UCS Manager controls multiple servers and manages resources for thousands of VMs.

Some management consoles also allow server over allocation. Over allocation is when multiple OS instances are installed, but their memory allocation exceeds the total amount of memory that a server has. For example, a server has 16 GB of RAM, but the administrator creates four OS instances with 10 GB of RAM allocated to each. This type of over allocation is a common practice because all four OS instances rarely require the full 10 GB of RAM at any one moment.

The Complexity of Network Virtualization - 13.3.3

Refer to **Online Course** for Illustration

Server virtualization hides server resources, such as the number and identity of physical servers, processors, and OSs from server users. This practice can create problems if the data center is using traditional network architectures.

For example, Virtual LANs (VLANs) used by VMs must be assigned to the same switch port as the physical server running the hypervisor. However, VMs are movable, and the network administrator must be able to add, drop, and change network resources and profiles. This process would be manual and time-consuming with traditional network switches.

Another problem is that traffic flows differ substantially from the traditional client-server model. Typically, a data center has a considerable amount of traffic being exchanged between virtual servers, such as the UCS servers shown in the figure. These flows are called East-West traffic and can change in location and intensity over time. North-South traffic occurs between the distribution and core layers and is typically traffic destined for offsite locations such as another data center, other cloud providers, or the internet.

Dynamic ever-changing traffic requires a flexible approach to network resource management. Existing network infrastructures can respond to changing requirements related to the management of traffic flows by using Quality of Service (QoS) and security level configurations for individual flows. However, in large enterprises using multivendor equipment, each time a new VM is enabled, the necessary reconfiguration can be very time-consuming.

The network infrastructure can also benefit from virtualization. Network functions can be virtualized. Each network device can be segmented into multiple virtual devices that operate as independent devices. Examples include subinterfaces, virtual interfaces, VLANs, and routing tables. Virtualized routing is called virtual routing and forwarding (VRF).

How is the network virtualized? The answer is found in how a networking device operates using a data plane and a control plane, as discussed in the next topic.

Go to the online course to take the quiz and exam.

Check Your Understanding - Virtual Network Infrastructure - 13.3.4

Software-Defined Networking - 13.4

Refer to **Video** in online course

Video - Software-Defined Networking - 13.4.1

Click Play to view a video on network programming, software-defined networking (SDN), and controllers.

Refer to **Interactive Graphic** in online course

Control Plane and Data Plane - 13.4.2

Refer to **Online Course** for Illustration

The previous topic explained virtual network infrastructure. This topic will cover Software-Defined Networking (SDN). SDN was explained in the previous video. We will cover more details here.

A network device contains the following planes:

- **Control plane** - This is typically regarded as the brains of a device. It is used to make forwarding decisions. The control plane contains Layer 2 and Layer 3 route forwarding mechanisms, such as routing protocol neighbor tables and topology tables, IPv4 and IPv6 routing tables, STP, and the ARP table. Information sent to the control plane is processed by the CPU.

- **Data plane** - Also called the forwarding plane, this plane is typically the switch fabric connecting the various network ports on a device. The data plane of each device is used to forward traffic flows. Routers and switches use information from the control plane to forward incoming traffic out the appropriate egress interface. Information in the data plane is typically processed by a special data plane processor without the CPU getting involved.

Click each button for an illustration and explanation of the difference between the operation of localized control on a Layer 3 switch and a centralized controller in SDN

Layer 3 Switch and CEF

The figure illustrates how Cisco Express Forwarding (CEF) uses the control plane and data plane to process packets.

CEF is an advanced, Layer 3 IP switching technology that enables forwarding of packets to occur at the data plane without consulting the control plane. In CEF, the control plane's routing table pre-populates the CEF Forwarding Information Base (FIB) table in the data plane. The control plane's ARP table pre-populates the adjacency table. Packets are then forwarded directly by the data plane based on the information contained in the FIB and adjacency table, without needing to consult the information in the control plane.

SDN and Central Controller

SDN is basically the separation of the control plane and data plane. The control plane function is removed from each device and is performed by a centralized controller, as shown in the figure. The centralized controller communicates control plane functions to each device. Each device can now focus on forwarding data while the centralized controller manages data flow, increases security, and provides other services.

Management Plane

Not shown in the figures is the management plane, which is responsible for managing a device through its connection to the network. Network administrators use applications such as Secure Shell (SSH), Trivial File Transfer Protocol (TFTP), Secure FTP, and Secure Hypertext Transfer Protocol (HTTPS) to access the management plane and configure a device. The management plane is how you have accessed and configured devices in your networking studies. In addition, protocols like Simple Network Management Protocol (SNMP), use the management plane.

Refer to
Online Course
for Illustration

Network Virtualization Technologies - 13.4.3

Over a decade ago, VMware developed a virtualizing technology that enabled a host OS to support one or more client OSs. Most virtualization technologies are now based on this technology. The transformation of dedicated servers to virtualized servers has been embraced and is rapidly being implemented in data center and enterprise networks.

Two major network architectures have been developed to support network virtualization:

- **Software-Defined Networking (SDN)** - A network architecture that virtualizes the network, offering a new approach to network administration and management that seeks to simplify and streamline the administration process.

- **Cisco Application Centric Infrastructure (ACI)** - A purpose-built hardware solution for integrating cloud computing and data center management.

Components of SDN may include the following:

- **OpenFlow** - This approach was developed at Stanford University to manage traffic between routers, switches, wireless access points, and a controller. The OpenFlow protocol is a basic element in building SDN solutions. Search for OpenFlow and the Open Networking Foundation for more information.

- **OpenStack** - This approach is a virtualization and orchestration platform designed to build scalable cloud environments and provide an IaaS solution. OpenStack is often used with Cisco ACI. Orchestration in networking is the process of automating the provisioning of network components such as servers, storage, switches, routers, and applications. Search for OpenStack for more information.

- **Other components** - Other components include Interface to the Routing System (I2RS), Transparent Interconnection of Lots of Links (TRILL), Cisco FabricPath (FP), and IEEE 802.1aq Shortest Path Bridging (SPB).

Refer to
Online Course
for Illustration

Traditional and SDN Architectures - 13.4.4

In a traditional router or switch architecture, the control plane and data plane functions occur in the same device. Routing decisions and packet forwarding are the responsibility of the device operating system. In SDN, management of the control plane is moved to a centralized SDN controller. The figure compares traditional and SDN architectures.

The SDN controller is a logical entity that enables network administrators to manage and dictate how the data plane of switches and routers should handle network traffic. It orchestrates, mediates, and facilitates communication between applications and network elements.

The complete SDN framework is shown in the figure. Note the use of Application Programming Interfaces (APIs) within the SDN framework. An API is a set of standardized requests that define the proper way for an application to request services from another application. The SDN controller uses northbound APIs to communicate with the upstream applications. These APIs help network administrators shape traffic and deploy services. The SDN controller also uses southbound APIs to define the behavior of the data planes on downstream switches and routers. OpenFlow is the original and widely implemented southbound API.

Go to the online
course to take the
quiz and exam.

Check Your Understanding - Software-Defined Networking - 13.4.5

Refer to
Online Course
for Illustration

Controllers - 13.5

SDN Controller and Operations - 13.5.1

The previous topic covered SDN. This topic will explain controllers.

The SDN controller defines the data flows between the centralized control plane and the data planes on individual routers and switches.

Each flow traveling through the network must first get permission from the SDN controller, which verifies that the communication is permissible according to the network policy. If the controller allows a flow, it computes a route for the flow to take and adds an entry for that flow in each of the switches along the path.

All complex functions are performed by the controller. The controller populates flow tables. Switches manage the flow tables. In the figure, an SDN controller communicates with OpenFlow-compatible switches using the OpenFlow protocol. This protocol uses Transport Layer Security (TLS) to securely send control plane communications over the network. Each OpenFlow switch connects to other OpenFlow switches. They can also connect to end-user devices that are part of a packet flow.

Within each switch, a series of tables implemented in hardware or firmware are used to manage the flows of packets through the switch. To the switch, a flow is a sequence of packets that matches a specific entry in a flow table.

The three tables types shown in the previous figure are as follows:

- **Flow Table** - This table matches incoming packets to a particular flow and specifies the functions that are to be performed on the packets. There may be multiple flow tables that operate in a pipeline fashion.

- **Group Table** - A flow table may direct a flow to a Group Table, which may trigger a variety of actions that affect one or more flows.

- **Meter Table** - This table triggers a variety of performance-related actions on a flow including the ability to rate-limit the traffic.

Refer to **Video** in online course

Video - Cisco ACI - 13.5.2

Very few organizations actually have the desire or skill to program the network using SDN tools. However, the majority of organizations want to automate the network, accelerate application deployments, and align their IT infrastructures to better meet business requirements. Cisco developed the Application Centric Infrastructure (ACI) to meet these objectives in more advanced and innovative ways than earlier SDN approaches.

Cisco ACI is a hardware solution for integrating cloud computing and data center management. At a high level, the policy element of the network is removed from the data plane. This simplifies the way data center networks are created.

Click Play to view a video about the evolution of SDN and ACI.

Refer to **Online Course** for Illustration

Core Components of ACI - 13.5.3

These are the three core components of the ACI architecture:

- **Application Network Profile (ANP)** - An ANP is a collection of end-point groups (EPG), their connections, and the policies that define those connections. The EPGs shown in the figure, such as VLANs, web services, and applications, are just examples. An ANP is often much more complex.

- **Application Policy Infrastructure Controller (APIC)** - The APIC is considered to be the brains of the ACI architecture. APIC is a centralized software controller that manages and operates a scalable ACI clustered fabric. It is designed for programmability and centralized management. It translates application policies into network programming.

- **Cisco Nexus 9000 Series switches** - These switches provide an application-aware switching fabric and work with an APIC to manage the virtual and physical network infrastructure.

The APIC is positioned between the APN and the ACI-enabled network infrastructure. The APIC translates the application requirements into a network configuration to meet those needs, as shown in the figure

Refer to **Online Course** for Illustration

Spine-Leaf Topology - 13.5.4

The Cisco ACI fabric is composed of the APIC and the Cisco Nexus 9000 series switches using two-tier spine-leaf topology, as shown in the figure. The leaf switches always attach to the spines, but they never attach to each other. Similarly, the spine switches only attach to the leaf and core switches (not shown). In this two-tier topology, everything is one hop from everything else.

The Cisco APICs and all other devices in the network physically attach to leaf switches.

When compared to SDN, the APIC controller does not manipulate the data path directly. Instead, the APIC centralizes the policy definition and programs the leaf switches to forward traffic based on the defined policies.

Refer to **Interactive Graphic** in online course

Refer to **Online Course** for Illustration

SDN Types - 13.5.5

The Cisco Application Policy Infrastructure Controller - Enterprise Module (APIC-EM) extends ACI aimed at enterprise and campus deployments. To better understand APIC-EM, it is helpful to take a broader look at the three types of SDN.

Click each SDN type to for more information.

Device-based SDN

In this type of SDN, the devices are programmable by applications running on the device itself or on a server in the network, as shown in the figure. Cisco OnePK is an example of a device-based SDN. It enables programmers to build applications using C, and Java with Python, to integrate and interact with Cisco devices.

Controller-based SDN

This type of SDN uses a centralized controller that has knowledge of all devices in the network, as shown in the figure. The applications can interface with the controller responsible for managing devices and manipulating traffic flows throughout the network. The Cisco Open SDN Controller is a commercial distribution of OpenDaylight.

Policy-based SDN

This type of SDN is similar to controller-based SDN where a centralized controller has a view of all devices in the network, as shown in the figure. Policy-based SDN includes an additional Policy layer that operates at a higher level of abstraction. It uses built-in applications that automate advanced configuration tasks via a guided workflow and user-friendly GUI. No programming skills are required. Cisco APIC-EM is an example of this type of SDN.

Refer to **Online Course** for Illustration

APIC-EM Features - 13.5.6

Each type of SDN has its own features and advantages. Policy-based SDN is the most robust, providing for a simple mechanism to control and manage policies across the entire network.

Cisco APIC-EM is an example of policy-based SDN. Cisco APIC-EM provides a single interface for network management including:

- discovering and accessing device and host inventories,
- viewing the topology (as shown in the figure),
- tracing a path between end points, and
- setting policies.

Refer to **Online Course** for Illustration

APIC-EM Path Trace - 13.5.7

The APIC-EM Path Trace tool allows the administrator to easily visualize traffic flows and discover any conflicting, duplicate, or shadowed ACL entries. This tool examines specific

ACLs on the path between two end nodes, displaying any potential issues. You can see where any ACLs along the path either permitted or denied your traffic, as shown in the figure. Notice how Branch-Router2 is permit all traffic. The network administrator can now make adjustments, if necessary, to better filter traffic.

Go to the online course to take the quiz and exam.

Check Your Understanding - Controllers - 13.5.8

Module Practice and Quiz - 13.6

Refer to **Lab Activity** for this chapter

Lab - Install Linux in a Virtual Machine and Explore the GUI - 13.6.1

In this lab, you will install a Linux OS in a virtual machine using a desktop virtualization application, such as VirtualBox. After completing the installation, you will explore the GUI interface.

What did I learn in this module? - 13.6.2

Cloud Computing

Cloud computing involves large numbers of computers connected through a network that can be physically located anywhere. Cloud computing can reduce operational costs by using resources more efficiently. Cloud computing addresses a variety of data management issues:

- It enables access to organizational data anywhere and at any time.

- It streamlines the organization's IT operations by subscribing only to needed services.

- It eliminates or reduces the need for onsite IT equipment, maintenance, and management.

- It reduces cost for equipment, energy, physical plant requirements, personnel training needs.

- It enables rapid responses to increasing data volume requirements.

The three main cloud computing services defined by the National Institute of Standards and Technology (NIST) are Software as a Service (SaaS), Platform as a Service (PaaS), and Infrastructure as a Service (IaaS). With SaaS, the cloud provider is responsible for access to applications and services, such as email, communication, and Office 365 that are delivered over the internet. With PaaS, the cloud provider is responsible for providing users access to the development tools and services used to deliver the applications. With IaaS, the cloud provider is responsible for giving IT managers access to the network equipment, virtualized network services, and supporting network infrastructure. The four types of clouds are public, private, hybrid, and community. Cloud-based applications and services offered in a public cloud are made available to the general population. Cloud-based applications and services offered in a private cloud are intended for a specific organization or entity, such as the government. A hybrid cloud is made up of two or more clouds (example: part private, part public), where each part remains a separate object, but both are connected using a single architecture. A community cloud is created for exclusive use by a specific community.

Virtualization

The terms "cloud computing" and "virtualization" are often used interchangeably; however, they mean different things. Virtualization is the foundation of cloud computing. Virtualization separates the operating system (OS) from the hardware. Historically, enterprise servers consisted of a server OS, such as Windows Server or Linux Server, installed on specific hardware. All of a server's RAM, processing power, and hard drive space were dedicated to the service. When a component fails, the service that is provided by this server becomes unavailable. This is known as a single point of failure. Another problem with dedicated servers is that they often sat idle for long periods of time, waiting until there was a need to deliver the specific service they provide. This wastes energy and resources (server sprawl). Virtualization reduces costs because less equipment is required, less energy is consumed, and less space is required. It provides for easier prototyping, faster server provisioning, increased server uptime, improved disaster recovery, and legacy support. A computer system consists of the following abstraction layers: services, OS, firmware, and hardware. With Type 1 hypervisors, the hypervisor is installed directly on the server or networking hardware. A Type 2 hypervisor is software that creates and runs VM instances. It can be installed on top of the OS or can be installed between the firmware and the OS. A Type 2 hypervisor is software that creates and runs VM instances.

Virtual Network Infrastructure

Type 1 hypervisors are also called the "bare metal" approach because the hypervisor is installed directly on the hardware. Type 1 hypervisors have direct access to the hardware resources and are more efficient than hosted architectures. They improve scalability, performance, and robustness. Type 1 hypervisors require a "management console" to manage the hypervisor. Management software is used to manage multiple servers using the same hypervisor. The management console can automatically consolidate servers and power on or off servers as required. The management console provides recovery from hardware failure. Some management consoles also allow server over allocation. Server virtualization hides server resources, such as the number and identity of physical servers, processors, and OSs from server users. This practice can create problems if the data center is using traditional network architectures. Another problem is that traffic flows differ substantially from the traditional client-server model. Typically, a data center has a considerable amount of traffic being exchanged between virtual servers. These flows are called East-West traffic and can change in location and intensity over time. North-South traffic occurs between the distribution and core layers and is typically traffic destined for offsite locations such as another data center, other cloud providers, or the internet.

Software-Defined Networking

Two major network architectures have been developed to support network virtualization: Software-Defined Networking (SDN) and Cisco Application Centric Infrastructure (ACI). SDN is an approach to networking where the network is software programmable remotely. Components of SDN may include OpenFlow, OpenStack, and other components. The SDN controller is a logical entity that enables network administrators to manage and dictate how the data plane of switches and routers should handle network traffic. A network device contains a control plane and a data plane. The control plane is regarded as the brains of a device. It is used to make forwarding decisions. The control plane contains Layer 2 and Layer 3 route forwarding mechanisms, such as routing protocol neighbor tables and topology tables, IPv4 and IPv6 routing tables, STP, and the ARP table. Information sent to the control plane, is processed by the CPU. The data plane, also called the forwarding plane, is typically the switch fabric connecting the various network ports on a device. The data plane of each device is used to forward traffic flows. Routers and switches use information from the control plane to forward incoming traffic out the appropriate egress interface. Information in the data plane is typically processed by a special data plane processor without the CPU getting involved. Cisco Express Forwarding (CEF) uses the control plane and data plane to process packets. CEF is an advanced, Layer 3 IP switching technology that enables forwarding of packets to occur at the data plane without consulting the control plane. SDN is basically the separation of the control plane and data plane. The control plane function is removed from each device and is performed by a centralized controller. The centralized controller communicates control plane functions to each device. The management plane is responsible for managing a device through its connection to the network. Network administrators use applications such as Secure Shell (SSH), Trivial File Transfer Protocol (TFTP), Secure FTP, and Secure Hypertext Transfer Protocol (HTTPS) to access the management plane and configure a device. Protocols like Simple Network Management Protocol (SNMP) use the management plane.

Controllers

The SDN controller is a logical entity that enables network administrators to manage and dictate how the data plane of switches and routers should handle network traffic. The SDN controller defines the data flows between the centralized control plane and the data planes on individual routers and switches. Each flow traveling through the network must first get permission from the SDN controller, which verifies that the communication is permissible according to the network policy. If the controller allows a flow, it computes a route for the flow to take and adds an entry for that flow in each of the switches along the path. The controller populates flow tables. Switches manage the flow tables. A flow table matches incoming packets to a particular flow and specifies the functions that are to be performed on the packets. There may be multiple flow tables that operate in a pipeline fashion. A flow table may direct a flow to a group table, which may trigger a variety of actions that affect one or more flows. A meter table triggers a variety of performance-related actions on a flow including the ability to rate-limit the traffic. Cisco developed the Application Centric Infrastructure (ACI) which is a more advanced and innovative way than earlier SDN approaches. Cisco ACI is a hardware solution for integrating cloud computing and data center management. At a high level, the policy element of the network is removed from the data plane. This simplifies the way data center networks are created. The three core components of the ACI architecture are Application Network Profile (ANP), Application Policy Infrastructure Controller (APIC), and Cisco Nexus 9000 Series switches. The Cisco ACI fabric is composed of the APIC and the Cisco Nexus 9000 series switches using two-tier spine-leaf topology. When compared to SDN, the APIC controller does not manipulate the data path directly. Instead, the APIC centralizes the policy definition and programs the leaf switches to forward traffic based on the defined policies. There are three types of SDN. Device-based SDN is when the devices are programmable by applications running on the device itself or on a server in the network. Controller-based SDN uses a centralized controller that has knowledge of all devices in the network. Policy based SDN is similar to controller-based SDN where a centralized controller has a view of all devices in the network. Policy-based SDN includes an additional Policy layer that operates at a higher level of abstraction. Policy-based SDN is the most robust, providing for a simple mechanism to control and manage policies across the entire network. Cisco APIC-EM is an example of policy-based SDN. Cisco APIC-EM provides a single interface for network management including discovering and accessing device and host inventories, viewing the topology, tracing a path between end points, and setting policies. The APIC-EM Path Trace tool allows the administrator to easily visualize traffic flows and discover any conflicting, duplicate, or shadowed ACL entries. This tool examines specific ACLs on the path between two end nodes, displaying any potential issues.

Go to the online course to take the quiz and exam.

Chapter Quiz - Network Virtualization

Your Chapter Notes

CHAPTER 14

Network Automation

Introduction - 14.0

Why should I take this module? - 14.0.1

Welcome to Network Automation!

Have you set up your home network? A small office network? Imagine doing those tasks for tens of thousands of end devices and thousands of routers, switches, and access points! Did you know that there is software that automates those tasks for an enterprise network? In fact, there is software that can automate the *design* of an enterprise network. It can automate all of the monitoring, operations and maintenance for your network. Interested? Get started!

What will I learn in this module? - 14.0.2

Module Title: Network Automation

Module Objective: Explain how network automation is enabled through RESTful APIs and configuration management tools.

Topic Title	Topic Objective
Automation Overview	Describe automation.
Data Formats	Compare JSON, YAML, and XML data formats.
APIs	Explain how APIs enable computer to computer communications.
REST	Explain how REST enables computer to computer communications.
Configuration Management	Compare the configuration management tools Puppet, Chef, Ansible, and SaltStack.
IBN and Cisco DNA Center	Explain how Cisco DNA center enables intent-based networking.

Automation Overview - 14.1

Video - Automation Everywhere - 14.1.1

Refer to **Video** in online course

We now see automation everywhere, from self-serve checkouts at stores and automatic building environmental controls, to autonomous cars and planes. How many automated systems do you encounter in a single day?

Click Play in the video to see examples of automation.

The Increase in Automation - 14.1.2

Automation is any process that is self-driven, that reduces and potentially eliminates, the need for human intervention.

Automation was once confined to the manufacturing industry. Highly repetitive tasks, such as automobile assembly, were turned over to machines and the modern assembly line was born. Machines excel at repeating the same task without fatigue and without the errors that humans are prone to make in such jobs.

These are some of the benefits of automation:

- Machines can work 24 hours a day without breaks, which results in greater output.

- Machines provide a more uniform product.

- Automation allows the collection of vast amounts of data that can be quickly analyzed to provide information which can help guide an event or process.

- Robots are used in dangerous conditions such as mining, firefighting, and cleaning up industrial accidents. This reduces the risk to humans.

- Under certain circumstances, smart devices can alter their behavior to reduce energy usage, make a medical diagnosis, and improve automobile driving safety.

Refer to
Online Course
for Illustration

Thinking Devices - 14.1.3

Can devices think? Can they learn from their environment? In this context, there are many definitions of the word "think". One possible definition is the ability to connect a series of related pieces of information together, and then use them to alter a course of action.

Many devices now incorporate smart technology to help to govern their behavior. This can be as simple as a smart appliance lowering its power consumption during periods of peak demand or as complex as a self-driving car.

Whenever a device takes a course of action based on an outside piece of information, then that device is referred to as a smart device. Many devices that we interact with now have the word smart in their names. This indicates that the device has the ability to alter its behavior depending on its environment.

In order for devices to "think", they need to be programmed using network automation tools.

Go to the online
course to take the
quiz and exam.

Check Your Understanding - Benefits of Automation - 14.1.4

Data Formats - 14.2

Refer to **Video**
in online course

Video - Data Formats - 14.2.1

Smart devices are, in fact, tiny computers. For a smart device, such as an actuator, to react to changing conditions, it must be able to receive and interpret information sent to it by another smart device, such as a sensor. These two smart devices must share a common 'language' which is called a data format. Shared data formats are also used by other devices in the network.

Click play in the video to learn about data formats.

Refer to
Online Course
for Illustration

The Data Formats Concept - 14.2.2

When sharing data with people, the possibilities for how to display that information are almost endless. For example, think of how a restaurant might format their menu. It could be text-only, a bulleted list, or photos with captions, or just photos. These are all different ways in which the restaurant can format the data that makes up the menu. A well-designed form is dictated by what makes the information the easiest for the intended audience to understand. This same principle applies to shared data between computers. A computer must put the data into a format that another computer can understand.

Data formats are simply a way to store and exchange data in a structured format. One such format is called Hypertext Markup Language (HTML). HTML is a standard markup language for describing the structure of web pages, as shown the figure.

These are some common data formats that are used in many applications including network automation and programmability:

■ JavaScript Object Notation (JSON)

■ eXtensible Markup Language (XML)

■ YAML Ain't Markup Language (YAML)

The data format that is selected will depend on the format that is used by the application, tool, or script that you are using. Many systems will be able to support more than one data format, which allows the user to choose their preferred one.

Data Format Rules - 14.2.3

Data formats have rules and structure similar to what we have with programming and written languages. Each data format will have specific characteristics:

■ Syntax, which includes the types of brackets used, such as [ ], ( ), { }, the use of white space, or indentation, quotes, commas, and more.

■ How objects are represented, such as characters, strings, lists, and arrays.

■ How key/value pairs are represented. The key is usually on the left side and it identifies or describes the data. The value on the right is the data itself and can be a character, string, number, list or another type of data.

Search the internet for "open notify ISS location now" to find a web site that tracks the current location of the International Space Station. At this web site you can see how data formats are used and some of the similarities between them. This web site includes a link for a simple Application Programming Interface (API) call to a server, which returns the current latitude and longitude of the space station along with a UNIX timestamp. The following example shows the information returned by the server using JavaScript Object Notation (JSON). The information is displayed in a raw format. This can make it difficult to understand the structure of the data.

```
{"message": "success", "timestamp": 1560789216, "iss_position":
{"latitude": "25.9990", "longitude": "-132.6992"}}
```

Search the internet to find the "JSONView" browser extension or any extension that will allow you to view JSON in a more readable format. Data objects are displayed in key/value pairs. The following output shows this same output using JSONView. The key/value pairs are much easier to interpret. In the example below, you can see the key **latitude** and its value **25.9990**.

```
{
        "message": "success",
        "timestamp": 1560789260,
        "iss_position": {
                "latitude": "25.9990",
                "longitude": "-132.6992"
        }
}
```

Note: JSONView may remove the quotation marks from the key. Quotation marks are required when coding JSON key/value pairs.

Compare Data Formats - 14.2.4

To see this same data formatted as XML or YAML, search the internet for a JSON conversion tool. At this point it is not important to understand the details of each data format, but notice how each data format makes use of syntax and how the key/value pairs are represented.

JSON Format

```
{
        "message": "success",
        "timestamp": 1560789260,
        "iss_position": {
                "latitude": "25.9990",
                "longitude": "-132.6992"
        }
}
```

YAML Format

```
message: success
timestamp: 1560789260
iss_position:
    latitude: '25.9990'
    longitude: '-132.6992'
```

XML Format

```
<?xml version="1.0" encoding="UTF-8" ?>

<root>

  <message>success</message>

  <timestamp>1560789260</timestamp>

  <iss_position>

    <latitude>25.9990</latitude>

    <longitude>-132.6992</longitude>

  </iss_position>

</root>
```

JSON Data Format - 14.2.5

JSON is a human readable data format used by applications for storing, transferring and reading data. JSON is a very popular format used by web services and APIs to provide public data. This is because it is easy to parse and can be used with most modern programming languages, including Python.

The following output shows an example of partial IOS output from a **show interface GigabitEthernet0/0/0** command on a router.

IOS Router Output

```
GigabitEthernet0/0/0 is up, line protocol is up (connected)

  Description: Wide Area Network

  Internet address is 172.16.0.2/24
```

This same information can be represented in JSON format. Notice that each object (each key/value pair) is a different piece of data about the interface including its name, a description, and whether the interface is enabled.

JSON Output

```
{

    "ietf-interfaces:interface": {

        "name": "GigabitEthernet0/0/0",

        "description": "Wide Area Network",

        "enabled": true,

        "ietf-ip:ipv4": {

            "address": [

                {

                    "ip": "172.16.0.2",

                    "netmask": "255.255.255.0"
```

```
                  }
               ]
            }
         }
      }
```

JSON Syntax Rules - 14.2.6

These are some of the characteristics of JSON:

- It uses a hierarchical structure and contains nested values.

- It uses braces { } to hold objects and square brackets [ ] hold arrays.

- Its data is written as key/value pairs.

In JSON, the data known as an object is one or more key/value pairs enclosed in braces { }. The syntax for a JSON object includes:

- Keys must be strings within double quotation marks " ".

- Values must be a valid JSON data type (string, number, array, Boolean, null, or another object).

- Keys and values are separated by a colon.

- Multiple key/value pairs within an object are separated by commas.

- Whitespace is not significant.

At times a key may contain more than one value. This is known as an array. An array in JSON is an ordered list of values. Characteristics of arrays in JSON include:

- The key followed by a colon and a list of values enclosed in square brackets [ ].

- The array is an ordered list of values.

- The array can contain multiple value types including a string, number, Boolean, object or another array inside the array.

- Each value in the array is separated by a comma.

For example, a list of IPv4 addresses might look like the following output. The key is "addresses". Each item in the list is a separate object, separated by braces { }. The objects are two key/value pairs: an IPv4 address ("ip") and a subnet mask ("netmask") separated by a comma. The array of objects in the list is also separated by a comma following the closing brace for each object.

JSON List of IPv4 Addresses

```
{
    "addresses": [
        {
```

```
    "ip": "172.16.0.2",

    "netmask": "255.255.255.0"

},

{

    "ip": "172.16.0.3",

    "netmask": "255.255.255.0"

},

{

    "ip": "172.16.0.4",

    "netmask": "255.255.255.0"

}

    ]

}
```

YAML Data Format - 14.2.7

YAML is another type of human readable data format used by applications for storing, transferring, and reading data. Some of the characteristic of YAML include:

- It is like JSON and is considered a superset of JSON.

- It has a minimalist format making it easy to both read and write.

- It uses indentation to define its structure, without the use of brackets or commas.

For example, look at this JSON output for a Gigabit Ethernet 2 interface.

JSON for GigabitEthernet2

```
{

    "ietf-interfaces:interface": {

        "name": "GigabitEthernet2",

        "description": "Wide Area Network",

        "enabled": true,

        "ietf-ip:ipv4": {

            "address": [

                {

                    "ip": "172.16.0.2",

                    "netmask": "255.255.255.0"

                },

                {

                    "ip": "172.16.0.3",
```

```
                        "netmask": "255.255.255.0"
                    },
                    {

                        "ip": "172.16.0.4",

                        "netmask": "255.255.255.0"

                    }

                ]

            }

        }

}
```

That same data in YAML format is easier to read. Similar to JSON, a YAML object is one or more key value pairs. Key value pairs are separated by a colon without the use of quotation marks. In YAML, a hyphen is used to separate each element in a list. This is shown for the three IPv4 addresses in the following output.

YAML for GigabitEthernet2

```
ietf-interfaces:interface:

    name: GigabitEthernet2

    description: Wide Area Network

    enabled: true

    ietf-ip:ipv4:

        address:

        - ip: 172.16.0.2

          netmask: 255.255.255.0

        - ip: 172.16.0.3

          netmask: 255.255.255.0

        - ip: 172.16.0.4

          netmask: 255.255.255.0
```

XML Data Format - 14.2.8

XML is one more type of human readable data format used to store, transfer, and read data by applications. Some of the characteristics of XML include:

- It is like HTML, which is the standardized markup language for creating web pages and web applications.

- It is self-descriptive. It encloses data within a related set of tags: **\<tag>data\</tag>**

- Unlike HTML, XML uses no predefined tags or document structure.

XML objects are one or more key/value pairs, with the beginning tag used as the name of the key: **\<key>value\</key>**

The following output shows the same data for GigabitEthernet2 formatted as an XML data structure. Notice how the values are enclosed within the object tags. In this example, each key/value pair is on a separate line and some lines are indented. This is not required but is done for readability. The list uses repeated instances of **<tag></tag>** for each element in the list. The elements within these repeated instances represent one or more key/value pairs.

XML for GigabitEthernet2

```
<?xml version="1.0" encoding="UTF-8" ?>

<ietf-interfaces:interface>

  <name>GigabitEthernet2</name>

  <description>Wide Area Network</description>

  <enabled>true</enabled>

  <ietf-ip:ipv4>

    <address>

      <ip>172.16.0.2</ip>

      <netmask>255.255.255.0</netmask>

    </address>

    <address>

      <ip>172.16.0.3</ip>

      <netmask>255.255.255.0</netmask>

    </address>

    <address>

      <ip>172.16.0.4</ip>

      <netmask>255.255.255.0</netmask>

    </address>

  </ietf-ip:ipv4>

</ietf-interfaces:interface>
```

Go to the online course to take the quiz and exam.

Check Your Understanding - Data Formats - 14.2.9

APIs - 14.3

Refer to **Video** in online course

Video - APIs - 14.3.1

Data formats shared between smart devices often use an Application Programming Interface (API). As you will learn in this topic, an API is software that allows other applications to access its data or services.

Click play in the video to learn about APIs.

Refer to
Online Course
for Illustration

The API Concept - 14.3.2

APIs are found almost everywhere. Amazon Web Services, Facebook, and home automation devices such as thermostats, refrigerators, and wireless lighting systems, all use APIs. They are also used for building programmable network automation.

An API is software that allows other applications to access its data or services. It is a set of rules describing how one application can interact with another, and the instructions to allow the interaction to occur. The user sends an API request to a server asking for specific information and receives an API response in return from the server along with the requested information.

An API is similar to a waiter in a restaurant, as shown in the following figure. A customer in a restaurant would like to have some food delivered to the table. The food is in the kitchen where it is cooked and prepared. The waiter is the messenger, similar to an API. The waiter (the API) is the person who takes the customer's order (the request) and tells the kitchen what to do. When the food is ready, the waiter will then deliver the food (the response) back to the customer.

Previously, you saw an API request to a server which returned the current latitude and longitude of the International Space Station. This was an API that Open Notify provides to access data from a web browser at National Aeronautics and Space Administration (NASA).

Refer to
Online Course
for Illustration

An API Example - 14.3.3

To really understand how APIs can be used to provide data and services, we will look at two options for booking airline reservations. The first option uses the web site of a specific airline, as shown in the figure. Using the airline's web site, the user enters the information to make a reservation request. The web site interacts directly with the airline's own database and provides the user with information matching the user's request.

Instead of using an individual airline web site which has direct access to its own information, there is a second option. Users can use a travel site to access this same information, not only from a specific airline but a variety of airlines. In this case, the user enters in similar reservation information. The travel service web site interacts with the various airline databases using APIs provided by each airline. The travel service uses each airline API to request information from that specific airline, and then it displays the information from all the airlines on the its web page, as shown in the figure.

The API acts as a kind of messenger between the requesting application and the application on the server that provides the data or service. The message from the requesting application to the server where the data resides is known as an API call.

Refer to
Online Course
for Illustration

Open, Internal, and Partner APIs - 14.3.4

An important consideration when developing an API is the distinction between open, internal, and partner APIs:

- **Open APIs or Public APIs** - These APIs are publicly available and can be used with no restrictions. The International Space Station API is an example of a Public API. Because these APIs are public, many API providers, such as Google Maps, require the user to get a free key, or token, prior to using the API. This is to help control the number of API requests they receive and process. Search the internet for a list of public APIs.

- **Internal or Private APIs** - These are APIs that are used by an organization or company to access data and services for internal use only. An example of an internal API is allowing authorized salespeople access to internal sales data on their mobile devices.

- **Partner APIs** - These are APIs that are used between a company and its business partners or contractors to facilitate business between them. The business partner must have a license or other form of permission to use the API. A travel service using an airline's API is an example of a partner API.

Types of Web Service APIs - 14.3.5

A web service is a service that is available over the internet, using the World Wide Web. There are four types of web service APIs:

- Simple Object Access Protocol (SOAP)

- Representational State Transfer (REST)

- eXtensible Markup Language-Remote Procedure Call (XML-RPC)

- JavaScript Object Notation-Remote Procedure Call (JSON-RPC)

Characteristic	SOAP	REST	XML-RPC	JSON-RPC
Data Format	XML	JSON, XML, YAML, and others	XML	JSON
First released	1998	2000	1998	2005
Strengths	Well-established	Flexible formatting and most widely used	Well-established, simplicity	Simplicity

SOAP is a messaging protocol for exchanging XML-structured information, most often over HTTP or Simple Mail Transfer Protocol (SMTP). Designed by Microsoft in 1998, SOAP APIs are considered slow to parse, complex, and rigid.

This led to the development of a simpler REST API framework which does not require XML. REST uses HTTP, is less verbose, and is easier to use than SOAP. REST refers to the style of software architecture and has become popular due to its performance, scalability, simplicity, and reliability.

REST is the most widely used web service API, accounting for over 80% of all the API types used. REST will be further discussed in this module.

RPC is when one system requests that another system executes some code and returns the information. This is done without having to understand the details of the network. This works much like a REST API but there are differences dealing with formatting and flexibility. XML-RPC is a protocol developed prior to SOAP, and later evolved into what became SOAP. JSON-RPC is a very simple protocol and similar to XML-RPC.

Go to the online course to take the quiz and exam.

Check Your Understanding - APIs - 14.3.6

REST - 14.4

Refer to **Video** in online course

Video - REST - 14.4.1

As you have just learned, REST is currently the most widely used API. This topic covers REST in more detail.

Click play in the video to learn about more about REST.

Refer to **Online Course** for Illustration

REST and RESTful API - 14.4.2

Web browsers use HTTP or HTTPS to request (GET) a web page. If successfully requested (HTTP status code 200), web servers respond to GET requests with an HTML coded web page, as shown in the figure.

REST is an architectural style for designing web service applications. It refers to a style of web architecture that has many underlying characteristics and governs the behavior of clients and servers. Simply stated, a REST API is an API that works on top of the HTTP protocol. It defines a set of functions developers can use to perform requests and receive responses via HTTP protocol such as GET and POST.

Conforming to the constraints of the REST architecture is generally referred to as being "RESTful". An API can be considered "RESTful" if it has the following features:

- **Client-Server** - The client handles the front end and the server handles the back end. Either can be replaced independently of the other.

- **Stateless** - No client data is stored on the server between requests. The session state is stored on the client.

- **Cacheable** - Clients can cache responses to improve performance.

Refer to **Online Course** for Illustration

RESTful Implementation - 14.4.3

A RESTful web service is implemented using HTTP. It is a collection of resources with four defined aspects:

- The base Uniform Resource Identifier (URI) for the web service, such as http://example.com/resources.

- The data format supported by the web service. This is often JSON, YAML, or XML but could be any other data format that is a valid hypertext standard.

- The set of operations supported by the web service using HTTP methods.

- The API must be hypertext driven.

RESTful APIs use common HTTP methods including POST, GET, PUT, PATCH and DELETE. As shown in the following table, these correspond to RESTful operations: Create, Read, Update, and Delete (or CRUD).

HTTP Method	RESTful Operation
POST	Create
GET	Read
PUT/PATCH	Update
DELETE	Delete

In the figure, the HTTP request asks for JSON-formatted data. If the request is successfully constructed according to the API documentation, the server will respond with JSON data. This JSON data can be used by a client's web application to display the data. For example, a smartphone mapping app show the location of San Jose, California.

URI, URN, and URL - 14.4.4

Refer to **Online Course** for Illustration

Web resources and web services such as RESTful APIs are identified using a URI. A URI is a string of characters that identifies a specific network resource. As shown in the figure, a URI has two specializations:

- **Uniform Resource Name (URN)** - identifies only the namespace of the resource (web page, document, image, etc.) without reference to the protocol.

- **Uniform Resource Locator (URL)** - defines the network location of a specific resource on the network. HTTP or HTTPS URLs are typically used with web browsers. Other protocols such as FTP, SFTP, SSH, and others can use a URL. A URL using SFTP might look like: sftp://sftp.example.com.

These are the parts of a URI, as shown in the figure:

- **Protocol/scheme** – HTTPS or other protocols such as FTP, SFTP, mailto, and NNTP
- **Hostname** - www.example.com
- **Path and file name** - /author/book.html
- **Fragment** - #page155

Anatomy of a RESTful Request - 14.4.5

Refer to **Online Course** for Illustration

In a RESTful Web service, a request made to a resource's URI will elicit a response. The response will be a payload typically formatted in JSON, but could be HTML, XML, or some other format. The figure shows the URI for the MapQuest directions API. The API request is for directions from San Jose, California to Monterey, California.

The figure shows part of the API response. In this example it is the MapQuest directions from San Jose to Monterey in JSON format.

These are the different parts of the API request:

- **API Server** - This is the URL for the server that answers REST requests. In this example it is the MapQuest API server.

- **Resources** - Specifies the API that is being requested. In this example it is the MapQuest directions API.

- **Query** - Specifies the data format and information the client is requesting from the API service. Queries can include:

  - **Format** – This is usually JSON but can be YAML or XML. In this example JSON is requested.

  - **Key** - The key is for authorization, if required. MapQuest requires a key for their directions API. In the above URI, you would need to replace "KEY" with a valid key to submit a valid request.

  - **Parameters** - Parameters are used to send information pertaining to the request. In this example, the query parameters include information about the directions that the API needs so it knows what directions to return: "from=San+Jose,Ca" and "to=Monterey,Ca".

Many RESTful APIs, including public APIs, require a key. The key is used to identify the source of the request. Here are some reasons why an API provider may require a key:

- To authenticate the source to make sure they are authorized to use the API.

- To limit the number of people using the API.

- To limit the number of requests per user.

- To better capture and track the data being requested by users.

- To gather information on the people using the API.

Note: If you wish to use the MapQuest API, the API does require a key. Search the internet for the URL to obtain a MapQuest key. Use the search parameters: developer.mapquest. You can also search the internet for the current URL that outlines the MapQuest privacy policy.

Refer to
Interactive Graphic
in online course

RESTful API Applications - 14.4.6

Many web sites and applications use APIs to access information and provide service for their customers. For example, when using a travel service web site, the travel service uses the API of various airlines to provide the user with airline, hotel and other information.

Some RESTful API requests can be made by typing in the URI from within a web browser. The MapQuest directions API is an example of this. A RESTful API request can also be made in other ways.

Click each API application scenario below for more information.

Developer Web Site

Developers often maintain web sites that include information about the API, parameter information, and usage examples. These sites may also allow the user to perform the API request within the developer web page by entering in the parameters and other information. The following figure shows an example of the MapQuest Directions API web page.

Postman

Postman is an application for testing and using REST APIs. It is available as a browser app or a standalone install. It contains everything required for constructing and sending REST API requests, including entering query parameters and keys. Postman allows you to collect and save frequently used API calls in history, or as collections. Postman is an excellent tool for learning how to construct API requests, and for analyzing the data that is returned from an API. The following figure shows an example of using the MapQuest API with Postman.

Python

APIs can also be called from within a Python program. This allows for possible automation, customization, and App integration of the API. The following figure shows an example of part of a Python program used to submit requests to the MapQuest API.

Network Operating Systems

Using protocols such as NETCONF (NET CONFiguration) and RESTCONF, network operating systems are beginning to provide an alternative method for configuration, monitoring, and management. For example, the following output might be the opening response from a router after the user has established a NETCONF session at the command line. However, working at the command line is not automating the network. Instead, a network administrator can use Python scripts or other automation tools, like Cisco DNA Center, to programmatically interact with the router.

```
$ ssh admin@192.168.0.1 -p 830 -s netconf

admin@192.168.0.1's password:

<hello xmlns="urn:ietf:params:xml:ns:netconf:base:1.0">

<capabilities>

  <capability>urn:ietf:params:netconf:base:1.1</capability>


<capability>urn:ietf:params:netconf:capability:candidate:1.0</capability>

  <capability>urn:ietf:params:xml:ns:yang:ietf-netconf-monitoring</
capability>

  <capability>urn:ietf:params:xml:ns:yang:ietf-interfaces</capability>

  [output omitted and edited for clarity]

</capabilities>

<session-id>19150</session-id></hello>
```

Go to the online course to take the quiz and exam.

Check Your Understanding - REST - 14.4.7

Configuration Management Tools - 14.5

Video - Configuration Management Tools - 14.5.1

Refer to **Video** in online course

As mentioned in the introduction to this module, setting up a network can be very time consuming. Configuration management tools can help you to automate the configuration of routers, switches, firewalls and many other aspects of your network.

Click play in the video to learn about configuration management tools.

Refer to
Online Course
for Illustration

Traditional Network Configuration - 14.5.2

Network devices such as router, switches, and firewalls have traditionally been configured by a network administrator using the CLI, as shown in the figure. Whenever there is a change or new feature, the necessary configuration commands must be manually entered on all of the appropriate devices. In many cases, this is not only time-consuming, but can also be prone to errors. This becomes a major issue on larger networks or with more complex configurations.

Simple Network Management Protocol (SNMP) was developed to allow administrators to manage nodes such as servers, workstations, routers, switches, and security appliances, on an IP network. Using a network management station (NMS), shown in the following figure, SNMP enables network administrators to monitor and manage network performance, find and solve network problems, and perform queries for statistics. SNMP works reasonably well for device monitoring. However, it is not typically used for configuration due to security concerns and difficulty in implementation. Although SNMP is widely available, it cannot serve as an automation tool for today's networks.

You can also use APIs to automate the deployment and management of network resources. Instead of the network administrator manually configuring ports, access lists, quality of service (QoS), and load balancing policies, they can use tools to automate configurations. These tools hook into network APIs to automate routine network provisioning tasks, enabling the administrator to select and deploy the network services they need. This can significantly reduce many repetitive and mundane tasks to free up time for network administrators to work on more important things.

Refer to
Online Course
for Illustration

Network Automation - 14.5.3

We are rapidly moving away from a world where a network administrator manages a few dozen network devices, to one where they are deploying and managing hundreds, thousands, and even tens of thousands of complex network devices (both physical and virtual) with the help of software. This transformation is quickly spreading from its beginnings in the data center, to all places in the network. There are new and different methods for network operators to automatically monitor, manage, and configure the network. As shown in the figure, these include protocols and technologies such as REST, Ansible, Puppet, Chef, Python, JSON, XML, and more.

Refer to
Online Course
for Illustration

Configuration Management Tools - 14.5.4

Configuration management tools make use of RESTful API requests to automate tasks and can scale across thousands of devices. Configuration management tools maintain the characteristics of a system, or network, for consistency. These are some characteristics of the network that administrators benefit from automating:

- Software and version control

- Device attributes such as names, addressing, and security

- Protocol configurations

- ACL configurations

Configuration management tools typically include automation and orchestration. Automation is when a tool automatically performs a task on a system. This might be configuring an interface or deploying a VLAN. Orchestration is the process of how all these automated activities need to happen, such as the order in which they must be done, what must be completed before another task is begun, etc. Orchestration is the arranging of the automated tasks that results in a coordinate process or workflow.

There are several tools available to make configuration management easier:

- Ansible

- Chef

- Puppet

- SaltStack

The goal of all of these tools is to reduce the complexity and time involved in configuring and maintaining a large-scale network infrastructure with hundreds, even thousands of devices. These same tools can benefit smaller networks as well.

Compare Ansible, Chef, Puppet, and SaltStack - 14.5.5

Ansible, Chef, Puppet, and SaltStack all come with API documentation for configuring RESTful API requests. All of them support JSON and YAML as well as other data formats. The following table shows a summary of a comparison of major characteristics of Ansible, Puppet, Chef, and SaltStack configuration management tools.

Characteristic	Ansible	Chef	Puppet	SaltStack
What programming language?	Python + YAML	Ruby	Ruby	Python
Agent-based or agentless?	Agentless	Agent-based	Supports both	Supports both
How are devices managed?	Any device can be "controller"	Chef Master	Puppet Master	Salt Master
What is created by the tool?	Playbook	Cookbook	Manifest	Pillar

- **What programming language?** - Ansible and SaltStack are both built on Python whereas Puppet and Chef are built on Ruby. Similar to Python, Ruby is an open-source programming language that is cross-platform. However, Ruby is typically considered a more difficult language to learn than Python.

- **Agent-based or agentless?** - Configuration management is either agent-based or agentless. Agent-based configuration management is "pull-based", meaning the agent on the managed device periodically connects with the master for its configuration information. Changes are done on the master and pulled down and executed by the device. Agentless configuration management is "push-based." A configuration script is run on the master. The master connects to the device and executes the tasks in the script. Of the four configuration tools in the table, only Ansible is agentless.

- **How are devices managed?** - This lies with a device called the Master in Puppet, Chef, and SaltStack. However, because Ansible is agentless, any computer can be the controller.

- **What is created by the tool?** - Network administrators use configuration management tools to create a set of instructions to be executed. Each tool has its own name for these instructions: Playbook, Cookbook, Manifest, and Pillar. Common to each of this is specification of a policy or a configuration that is to be applied to devices. Each device type might have its own policy. For example, all Linux servers might get the same basic configuration and security policy.

Go to the online course to take the quiz and exam.

Check Your Understanding - Configuration Management - 14.5.6

IBN and Cisco DNA Center - 14.6

Refer to **Video** in online course

Video - Intent-Based Networking - 14.6.1

You have learned of the many tools and software that can help you automate your network. Intent-Based Networking (IBN) and Cisco Digital Network Architecture (DNA) Center can help you bring it all together to create an automated network.

Click Play in the figure to view a video by Cisco's John Apostolopoulos and Anand Oswal explaining how artificial intelligence and intent-based networking (IBN) can improve networks.

Refer to **Online Course** for Illustration

Intent-Based Networking Overview - 14.6.2

IBN is the emerging industry model for the next generation of networking. IBN builds on Software-Defined Networking (SDN), transforming a hardware-centric and manual approach to designing and operating networks to one that is software-centric and fully automated.

Business objectives for the network are expressed as intent. IBN captures business intent and uses analytics, machine learning, and automation to align the network continuously and dynamically as business needs change.

IBN captures and translates business intent into network policies that can be automated and applied consistently across the network.

Cisco views IBN as having three essential functions: translation, activation, and assurance. These functions interact with the underlying physical and virtual infrastructure, as shown in the figure.

Refer to **Online Course** for Illustration

Network Infrastructure as Fabric - 14.6.3

From the perspective of IBN, the physical and virtual network infrastructure is a fabric. Fabric is a term used to describe an overlay that represents the logical topology used to virtually connect to devices, as shown in the figure. The overlay limits the number of devices the network administrator must program. It also provides services and alternative forwarding methods not controlled by the underlying physical devices. For example, the overlay is where encapsulation protocols like IP security (IPsec) and Control and Provisioning of Wireless Access Points (CAPWAP) occur. Using an IBN solution, the network administrator can specify through policies exactly what happens in the overlay control plane. Notice that how the switches are physically connected is not a concern of the overlay.

The underlay network is the physical topology that includes all hardware required to meet business objectives. The underlay reveals additional devices and specifies how these devices are connected, as shown in the figure. End points, such as the servers in the figure, access the network through the Layer 2 devices. The underlay control plane is responsible for simple forwarding tasks.

Refer to **Online Course** for Illustration

Cisco Digital Network Architecture (DNA) - 14.6.4

Cisco implements the IBN fabric using Cisco DNA. As displayed in the figure, the business intent is securely deployed into the network infrastructure (the fabric). Cisco DNA then continuously gathers data from a multitude of sources (devices and applications) to provide a rich context of information. This information can then be analyzed to make sure the network is performing securely at its optimal level and in accordance with business intent and network policies.

Cisco DNA is a system that is constantly learning, adapting to support the business needs. The table lists some Cisco DNA products and solutions.

Cisco DNA Solution	Description	Benefits
SD-Access	• First intent-based enterprise networking solution built using Cisco DNA. • It uses a single network fabric across LAN and WLAN to create a consistent, highly secure user experience. • It segments user, device, and application traffic and automates user-access policies to establish the right policy for any user or device, with any application, across a network.	Enables network access in minutes for any user or device to any application without compromising security.
SD-WAN	• It uses a secure cloud-delivered architecture to centrally manage WAN connections. • It simplifies and accelerates delivery of secure, flexible and rich WAN services to connect data centers, branches, campuses, and colocation facilities.	• Delivers better user experiences for applications residing on-premise or in the cloud. • Achieve greater agility and cost savings through easier deployments and transport independence.
Cisco DNA Assurance	• Used to troubleshoot and increase IT productivity. • It applies advanced analytics and machine learning to improve performance and issue resolution, and predict to assure network performance. • It provides real-time notification for network conditions that require attention.	• Allows you to identify root causes and provides suggested remediation for faster troubleshooting. • The Cisco DNA Center provides an easy-to-use single dashboard with insights and drill-down capabilities. • Machine learning continually improves network intelligence to predict problems before they occur.

Cisco DNA Solution	Description	Benefits
Cisco DNA Security	• Used to provide visibility by using the network as a sensor for real-time analysis and intelligence. • It provides increased granular control to enforce policy and contain threats across the network.	• Reduce risk and protect your organization against threats - even in encrypted traffic. • Gain 360-degree visibility through real-time analytics for deep intelligence across the network. • Lower complexity with end-to-end security.

These solutions are not mutually exclusive. For example, all four solutions could be deployed by an organization.

Many of these solutions are implemented using the Cisco DNA Center which provides a software dashboard for managing an enterprise network.

Refer to **Online Course** for Illustration

Cisco DNA Center - 14.6.5

Cisco DNA Center is the foundational controller and analytics platform at the heart of Cisco DNA. It supports the expression of intent for multiple use cases, including basic automation capabilities, fabric provisioning, and policy-based segmentation in the enterprise network. Cisco DNA Center is a network management and command center for provisioning and configuring network devices. It is a hardware and software platform providing a 'single-pane-of-glass' (single interface) that focuses on assurance, analytics, and automation.

The DNA Center interface launch page gives you an overall health summary and network snapshot, as shown in the figure. From here, the network administrator can quickly drill down into areas of interest.

At the top, menus provide you access to DNA Center's five main areas. As shown in the figure, these are

- **Design** - Model your entire network, from sites and buildings to devices and links, both physical and virtual, across campus, branch, WAN and cloud.

- **Policy** - Use policies to automate and simplify network management, reducing cost and risk while speeding rollout of new and enhanced services.

- **Provision** - Provide new services to users with ease, speed, and security across your enterprise network, regardless of network size and complexity.

- **Assurance** - Use proactive monitoring and insights from the network, devices, and applications to predict problems faster and ensure that policy and configuration changes achieve the business intent and the user experience you want.

- **Platform** - Use APIs to integrate with your preferred IT systems to create end-to-end solutions and add support for multi-vendor devices.

Refer to **Video** in online course

Video - DNA Center Overview and Platform APIs - 14.6.6

This is Part One of a four-part series demonstrating the Cisco DNA Center.

Part One is an overview of the Cisco DNA Center GUI. It includes design, policy, provision, and assurance tools used to control multiple sites and multiple devices.

Click Play in the figure to view the video.

Refer to **Video** in online course

Video - DNA Center Design and Provision - 14.6.7

This is Part Two of a four-part series demonstrating the Cisco DNA Center.

Part Two is an overview of the Cisco DNA Center design and provision areas.

Click Play in the figure to view the video.

Refer to **Video** in online course

Video - DNA Center Policy and Assurance - 14.6.8

This is Part Three of a four-part series demonstrating the Cisco DNA Center.

Part Three explains the Cisco DNA Center policy and assurance areas.

Click Play in the figure to view the video.

Refer to **Video** in online course

Video - DNA Center Troubleshooting User Connectivity - 14.6.9

This is Part Four of a four-part series demonstrating the Cisco DNA Center.

Part Four explains how to use Cisco DNA Center to troubleshoot devices.

Click Play in the figure to view the video.

Go to the online course to take the quiz and exam.

Check Your Understanding - IBN and Cisco DNA Center - 14.6.10

Module Practice and Quiz - 14.7

What did I learn in this module? - 14.7.1

Automation is any process that is self-driven, reducing and potentially eliminating, the need for human intervention. Whenever a course of action is taken by a device based on an outside piece of information, then that device is a smart device. For smart devices to "think", they need to be programmed using network automation tools.

Data formats are simply a way to store and interchange data in a structured format. One such format is called Hypertext Markup Language (HTML). Common data formats that are used in many applications including network automation and programmability are JavaScript Object Notation (JSON), eXtensible Markup Language (XML), and YAML Ain't Markup Language (YAML). Data formats have rules and structure similar to what we have with programming and written languages.

An API is a set of rules describing how one application can interact with another, and the instructions to allow the interaction to occur. Open/Public APIs are, as the name suggests, publicly available. Internal/Private APIs are used only within an organization. Partner APIs are used between a company and its business partners. There are four types of web service APIs: Simple Object Access Protocol (SOAP), Representational State Transfer (REST), eXtensible Markup Language-Remote Procedure Call (XML-RPC), and JavaScript Object Notation-Remote Procedure Call (JSON-RPC).

A REST API defines a set of functions developers can use to perform requests and receive responses via HTTP protocol such as GET and POST. Conforming to the constraints of the REST architecture is generally referred to as being "RESTful". RESTful APIs use common HTTP methods including POST, GET, PUT, PATCH and DELETE. These methods correspond to RESTful operations: Create, Read, Update, and Delete (or CRUD). Web resources and web services such as RESTful APIs are identified using a URI. A URI has two specializations, Uniform Resource Name (URN) and Uniform Resource Locator (URL). In a RESTful Web service, a request made to a resource's URI will elicit a response. The response will be a payload typically formatted in JSON. The different parts of the API request are API server, Resources, and Query. Queries can include format, key, and parameters.

There are now new and different methods for network operators to automatically monitor, manage, and configure the network. These include protocols and technologies such as REST, Ansible, Puppet, Chef, Python, JSON, XML, and more. Configuration management tools use RESTful API requests to automate tasks and scale across thousands of devices. Characteristics of the network that benefit from automation include software and version control, device attributes such as names, addressing, and security, protocol configurations, and ACL configurations. Configuration management tools typically include automation and orchestration. Orchestration is the arranging of the automated tasks that results in a coordinate process or workflow. Ansible, Chef, Puppet, and SaltStack all come with API documentation for configuring RESTful API requests.

IBN builds on SDN, taking a software-centric, fully automated approach to designing and operating networks. Cisco views IBN as having three essential functions: translation, activation, and assurance. The physical and virtual network infrastructure is a fabric. The term fabric describes an overlay that represents the logical topology used to virtually connect to devices. The underlay network is the physical topology that includes all hardware required to meet business objectives. Cisco implements the IBN fabric using Cisco DNA. The business intent is securely deployed into the network infrastructure (the fabric). Cisco DNA then continuously gathers data from a multitude of sources (devices and applications) to provide a rich context of information. Cisco DNA Center is the foundational controller and analytics platform at the heart of Cisco DNA. Cisco DNA Center is a network management and command center for provisioning and configuring network devices. It is a single interface hardware and software platform that focuses on assurance, analytics, and automation.

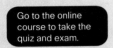
Go to the online course to take the quiz and exam.

Chapter Quiz - Network Automation

Your Chapter Notes